KRASNER HYPERRING THEORY

KRASNER
HYPERRING
THEORY

Bijan Davvaz
Yazd University, Iran

Violeta Leoreanu-Fotea
Alexandru Ioan Cuza University of Iasi, Romania

W😊 World Scientific

NEW JERSEY · LONDON · SINGAPORE · BEIJING · SHANGHAI · HONG KONG · TAIPEI · CHENNAI · TOKYO

Published by

World Scientific Publishing Co. Pte. Ltd.
5 Toh Tuck Link, Singapore 596224
USA office: 27 Warren Street, Suite 401-402, Hackensack, NJ 07601
UK office: 57 Shelton Street, Covent Garden, London WC2H 9HE

Library of Congress Control Number: 2023047505

British Library Cataloguing-in-Publication Data
A catalogue record for this book is available from the British Library.

KRASNER HYPERRING THEORY

ISBN 978-981-12-8540-0 (hardcover)
ISBN 978-981-12-8541-7 (ebook for institutions)
ISBN 978-981-12-8542-4 (ebook for individuals)

For any available supplementary material, please visit
https://www.worldscientific.com/worldscibooks/10.1142/13652#t=suppl

Desk Editors: Aanand Jayaraman/Ana Ovey

Typeset by Stallion Press
Email: enquiries@stallionpress.com

Preface

The theory of algebraic hyperstructures, in particular the theory of Krasner hyperrings, has seen a spectacular development in the past 20 years, which is why we have dedicated this book to the study of this topic.

Marc Krasner (1912–1985, Paris) was a Russian Empire–born French mathematician who worked on algebraic number theory. He received his PhD in 1935 from the University of Paris under the guidance of Jacques Hadamard, with a thesis titled *Sur la théorie de la ramification des idéaux de corps non-galoisiens de nombres algébriques*. Krasner did research on p-adic analysis. He introduced the concept of ultrametric spaces and, alongside Lev Kaluznin, proved the Krasner–Kaloujnine universal embedding theorem, which states that every extension of one group by another is isomorphic to a subgroup of the wreath product. The well-known Krasner's lemma relies on the topological structure and the algebraic structure of vector spaces over local fields.

Krasner hyperrings are a generalization of hyperfields, introduced by Krasner himself in order to study complete valued fields. A Krasner hyperring $(R, +, \cdot)$ is an algebraic structure, where $(R, +)$ is a canonical hypergroup, $(R, \cdot)$ is a semigroup with zero as a bilaterally absorbing element, and the multiplication is distributive with respect to the hyperoperation $+$.

The current book is structured in 10 chapters, presenting an elaborate study of the theme and being rich in examples. The book not only contains the results of the authors but also those of other researchers in the field, focusing especially on recent research.

The following is a brief summary of the chapters.

Since the additive structure of a Krasner hyperring is a canonical hypergroup, Chapter 1 is dedicated to this topic and contains some recent results. It also connects with another important algebraic hyperstructure, namely the join space. Special attention is dedicated to the class of canonical hypergroups C_n.

The basic notions of Krasner hyperrings are then presented. Various types of hyperideals are introduced and analyzed, as well as their generalizations, in Chapter 4: 2-absorbing hyperideals, r hyperideals, and the generalizations of prime and primary hyperideals. The existence of non-quotient Krasner hyperrings, as well as hypervaluations and semigroups that admit Krasner hyperring structures, are analyzed.

This is followed by a chapter dedicated to homomorphisms and isomorphisms, in which the hyperrings of fractions are also presented. Chapter 5 contains lower and upper approximations in Krasner hyperrings, with a preamble on Pawlak approximations, followed by rough hyperideals and rough sets in a quotient hyperring.

Chapter 6 contains hyperstructures derived from hyperconics. In Chapter 7, fundamental relations in Krasner hyperrings are analyzed, in particular the relations γ^* and α^*. Then, some special hyperrings are presented: graded hyperrings and hyperring extensions of the Krasner hyperfield.

Chapter 9 is dedicated to differential Krasner hyperrings. Several types of derivations are defined and analyzed.

The last chapter contains a study of ordered Krasner hyperrings, especially of hyperideals, bi-hyperideals, prime bi-hyperideals, interior hyperideals of ordered Krasner hyperrings, and clean ordered Krasner hyperrings and hyperringoids associated with a Krasner hyperring.

This book is especially addressed to doctoral students or researchers in the field, as well as to all those interested in this fascinating part of algebra with applications to other fields.

Bijan Davvaz
Department of Mathematical Sciences,
Yazd University, Yazd, Iran

Violeta Leoreanu-Fotea
Faculty of Mathematics,
Alexandru Ioan Cuza University of Iasi, Romania

About the Authors

Bijan Davvaz is a professor at the Department of Mathematical Sciences, Yazd University, Iran. He earned his PhD in mathematics with a thesis on "Topics in Algebraic Hyperstructures" from Tarbiat Modares University, Iran, and completed his MSc in mathematics from the University of Tehran, Iran. He has also served as the head of the Department of Mathematics (1998–2002), the chairman of the Faculty of Science (2004–2006), and the vice president for research (2006–2008) at Yazd University. His areas of interest include algebra, algebraic hyperstructures, rough sets, and fuzzy logic. He is on the editorial boards of 25 mathematical journals. Prof. Davvaz has authored 12 books and over 750 research papers, especially on algebra, fuzzy logic, algebraic hyperstructures, and their applications.

Violeta Leoreanu-Fotea is a professor at the Faculty of Mathematics of Al. I. Cuza University in Iaşi, Romania. She earned her PhD in mathematics in 1999 from Babes-Bolyai University in Cluj-Napoca, Romania, with the doctoral thesis, "Contributions to the Study of the Heart of a Hypergroup." In 2011, she was conferred the Grigore Moisil Award by the Romanian Academy. She has delivered lectures in algebra in various European countries, Canada, Thailand, China, Iran, and Turkey. Prof. Leoreanu-Fotea is the author of nine algebra books and over 100 research papers about algebraic hyperstructures, fuzzy and soft algebra, categories, and their applications.

Contents

Chapter 1

Canonical Hypergroups

1.1 The Concept of Canonical Hypergroups

Canonical hypergroups are an important class of hypergroups, which are additive structures of Krasner hyperrings and hyperfields. They were introduced and analyzed especially by Mittas [64–67].

Several other classes of hyperstructures connected with canonical hypergroups have been considered. Corsini [19–21] studied a particular class of canonical hypergroups, called sd-hypergroups. Roth [90,91] used canonical hypergroups in order to prove theorems in the finite group character theory.

Moreover, Prenowitz and Jantosciak [86] emphasized the role of canonical hypergroups in geometry. McMullen and Price [56,57] used a generalization of them in harmonic analysis and particle physics. Serafimidis et al. [92,93] and Corsini [17,18] analyzed strongly canonical and i.p.s. hypergroups.

Quasi-canonical hypergroups, also called polygroups, were investigated especially by Comer [13–15] in connection with graphs, relations, and Boole and cylindric algebras. Feebly canonical hypergroups were studied by Corsini [20] and De Salvo [31]. Massouros and Massouros studied canonical hypergroups and join spaces, and they introduced fortified join hypergroups in connection with the theory of languages and automata [61].

Canonical hypergroups can be characterized using another important class of hypergroups, called join spaces, which are very useful especially for their applications in geometry [86].

Let us see now what canonical hypergroups and join spaces are.

Definition 1.1. A commutative hypergroup $(H, \circ)$ is called *canonical* if:

1°. it has a scalar identity, which means that $\exists e \in H, \forall x \in H$,
   $e \circ x = x \circ e = \{x\}$;
2°. for all $x \in H$, there exists a unique $x' \in H$ such that $e \in x \circ x'$;
3°. it is reversible, which means that from $x \in y \circ z$, it follows that
   $y \in x \circ z'$, where z' is the inverse of z.

Example 1.1. Let S be a set of elements, called points, and T be a set lines, i.e., sets of points, which satisfy the following conditions:

- Any line contains at least three points.
- Two distinct points a, b are contained in a unique line, which we denote by $L(a, b)$.
- If a, b, c, d are distinct points of S and $L(a, b) \cap L(c, d) \neq \emptyset$, then

$$L(a, c) \cap L(b, d) \neq \emptyset.$$

Consider $S' = S \cup \{e\}$, where e does not belong to S. We define the following hyperoperation in S':

- $\forall (a, b) \in S^2, a \neq b, \ a \circ b = L(a, b) - \{a, b\}$.
- If $a \in S$ and each line contains exactly three points, let $a \circ a = \{e\}$; otherwise, $a \circ a = \{a, e\}$.
- $\forall a \in S', \ e \circ a = a \circ e = a$.

Then, $(S', \circ)$ is a canonical hypergroup.

Remark 1.1. Not all subhypergroups of a canonical hypergroup are canonical.

Indeed, this follows from the following theorem (see [19, Th. 200]).

Theorem 1.1. *Let A be a commutative hypergroup such that for all elements $x, y, z \in A$, the following conditions hold:*

(1) $(x/y) \circ z = (x \circ z)/y$;
(2) $x \in (y/y) \circ x$;
(3) $x/(y/z) \subseteq (x \circ z)/y$.

Then, there exists a canonical hypergroup H such that A is its non-canonical subhypergroup.

Proof. Consider A' to be a set equipotent to and disjoint with A, and let $f : A \to A'$ denote a bijection. We denote $f(x)$ by x', for all $x \in A$. Moreover, if $B \subseteq A$, then we denote $f(B)$ by B'.

Denote $x \circ y$ by $x \circ_A y$, for all elements $x, y \in A$.

If for all $x, y \in A$, we define $x' \circ_{A'} y' = f(x \circ y)$, then $(A', \circ_{A'})$ is a hypergroup isomorphic to $(A, \circ_A)$. Let e be such that $e \notin A \cup A'$.

We shall define a canonical hypergroup structure on the set $H = A \cup A' \cup \{e\}$ such that A and A' are non-canonical subhypergroups of it.

Recall the usual notation $(x/y)_A = \{z \in H \mid x \in y \circ_A z\}$. Similarly, for $(x/y)_{A'}, (x/y)_H$.

We define $\forall x \in H$, $e \circ_H x = x \circ_H e = x$.

If $\forall x, y \in A$, $x \neq y$, we define

$$x \circ_H y' = (x/y)_A \cup (y'/x')_{A'}, \quad x \circ_H x' = (x/x)_A \cup (x'/x')_{A'} \cup \{e\},$$

then we have $x \circ_H y = x \circ_A y$, $x' \circ_H y' = x' \circ_{A'} y'$.

Hence, e is a bilateral scalar identity, and for all $x \in A$, x' is the inverse of x. Moreover, since $A \cap A' = \emptyset$, it follows that:

- for all $x, y \in A$, we have
  $(x/y')_H = \{h \in H \mid x \in h \circ_H y'\} = \{a \in A \mid x \in a \circ_H y'\} \cup \{b' \in A' \mid x \in b' \circ_H y'\} = (x/y')_A$.
- $\forall x, y \in A$, $(x/y')_A = \{a \in A \mid x \in a \circ y'\} = \{a \in A \mid x \in a/y\} = x \circ_A y$.

Hence,

$$\forall x, y \in A, \quad x \circ_H y = (x/y')_H.$$

Note also that $\forall (x', y) \in A' \times A$, $x' \circ_H y \neq \emptyset$.

Let us check now that H is reproducible. Let $x, y \in H$. We show that there exists $z \in H$ such that $x \in y \circ_H z$. If $x, y \in A$ or $x, y \in A'$, then it follows from the fact that A and A' are hypergroups. Suppose that $x \in A$, $y' \in A'$. Let $v' \in y' \circ x'$, then $y' \in (v'/x')_{A'} \subseteq v' \circ x$, whence $A' \subseteq H \circ x$. Moreover, if $w \in x \circ y$, then $x \in w/y \subseteq w \circ y'$; hence, $x \in H \circ y'$.

In order to show that H is a canonical hypergroup, we shall check the associativity law and the fact that H is reversible. First, we check the associativity law.

Let $(x, y', z') \in A \times A' \times A'$.

(a) Suppose that $x \neq y$.

    (i) Let $x' \notin y' \circ z'$. We have

$$x \circ (y' \circ z') = \left(\bigcup_{h' \in y' \circ z'} h'/x' \right) \cup \left(\bigcup_{h \in y \circ z} x/h \right).$$

On the other hand, we have $(x \circ y') \circ z' = ((y'/x')_{A'} \circ z') \cup ((x/y)_A \circ z')$. However, $((y'/x')_{A'} \circ z')_H = ((y'/x')_{A'} \circ z')_{A'} = ((y' \circ z')/x')_{A'}$, by hypothesis (1). Moreover, $((x/y)_A \circ z')_H = ((x/y)_A/z)_A \cup (z'/(x/y)'_{A'})_{A'}$. Since for $a, b \in A$, we have $x \in h \circ y \Leftrightarrow x' \in h' \circ y'$, it follows that $(x/y)'_{A'} = (x'/y')_{A'}$. On the other hand, in a commutative hypergroup H, we have $(x/y)/z = x/(y \circ z)$, for all elements $x, y, z \in H$. If the hypergroup H is regular, such that any element x has a unique bilateral inverse x', then H is reversible on the right if and only if for all elements $x, y \in H$, we have $x \circ y' = x/y$. We have $((x/y) \circ z')_H = (x/(y \circ z))_A \cup (z'/(x'/y')_{A'})_{A'}$, and from hypothesis (3), $(z'/(x'/y')_{A'})_{A'} \subseteq ((z' \circ y')/x')_{A'}$.

    (ii) If $x' \in y' \circ z'$, then $x \circ (y' \circ z') = R \cup \{e\}$. On the other hand $x' \in y' \circ z' \Leftrightarrow x \in y \circ z \Leftrightarrow z \in x/y$; hence, $e \in (x \circ y') \circ z'$.

(b) Suppose that $x = y \neq z$.

    (i) Let $x' \notin x' \circ z'$. We have $(x \circ x') \circ z' = ((x/x)_A \circ z') \cup ((x'/x')_{A'} \circ z') \cup (e \circ z')$. Furthermore,

$$x \circ (x' \circ z') = \left(\bigcup_{h' \in x' \circ z'} (h'/x')_{A'} \right) \cup \left(\bigcup_{h \in x \circ z} (x/h)_A \right).$$

We have $((x'/x')_{A'} \circ_H z' = ((x'/x')_{A'} \circ_{A'} z') = ((x' \circ z')/x')_A$. On the other hand, $((x/x)_A \circ_H z' = ((x/x)_A/z)_A \cup (z'/(x'/x')_{A'})_{A'} = (x/(x \circ z))_A \cup (z'/(x'/x')_{A'})_{A'}$. According to (3), we have $(z'/(x'/x')_{A'})_{A'} \subseteq ((x' \circ z')/x')_{A'}$. It remains to be shown that $z' \in x \circ (x' \circ z')$. By (2) and (1), we have $z' \in ((x'/x')_{A'} \circ z')_{A'} = ((x' \circ z')/x')_{A'}$, which is contained in $x \circ (x' \circ z')$.

    (ii) If $x' \in x' \circ z'$, then $e \in x \circ (x' \circ z')$; however, $x \in x \circ z$ implies $z \in x/x$ and hence $e \in (x \circ x') \circ z'$.

(c) Case $x = y = z$.

    (i) Let $x' \notin x' \circ x'$. Then,

$$x \circ (x' \circ x') = \left(\bigcup_{h' \in x' \circ x'} (h'/x')_{A'} \right) \cup \left(\bigcup_{h \in x \circ x} (x/h)_A \right).$$

        We have $(x \circ x') \circ x' = ((x/x)_A \circ_H x') \cup ((x'/x')_A \circ_H x')$ $\cup \{x'\}$ and $((x'/x')_{A'} \circ_H x' = ((x'/x')_{A'} \circ_{A'} x')$, $((x/x)_A \circ_H x' = ((x/x)_A/x)_A \cup (x'/(x/x)'_{A'})_{A'}$. On the other hand, $((x/x)_A/x)_A = (x/(x \circ x))_{A'}(x'/(x/x)'_{A'})_{A'} = (x'/(x'/x')_{A'})_{A'} \subseteq ((x' \circ x')/x')_{A'} = (x'/x')_{A'} \circ x'$. Similarly, as in case (b), posing $x = z$, we obtain $x' \in x \circ (x' \circ x')$.

    (ii) Consider now $x' \in x' \circ x'$. Then, $e \in x \circ (x' \circ x')$. From $x' \in x'/x'$, it follows that $x \in x/x$; hence, $e \in (x \circ x') \circ x'$. Let us now consider $(x, y', z) \in A \times A' \times A$.

(d) Suppose $x \neq y \neq z$.

    (i) Let $x' \notin y' \circ z$. Hence, $x \circ (y' \circ z) = ((y'/z')_{A'} \circ_H x \cup ((z/y)_A \circ_A x$ and $(x \circ y') \circ z = ((y'/x')_{A'} \circ_H z \cup ((x/y)_A \circ_H z$. By condition (2), we have $((z/y)_A \circ_A x = ((z \circ x)/y)_A = ((x/y)_A \circ z)_A = ((x/y)_A \circ_H z$. Moreover, $((y'/x')_{A'} \circ_H z = ((y'/x')_{A'}/z')_{A'} \cup (z/(y'/x')'_{A'})_A$, whence $((y'/x')_{A'} = (y'/(x' \circ z'))_{A'}$. Hence, $((y'/x')_{A'} \circ_H z = (y'/(x' \circ z'))_{A'} \cup ((z/(y/z)_A)_A$. According to (3), it follows that $((z/(y/x)_A)_A \subseteq ((z \circ x)/y)_A$. On the other hand, $((y'/z')_{A'} \circ_H x = ((y'/z')_{A'}/x')_{A'} \cup ((x/(y/z)_A)_A$, and by (3) $((x/(y/z)_A) \subseteq ((x \circ z)/y)_A$, whence $(y'/z')_{A'}/x')_{A'} = (y'/(x' \circ z'))_A = ((y'/x')_{A'}/z')_{A'}$. Thus, we obtain the associativity law in this case.

    (ii) Suppose now that $x' \in y' \circ z$, then $e \in x \circ (y' \circ z)$. We also have $z \in x'/y'$, whence $z' \in x/y$ and so, $e \in (x \circ y') \circ z$.

(e) Case $x = y \neq z$.

    (i) Let $x' \notin x' \circ z$. We have $x \circ (x' \circ z) = (x \circ_H (x'/z')_{A'}) \cup (x \circ_H (z/x)_A)$, $(x \circ x') \circ z = ((x/x)_A \circ_H z) \cup ((x'/x')_{A'} \circ z) \cup \{z\}$. However, $x \circ_H (z/x)_A = x \circ_A (z/x)_A = (x/x)_A \circ z = (x/x)_A \circ_H z$. Furthermore, $x \circ_H (x'/z')_{A'} = ((x'/z')_{A'}/x')_{A'} \cup ((x/(x'/z')'_A)_A = ((x'/z')_{A'}/x')_{A'} \cup ((x/(x/z)_A)_A$. But $(x'/x')_{A'} \circ_H z = ((x'/x')_{A'}/z')_{A'} \cup ((z/(x'/x')'_A)_A = ((x'/x')_{A'}/z')_{A'} \cup ((z/(x/x)_A)_A$. We also have $((x'/x')_{A'}/z')_{A'} = (x'/(x' \circ z'))_{A'} = ((x'/z')_{A'}/x')$.

The remaining terms are $((x/(x/z)_A)_A$ and $((z/(x/x)_A)_A$, which are both contained in $((x \circ z)/x)$, which is common to $(x \circ x') \circ z$ and $x \circ (x' \circ z)$. We still have to prove that $z \in x \circ (x' \circ z)$; however, this follows that in (b) and (c) from conditions (1) and (2).

(ii) Suppose that $x' \in x' \circ z$, then $e \in x \circ (x' \circ z)$; however, we also have $z \in x'/x'$, whence $z' \in x/x$, so $e \in (x \circ x') \circ z$.

(f) Let $x = z$. We have $(x \circ y') \circ x = (y' \circ x) \circ x = x \circ (y' \circ x)$.
Let us now check that H is reversible. We consider the following cases:

(i) Let $x, a, y \in A$. Then, $x \in a \circ y \Rightarrow y \in x/a \subseteq x \circ a'$.
(ii) Let $(x, a', y) \in A \times A' \times A$. We have $x \in a' \circ y \Rightarrow y \in x/a' = x \circ a$ and $x \in a' \circ y \Rightarrow a' \in x/y \subseteq x \circ y'$.

Therefore, $(H, \circ)$ is a canonical hypergroup. □

Let $(H, +)$ be a canonical hypergroup and $x \in H$. For any $n \in \mathbb{Z}$, we define

$$nx = \begin{cases} \underbrace{x + x + \cdots + x}_{n \text{ times}}, & n > 0 \\ 0, & n = 0 \\ \underbrace{(-x) + \cdots + (-x)}_{(-n) \text{ times}}, & n < 0, \end{cases}$$

where $\forall x \in H$, we denote by "$-x$" the inverse of x.
We have

$$mx + nx = \begin{cases} (m+n)x, & mn \geq 0 \\ (m+n)x + \min\{|m|, |n|\} \cdot (x - x), & mn < 0. \end{cases}$$

Definition 1.2. Let $(H, +)$ be a canonical hypergroup and $x \in H$. We say that the *order of x is infinite* $(o(x) = \infty)$ if $\forall (h, k) \in \mathbb{Z}^2$, where $h \neq 0$, we have $0 \notin hx + k(x - x)$.

Theorem 1.2. *Let $(H, +)$ be a canonical hypergroup and $x \in H$. Then, $o(x) = \infty$ if and only if $\forall (m, n) \in \mathbb{Z}^2$, $m \neq n$, we have $mx \cap nx \neq \emptyset$.*

Proof. Suppose that $m > n$ and $mx \cap nx \neq \emptyset$. We obtain that $(mx - nx) \cap (nx - nx) \neq \emptyset$, whence $((m-n)x + n(x-x)) \cap n(x-x) \neq \emptyset$.

Let $a \in ((m-n)x + n(x-x)) \cap n(x-x)$, then $-a \in n(x-x)$; hence, if $b \in (m-n)x$, $a' \in n(x-x)$ are such that $a \in b + a'$, we have $0 \in a - a \subseteq b + a' - a \subseteq (m-n)x + 2n(x-x)$, against the hypothesis.

Let us check now the opposite implication. Suppose that $0 \in mx + n(x-x)$, where $m \neq 0$. Since we have $b \in mx$ and $-b \in n(x-x)$, there are $p, q \in nx$ such that $b \in p - q$. Then, $p \in b + q$, so $p \in (mx + nx) \cap nx = (m+n)x \cap nx$, against the hypothesis. Hence, $o(x) = \infty$. $\qquad\square$

Definition 1.3. Let $(H, +)$ be a canonical hypergroup. Let us suppose that there exists $(m, n) \in \mathbb{Z} \times \mathbb{N}$, $m \neq 0$, such that $0 \in mx + n(x-x)$.

Let $h = \min\{r \in \mathbb{N}^* \mid \exists n \in \mathbb{N} : o \in rx + n(x-x)\}$.

The number h is called *the principal order of x*.

Theorem 1.3. *Let $(H, +)$ be a canonical hypergroup and $x \in H$. We have $0 \in mx + n(x-x)$ if and only if h divides m.*

Proof. "$\Rightarrow$" Let β be the fundamental equivalence relation, which means that H/β is a hypergroup. Recall that $x\beta y$ if and only if there are $a_1, \ldots, a_n \in H$, such that $\{x, y\} \subseteq \prod_{i=1}^{n} a_i$. Let $\bar{x}$ be the equivalence class of x mod β. From $0 \in mx + n(x-x)$, it follows that $m\bar{x} = 0$. Hence, $h|m$ since h is the order of the cyclic subgroup of H/β generated by $\bar{x}$.

"$\Leftarrow$" Let $m = kh$. Then, from $0 \in hx + n(x-x)$, it follows that $0 \in khx + kn(x-x)$, which completes the proof. $\qquad\square$

Definition 1.4. Let h divide m and

$$q = \min\{s \in \mathbb{N}^* \mid 0 \in mx + s(x-x)\}.$$

The couple (h, q) is called the *order of x*.

1.2 Connections with Join Spaces

Join spaces were introduced in the 1940s by Prenowitz and were utilized by him, and later with Jantosciak, to reconstruct several kinds of geometry. Join spaces had already many other applications, such as in graphs (Nieminen, Rosenberg, Bandelt, Mulder, and Corsini), median algebras (Bandelt and Hedlikova), hypergraphs

(Corsini and Leoreanu), and binary relations (Chvalina, Rosenberg, Corsini, Leoreanu, De Salvo, and Lo Faro). Transposition hypergroups, which are non-commutative join spaces, were introduced by Jantosciak, while fortified join hypergroups were introduced by G. G. Massouros, Ch. Massouros, and Mittas.

Let us see what a join space is.

For any x, y of H, we denote

$$x/y = \{u \mid x \in u \circ y\}.$$

Definition 1.5. A commutative hypergroup $(H, \circ)$ is called *a join space* if for all x, y, z, v of H, the following implication holds:

$$x/y \cap z/v \neq \emptyset \Rightarrow x \circ v \cap z \circ y \neq \emptyset.$$

Theorem 1.4 ([19]). *The class of canonical hypergroups coincide with that class of join spaces, which have a unique scalar identity.*

Proof. If $(H, \circ)$ is a canonical hypergroup, then for all $a, b \in H$, we have $a/b = a \circ b^{-1}$. Now, note that $a \circ b \cap c \circ d \neq \emptyset \Rightarrow a \circ c^{-1} \cap d \circ b^{-1} \neq \emptyset$. Indeed, let $u \in a \circ b \cap c \circ d$. Since H is reversible, it follows that $u^{-1} \in c^{-1} \circ d^{-1}$, whence $u \circ u^{-1} \subseteq a \circ b \circ c^{-1} \circ d^{-1}$. If e is the identity of H, then $e \in (a \circ c^{-1}) \circ (d \circ b^{-1})^{-1}$. Hence, there is an element $v \in (a \circ c^{-1}) \circ (d \circ b^{-1})$, which means that H is a join space.

Conversely, we check that the inverse of an element is unique. Denote by e the scalar identity. If $e \in a \circ b \cap a \circ c$, then $a \in e/b \cap e/c$. Hence, $e \circ c \cap e \circ b \neq \emptyset$, so $b = c = a^{-1}$. Let us check now the reversibility of H. We have $a \in b \circ c$ if and only if $b \in a/c$. From $e \in b \circ b^{-1}$, we obtain $b \in e/b^{-1}$, $a \circ b^{-1} \cap e \circ c \neq \emptyset$, which means $c \in a \circ b^1$. Therefore, H is canonical.   □

There are join spaces that are not canonical hypergroups, as we can see from the following example (see [72]).

Example 1.2. The complete combustion reaction of room-temperature gaseous alkanes is modeled using join spaces. We present here in detail this join space.

Alkanes, also known as paraffins, are hydrocarbons which contain only single bonds (e.g., C–H or C–C). Because there is hydrogen in every possible location of alkanes, they contain the maximum number of hydrogen atoms relative to the number of carbon atoms, and thus, alkanes are called saturated hydrocarbons [87, 88].

Table 1.1. The names and molecular formulas of the first 12 alkanes [87].

NCALCC	NA	MF	SF
1	methane	CH_4	CH_4
2	ethane	C_2H_6	CH_3CH_3
3	propane	C_3H_8	$CH_3CH_2CH_3$
4	butane	C_4H_{10}	$CH_3CH_2CH_2CH_3$
5	pentane	C_5H_{12}	$CH_3CH_2CH_2CH_2CH_3$
6	hexane	C_6H_{14}	$CH_3CH_2CH_2CH_2CH_2CH_3$
7	heptane	C_7H_{16}	$CH_3CH_2CH_2CH_2CH_2CH_2CH_3$
8	octane	C_8H_{18}	$CH_3CH_2CH_2CH_2CH_2CH_2CH_2CH_3$
9	nonane	C_9H_{20}	$CH_3CH_2CH_2CH_2CH_2CH_2CH_2CH_2CH_3$
10	decane	$C_{10}H_{22}$	$CH_3CH_2CH_2CH_2CH_2CH_2CH_2CH_2CH_2CH_3$
11	undecane	$C_{11}H_{24}$	$CH_3CH_2CH_2CH_2CH_2CH_2CH_2CH_2CH_2CH_2CH_3$
12	dodecane	$C_{12}H_{26}$	$CH_3CH_2CH_2CH_2CH_2CH_2CH_2CH_2CH_2CH_2CH_2CH_3$

Alkanes are one of the homologous series of organic compounds, and their general formula is C_nH_{2n+2}, where n is the number of carbon atoms in the molecule.

They can be subdivided into the following three groups: linear straight-chain alkanes, branched alkanes, and cycloalkanes. Alkanes have similar chemical properties, but their physical properties gradually change as the length of the carbon chain increases. The names of alkanes are based on the number of carbon atoms in the longest continuous chain in the compound [87].

The names and molecular formulas of the first 12 alkanes are given in Table 1.1 [87], where the number of carbon atoms in the longest continuous chain, names, molecular formulas, and structural formulas of the alkanes are denoted by NCALCC, NA, MF, and SF, respectively.

The principal sources of alkanes are crude oil and natural gas, and they are the major components of many fuels and solvents derived from petroleum [85, 88, 106].

The alkanes are highly combustible and valuable as clean fuels. The burning of alkanes forms water and carbon dioxide. Methane, ethane, propane, and butane are gaseous and used directly as fuels.

Alkanes from pentane up to around $C_{17}H_{36}$ are liquids. Gasoline is a mixture of alkanes, from pentane up to about decane. Kerosene contains alkanes from about $n = 10$ to $n = 16$. Above $n = 17$, they are solids at room temperature.

Alkanes with higher values of n are found in diesel fuel, fuel oil, petroleum jelly, paraffin wax, and motor oils, and those with the highest values of n are found in asphalt.

Alkanes are highly flammable compounds at high temperatures. The complete combustion of gaseous alkanes in the presence of sufficient oxygen results in the production of steam $(H_2O(g))$ and carbon dioxide $(CO_2(g))$. When they undergo this type of combustion, there is a blue, smokeless flame.

For example, when methane is burnt in excess oxygen, carbon dioxide and steam are produced. The chemical equation of the complete combustion of room-temperature gaseous alkanes (methane, ethane, propane, and butane) are as follows:

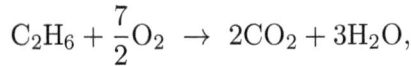

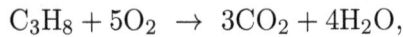

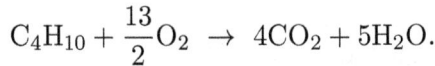

$$CH_4 + 2O_2 \;\rightarrow\; CO_2 + 2H_2O,$$

$$C_2H_6 + \frac{7}{2}O_2 \;\rightarrow\; 2CO_2 + 3H_2O,$$

$$C_3H_8 + 5O_2 \;\rightarrow\; 3CO_2 + 4H_2O,$$

$$C_4H_{10} + \frac{13}{2}O_2 \;\rightarrow\; 4CO_2 + 5H_2O.$$

Based on the above equations, the balanced chemical equation for the complete combustion of gaseous alkanes can be written as follows:

$$C_nH_{2n+2} + \frac{3n+1}{2}O_2 \rightarrow nCO_2 + (n+1)H_2O, \qquad (1.1)$$

where $n = 1, 2, 3$, and 4, respectively, gives methane, ethane, propane, and butane in C_nH_{2n+2}.

Now, consider reaction (1) for alkanes. Set

$$\mathbb{A} = \left\{ C_nH_{2n+2}, \frac{3n+1}{2}O_2, nCO_2, (n+1)H_2O \right\},$$

and rename the elements of $\mathbb{A}$ as

$$x := C_nH_{2n+2}, \quad y := \frac{3n+1}{2}O_2, \quad z := nCO_2, \quad w := (n+1)H_2O.$$

Define the hyperoperation "+" on $\mathbb{A}$ as follows:

For all $a, b \in \mathbb{A}$, the set $a + b$ is the result of a possible reaction between a and b in the complete combustion of gaseous alkanes, where if a and b do not react in the complete combustion, then we put $a + b = \{a, b\}$.

Therefore, all possible combinations for the set $\mathbb{A} = \{x, y, z, w\}$ under the complete combustion of gaseous alkanes can be displayed as the following table:

$+$	x	y	z	w
x	x	$\mathbb{A}$	$\{x, z\}$	$\{x, w\}$
y	$\mathbb{A}$	y	$\{y, z\}$	$\{y, w\}$
z	$\{x, z\}$	$\{y, z\}$	z	$\{z, w\}$
w	$\{x, w\}$	$\{y, w\}$	$\{z, w\}$	w

Clearly, $+$ is commutative. In the following, we investigate the associativity of $+$:

$(x + x) + x = x$	$x + (x + x) = x$
$(x + x) + y = x + y = \mathbb{A}$	$x + (x + y) = x + \mathbb{A} = \mathbb{A}$
$(x + x) + z = x + z = \{x, z\}$	$x + (x + z) = x + \{x, z\} = \{x, z\}$
$(x + x) + w = x + w = \{x, w\}$	$x + (x + w) = x + \{x, w\} = \{x, w\}$
$(x + y) + y = \mathbb{A} + y = \mathbb{A}$	$x + (y + y) = x + y = \mathbb{A}$
$(x + y) + z = \mathbb{A} + z = \mathbb{A}$	$x + (y + z) = x + \{y, z\} = \mathbb{A}$
$(x + y) + w = \mathbb{A} + w = \mathbb{A}$	$x + (y + w) = x + \{y, w\} = \mathbb{A}$
$(x + z) + y = \{x, z\} + y = \mathbb{A}$	$x + (z + y) = \mathbb{A}$
$(x + z) + z = \{x, z\} + z = \{x, z\}$	$x + (z + z) = x + z = \{x, z\}$
$(x + z) + w = \{x, z\} + w = \{x, z, w\}$	$x + (z + w) = x + \{z, w\} = \{x, z, w\}$
$(x + w) + y = \{x, w\} + y = \mathbb{A}$	$x + (w + y) = x + \{y, w\} = \mathbb{A}$
$(x + w) + z = \{x, w\} + z = \{x, z, w\}$	$x + (w + z) = x + \{z, w\} = \{x, z, w\}$
$(x + w) + w = \{x, w\} + w = \{x, w\}$	$x + (w + w) = x + w = \{x, w\}$
$(y + y) + y = y$	$y + (y + y) = y$
$(y + y) + z = y + z = \{y, z\}$	$y + (y + z) = y + \{y, z\} = \{y, z\}$
$(y + y) + w = y + w = \{y, w\}$	$y + (y + w) = y + \{y, w\} = \{y, w\}$
$(y + z) + z = \{y, z\} + z = \{y, z\}$	$y + (z + z) = y + z = \{y, z\}$
$(y + z) + w = \{y, z\} + w = \{y, z, w\}$	$y + (z + w) = y + \{z, w\} = \{y, z, w\}$
$(y + w) + z = \{y, w\} + z = \{y, z, w\}$	$y + (w + z) = y + \{z, w\} = \{y, z, w\}$
$(y + w) + w = \{y, w\} + w = \{y, w\}$	$y + (w + w) = y + w = \{y, w\}$
$(z + z) + z = z$	$z + (z + z) = z$
$(z + z) + w = z + w = \{z, w\}$	$z + (z + w) = z + \{z, w\} = \{z, w\}$
$(y + w) + w = \{y, w\} + w = \{y, w\}$	$y + (w + w) = y + w = \{y, w\}$
$(w + w) + w = w$	$w + (w + w) = w$

Hence, $(\mathbb{A}, +)$ is a semihypergroup. Also, $(\mathbb{A}, +)$ is a quasihypergroup since $x + \mathbb{A} = y + \mathbb{A} = z + \mathbb{A} = w + \mathbb{A} = \mathbb{A}$. Then, we have the following theorem.

Theorem 1.5. $(\mathbb{A},+)$ *is a commutative hypergroup.*

It can be seen that $A_1 = \{x\}$, $A_2 = \{y\}$, $A_3 = \{z\}$, $A_4 = \{w\}$, $A_5 = \{x,z\}$, $A_6 = \{x,w\}$, $A_7 = \{y,z\}$, $A_8 = \{y,w\}$, $A_9 = \{z,w\}$, $A_{10} = \{x,z,w\}$, and $A_{11} = \{y,z,w\}$ are subhypergroups of $(\mathbb{A},+)$. Also, we have $A_1 \cong A_2 \cong A_3 \cong A_4$, $A_5 \cong A_6 \cong A_7 \cong A_8 \cong A_9$, and $A_{10} \cong A_{11}$. Hence, we have the following corollary.

Corollary 1.1. *There are only three subhypergroups of $(\mathbb{A},+)$ up to isomorphism.*

Furthermore, the pair $(\mathbb{A},+)$ constructs a join space. Based on the concept of $a/b = \{t \in \mathbb{A} \mid a \in t + b\}$ for $a,b \in \mathbb{A}$, we have $x/x = \mathbb{A}$, $x/y = \{x\}$, $x/z = \{x\}$, $x/w = \{x\}$, $y/x = \{y\}$, $y/y = \mathbb{A}$, $y/z = \{y\}$, $y/w = \{y\}$, $z/x = \{y,z\}$, $z/y = \{x,z\}$, $z/z = \mathbb{A}$, $z/w = \{z\}$, $w/x = \{y,w\}$, $w/y = \{x,w\}$, $w/z = \{w\}$, and $w/w = \mathbb{A}$. Hence, we can see the implication $(a/b \cap c/d \neq \emptyset \ \rightarrow \ a+d \cap b+c \neq \emptyset)$ is valid for all $a,b \in \mathbb{A}$ through the following table:

$(x+x) \cap (y+y) = \emptyset \Rightarrow x/y \cap y/x = \emptyset$	$(z+x) \cap (w+w) = \emptyset \Rightarrow z/w \cap w/x = \emptyset$
$(x+x) \cap (z+z) = \emptyset \Rightarrow x/z \cap z/x = \emptyset$	$(z+y) \cap (x+w) = \emptyset \Rightarrow z/x \cap w/y = \emptyset$
$(x+x) \cap (w+w) = \emptyset \Rightarrow x/w \cap w/x = \emptyset$	$(z+y) \cap (w+x) = \emptyset \Rightarrow z/w \cap x/y = \emptyset$
$(y+y) \cap (z+z) = \emptyset \Rightarrow y/z \cap z/y = \emptyset$	$(z+y) \cap (w+w) = \emptyset \Rightarrow z/w \cap w/y = \emptyset$
$(y+y) \cap (w+w) = \emptyset \Rightarrow y/w \cap w/y = \emptyset$	$(z+z) \cap (x+w) = \emptyset \Rightarrow z/x \cap w/z = \emptyset$
$(z+z) \cap (w+w) = \emptyset \Rightarrow z/w \cap w/z = \emptyset$	$(z+z) \cap (w+x) = \emptyset \Rightarrow z/w \cap x/z = \emptyset$
$(x+z) \cap (y+y) = \emptyset \Rightarrow x/y \cap y/z = \emptyset$	$(z+z) \cap (y+w) = \emptyset \Rightarrow z/y \cap w/z = \emptyset$
$(x+z) \cap (y+w) = \emptyset \Rightarrow x/y \cap w/z = \emptyset$	$(z+z) \cap (w+y) = \emptyset \Rightarrow z/w \cap y/z = \emptyset$
$(x+z) \cap (w+y) = \emptyset \Rightarrow x/w \cap y/z = \emptyset$	$(z+w) \cap (x+x) = \emptyset \Rightarrow z/x \cap x/w = \emptyset$
$(x+z) \cap (w+w) = \emptyset \Rightarrow x/w \cap w/z = \emptyset$	$(z+w) \cap (y+y) = \emptyset \Rightarrow z/y \cap y/w = \emptyset$
$(x+w) \cap (y+y) = \emptyset \Rightarrow x/y \cap y/w = \emptyset$	$(w+x) \cap (y+y) = \emptyset \Rightarrow w/y \cap y/x = \emptyset$
$(x+w) \cap (y+z) = \emptyset \Rightarrow x/y \cap z/w = \emptyset$	$(w+x) \cap (y+z) = \emptyset \Rightarrow w/y \cap z/x = \emptyset$
$(x+w) \cap (z+y) = \emptyset \Rightarrow x/z \cap y/w = \emptyset$	$(w+x) \cap (z+y) = \emptyset \Rightarrow w/z \cap y/x = \emptyset$
$(x+w) \cap (z+z) = \emptyset \Rightarrow x/z \cap z/w = \emptyset$	$(w+x) \cap (z+z) = \emptyset \Rightarrow w/z \cap z/x = \emptyset$
$(y+z) \cap (x+w) = \emptyset \Rightarrow y/x \cap w/z = \emptyset$	$(w+y) \cap (x+x) = \emptyset \Rightarrow w/x \cap x/y = \emptyset$
$(y+z) \cap (w+x) = \emptyset \Rightarrow y/w \cap x/z = \emptyset$	$(w+y) \cap (x+z) = \emptyset \Rightarrow w/x \cap z/y = \emptyset$
$(y+z) \cap (w+w) = \emptyset \Rightarrow y/w \cap w/z = \emptyset$	$(w+y) \cap (z+x) = \emptyset \Rightarrow w/z \cap x/y = \emptyset$
$(y+w) \cap (x+z) = \emptyset \Rightarrow y/x \cap z/w = \emptyset$	$(w+y) \cap (z+z) = \emptyset \Rightarrow w/z \cap z/y = \emptyset$
$(y+w) \cap (z+x) = \emptyset \Rightarrow y/z \cap x/w = \emptyset$	$(w+z) \cap (x+x) = \emptyset \Rightarrow w/x \cap x/z = \emptyset$
$(y+w) \cap (x+x) = \emptyset \Rightarrow y/x \cap x/w = \emptyset$	$(w+z) \cap (y+y) = \emptyset \Rightarrow w/y \cap y/z = \emptyset$
$(y+w) \cap (z+z) = \emptyset \Rightarrow y/z \cap z/w = \emptyset$	$(w+w) \cap (x+z) = \emptyset \Rightarrow w/x \cap z/w = \emptyset$
$(z+x) \cap (y+y) = \emptyset \Rightarrow z/y \cap y/x = \emptyset$	$(w+w) \cap (z+x) = \emptyset \Rightarrow w/z \cap x/w = \emptyset$
$(z+x) \cap (y+w) = \emptyset \Rightarrow z/y \cap w/x = \emptyset$	$(w+w) \cap (y+z) = \emptyset \Rightarrow w/y \cap z/w = \emptyset$
$(z+x) \cap (w+y) = \emptyset \Rightarrow z/w \cap y/x = \emptyset$	$(w+w) \cap (z+y) = \emptyset \Rightarrow w/z \cap y/w = \emptyset$

Hence, we have the following theorem.

Theorem 1.6. $(\mathbb{A}, +)$ *is a join space, which is not a canonical hypergroup.*

Now, let $(H, \circ)$ be a hypergroup. An element $e \in H$ is said to be an identity if $a \in (a \circ e) \cap (e \circ a)$, for all $a \in H$. We denote the set of all identities of H by $Id(H)$. Also, $a' \in H$ is called an inverse of $a \in H$ whenever there exists at least an identity $e \in H$ such that $e \in (a \circ a') \cap (a' \circ a)$. A hypergroup $(H, \circ)$ is called *regular* if H has at least an identity and any element has an inverse.

Theorem 1.7. $(\mathbb{A}, +)$ *is a regular hypergroup.*

Proof. According to Theorem 1.5, $(\mathbb{A}, +)$ is a commutative hypergroup. It can be seen that $Id(\mathbb{A}) = \{e \in \mathbb{A} \mid \forall a \in \mathbb{A}, \ a \in a + e\} = \mathbb{A}$ and

$$\{x, x\} \subseteq x + x, \quad \{x, y\} \subseteq x + y, \quad \{x, z\} \subseteq x + z, \text{ and } \{x, w\} \subseteq x + w.$$

It shows that any element of $\mathbb{A}$ has an inverse. This completes the proof. □

Remark 1.2. An element a in a hypergroup $(H, \circ)$ is called *idempotent* if $a \circ a = \{a\}$. Every element in $(\mathbb{A}, +)$ is an idempotent. From a chemical point of view, this means that no reaction occurs between two identical elements, i.e., the element is unchanged.

Remark 1.3. According to Ref. [40, Proposition 5.1], every regular hypergroup is a completely simple semihypergroup. Therefore, $(\mathbb{A}, +)$ is a completely simple semihypergroup. Hence, we can apply the obtained results about completely simple semihypergroups to model on $\mathbb{A}$. For more details about completely simple semihypergroups, refer to Ref. [40].

1.3 The Canonical Hypergroup $(C_n, \circ)$

The following is an example studied in several papers:

Example 1.3. Let $(C_n, \circ)$ be the following hypergroup, where $C_n = \{e_0, e_1, \ldots, e_{k(n)}\}$, $n \in \mathbb{N}^*$, and

$$k(n) = \begin{cases} \frac{n}{2}, & n \in 2\mathbb{N}; \\ \frac{n-1}{2}, & n \in 2\mathbb{N} + 1. \end{cases}$$

For a pair (s, t) such that $s \leq k(n)$, $t \leq k(n)$, we consider

$$e_s \circ e_t = \{e_v, e_p\},$$

where $v = \min\{s + t, n - (s + t)\}$ and $p = |s - t|$.

The hypergroup $(C_n, \circ)$ is canonical was introduced by Corsini [19]. Many of its results were given by Leoreanu. She studied the subhypergroups of $(C_n, \circ)$ and established the conditions under which Lagrange's theorem could be applied.

Sonea and Davvaz proved the following results about Euler's function φ associated with the canonical hypergroup C_n in Ref. [95]. Euler's function in the context of canonical hypergroups represents the number of elements with periodicity equal to the exponent of the hypergroup. It is worth mentioning that the exponent of a hypergroup H, denoted by $exp(H)$, is the least common multiple of the orders of all elements of the hypergroup H.

Theorem 1.8.

(i) *If $n = 2p$, p is an odd number, then $\varphi(C_n) = |\omega_{C_n}|$.*
(ii) *If $n = 2p$ and p is an even number, then $\varphi(C_n) = |C_n| - |\omega_{C_n}|$.*
(iii) *If n is an odd number, then $\varphi(C_n) = |C_n| = |\omega_{C_n}|$,*

where ω_{C_n} is the heart of C_n.

Proof.   (i) We have

$$\varphi(C_n) = |\{e_j \in C_n : p(e_j) = \exp(C_n)\}|,$$

where

$$p(e_j) = \min\{k \in \mathbb{N}^* : e_j^k \subseteq \omega_{C_n}, \quad j \in \{0, 1, \ldots, p\}\}.$$

It is known that ω_{C_n} is a subhypergroup of the hypergroup C_n, and we show that $\omega_{C_n} = S_2$.

Note that e_0 is the identity of the canonical hypergroup C_n because

$$e_s \in e_s \circ e_0 \cap e_0 \circ e_s,$$

$\forall s \in \{0, 1, \ldots, p\}$ is equivalent to $e_s \in \{e_v, e_p\} \cap \{e_p, e_v\}$,

for any $s \in \{0, 1, \ldots, p\}$, where $v = \min\{s, n - s\}$, $p = |s - 0| = s$. We note that $s \leq \frac{n}{2} = p$, which implies that $v = \min\{s, n - s\} = s$. So,

$e_s \circ e_0 = e_0 \circ e_s = e_s$, for any $s \in \{0, 1, \ldots, p\}$. Using the definition of the heart of a hypergroup, we have $\omega_{C_n} = \beta^*(e_0) = \{e_j \in C_n |\ e_j \beta e_0\}$. We wish to check that $\omega_{C_n} = S_2$. So, we show that $e_{2i} \in \omega_{C_n}$, $i \in \left\{0, 1, 2, \ldots, \frac{p-1}{2}\right\}$ and $e_p \notin \omega_{C_n}$. Hence,

$$e_{2i} \in \omega_{C_n} \Leftrightarrow e_{2i} \beta e_0 \Leftrightarrow \exists m \in \mathbb{N}^*, \quad \exists\ e_{j_1}, e_{j_2}, \ldots, e_{j_m} \in C_n : \{e_{2i}, e_0\} \subseteq \prod_{s=1}^{m} e_{j_s}.$$

We consider $e_{j_1} = e_{j_2} = e_i$, then we obtain

$$e_i \circ e_i = \{e_v, e_p\}, \quad v = \min\{i + i, n - (i + i)\}, \quad p = |i - i| = 0.$$
$$v = \min\{2i, n - 2i\}, \quad p = 0.$$

We note that $\min\{2i, n - 2i\} = 2i$ if and only if $2i \leq n - 2i$, which is equivalent to $i \leq \frac{p}{2}$, which is true because $i \leq \frac{p-1}{2} \leq \frac{p}{2}$. Therefore,

$$e_i \circ e_i = \{e_{2i}, e_0\}, \quad i \in \left\{0, 1, 2, \ldots, \frac{p-1}{2}\right\},$$

i.e., $e_{2i} \in \omega_{C_n}$, $i \in \left\{0, 1, 2, \ldots, \frac{p-1}{2}\right\}$, whence $S_2 \subseteq \omega_{C_n}$.

In what follows, we want to prove that the element e_p, where p is an odd number and $p > 2$, does not belong to the heart of the canonical hypergroup C_n. So, we suppose that $e_p \in \omega_{C_n}$, and it follows that $m \in \mathbb{N}^*$, $e_{i_1}, e_{i_2}, \ldots, e_{i_m} \in C_n$ such that

$$\{e_0, e_p\} \subseteq \prod_{j=1}^{m} e_{i_j}.$$

To begin, we consider $m = 2$. So, we get $e_0 \in e_{i_1} \circ e_{i_2}$ and $e_p \in e_{i_1} \circ e_{i_2}$, where $e_{i_1} \circ e_{i_2} = \{e_v, e_t\}$, $v = \min\{i_1 + i_2, n - (i_1 + i_2)\}$, and $t = e_{|i_1 - i_2|}$. We note that if i_1 and i_2 have the same parity, then v and t are even, and if i_1 and i_2 have different parities, then v and t are odd. The set $\{e_v, e_t\}$ contains elements with the same parity, which leads to e_0 and e_p not belonging to $e_{i_1} \circ e_{i_2}$ because p is an odd number. Now, if $m = 3$, we have $\{e_0, e_p\} \subseteq e_{i_1} \circ e_{i_2} \circ e_{i_3}$. Denote $e_{i_1} \circ e_{i_2} = A$,

$$e_{i_1} \circ e_{i_2} \circ e_{i_3} = A \circ e_{i_3} = \bigcup_{e_z \in A} (e_z \circ e_{i_3}).$$

So, $\{e_0, e_p\} \subseteq \in e_{i_1} \circ e_{i_2} \circ e_{i_3}$ implies that there exist $e_z, e_s \in A$ such that $e_0 \in e_z \circ e_{i_3}$ and $e_p \in e_s \circ e_{i_3}$. To obtain $e_0 \in e_z \circ e_{i_3}$, it follows that either $z = i_3$ or $n = z + i_3$. If we have $z = i_3$, then $e_p \in e_s \circ e_z$. Then, the elements in A have the same parity, so we can affirm that $e_s \circ e_z$ contains elements with an even number of indexes. Then, $e_p \notin e_s \circ e_z$. Now, we analyze the case of $n = z + i_3$. We know that n is an even number, so z and i_3 are both even or odd. This implies that s, z, and i_3 have the same parity. Also, in this case, $e_p \notin e_s \circ e_{i_3}$. For $m > 3$, $\prod_{j=1}^{m} e_{i_j} = B \circ e_{i_m}$, where $B = e_{i_1} \circ \cdots \circ e_{i_{m-1}}$, and the process is the same as in case of $m = 3$. In conclusion, for $n = 2p$, with p being a prime number and $p > 3$, $e_p \notin \omega_{C_n}$. Therefore,

$$\omega_{C_n} = S_2 \Rightarrow |\omega_{C_n}| = |S_2| = \frac{p+1}{2}.$$

Now, we check that all elements with odd-numbered indexes have a periodicity of 2. Let $e_{2j+1} \in C_n$, $j \in \left\{0, 1, 2, \ldots, \frac{p-1}{2}\right\}$ and

$$p\left(e_{2j+1}\right) = \min\left\{k \in \mathbb{N}^* : e_{2j+1}^k \subseteq \omega_{C_n}\right\}.$$

We note that $e_{2j+1} \circ e_{2j+1} = \left\{e_{\min\{4j+2,\, n-(4j+2)},\, e_0\right\}$; however, n is even number, so we can say that $4j + 2$ and $n - (4j + 2)$ are even numbers. Thus,

$$e_{2j+1} \circ e_{2j+1} \subseteq \omega_{C_n}, \quad \text{for any } j \in \left\{0, 1, 2, \ldots, \frac{p-1}{2}\right\}.$$

In conclusion, all the elements with even-numbered indexes have periodicity equal to 2, which leads to $exp(C_n) = 2$ and

$$\varphi\left(C_n\right) = \frac{p+1}{2} = |\omega_{C_n}|.$$

(ii) Note that the elements with odd-numbered indexes do not belong to the heart of the canonical hypergroup C_n and the elements with even-numbered indexes satisfy the condition $\{e_{2i}, e_0\} \subseteq e_i \circ e_i$, which leads to $e_{2i} \in \omega_{C_n}$, $i \in \left\{0, 1, \ldots, \left[\frac{p}{2}\right]\right\}$, where $[x]$ represents the integer part of x. Also, the periodicity of the elements with odd-numbered indexes are the same: $e_{2j+1} \circ e_{2j+1} \subseteq \omega_{C_n}$, for any $j \in \left\{0, 1, 2, \ldots, \left[\frac{p-1}{2}\right]\right\}$. Hence, if $n = 2p$, p is an even number, then $\varphi\left(C_n\right) = |C_n| - |\omega_{C_n}|$.

(iii) Suppose that $n = 2p + 1$, p is a natural number, $p \geq 1$. Then, $C_n = \{e_0, e_1, \ldots, e_{k(n)}\}$, $k(n) = \frac{n-1}{2}$. This yields $C_n = \{e_0, e_1, \ldots, e_p\}$. We consider the following sets:

$$E_0 = \left\{ e_{2i}, \ i \in \left\{ 0, 1, \ldots, \left[\frac{p}{2}\right] \right\} \right\},$$

$$E_1 = \left\{ e_{2j+1}, \ j \in \left\{ 0, 1, \ldots, \left[\frac{p-1}{2}\right] \right\} \right\}.$$

We note that $C_n = E_0 \cup E_1$, and we want to prove that all the elements belong to w_{C_n}. For the elements from E_0, we can write

$$e_i \circ e_i = \{e_{2i}, e_0\} \quad \Rightarrow \quad e_{2i} \in w_{C_n}, \quad i \in \left\{ 0, 1, 2, \ldots, \left[\frac{p}{2}\right] \right\}.$$

We have $E_0 \subseteq w_{C_n}$. For the elements which are in the set E_1, we show that the following relation holds:

$$\{e_{2j+1}, e_0\} = e_{p-j} \circ e_{p-j}, \ j \in \left\{ 0, 1, \ldots, \left[\frac{p-1}{2}\right] \right\}$$

$$e_{p-j} \circ e_{p-j} = \{e_{\min\{2p-2j, n-(2p-2j)\}}, e_{|p-j-p+j|}\}$$
$$= \{e_{\min\{2p-2j, 2p+1-(2p-2j)\}}, e_0\}$$
$$= \{e_{\min\{2p-2j, 1+2j\}}, e_0\}.$$

We suppose that $\min\{2p-2j, 1+2j\} = 2p-2j$, which is equivalent to the inequality $2p - 2j \leq 1 + 2j$, i.e., $j \geq \frac{2p-1}{4}$. It is known that $j \leq \left[\frac{p-1}{2}\right] \leq \frac{p-1}{2}$. Then,

$$j \leq \frac{p-1}{2} < \frac{2p-1}{4}.$$

We obtain $\min\{2p - 2j, 1 + 2j\} = 2j + 1$. In conclusion,

$$\{e_{2j+1}, e_0\} = e_{p-j} \circ e_{p-j}, \quad j \in \left\{ 0, 1, \ldots, \left[\frac{p-1}{2}\right] \right\} \Rightarrow E_1 \subseteq w_{C_n}.$$

Therefore, $C_n = E_0 \cup E_1 \subseteq w_{C_n}$, but w_{C_n} is a subhypergroup of the canonical hypergroup C_n. So, $C_n = w_{C_n}$. We can state that all the elements have periodicity equal to one, which leads to $\varphi(C_n) = |C_n| = |w_{C_n}|$. $\square$

1.4   Constructions of Canonical Hypergroups

We present here a method to construct canonical hypergroups using extensions (see [95]). Also, in the same paper [95], the commutativity degree of this extension is calculated.

Let $\mathcal{H}_1 = <H_1, \cdot, e, ^{-1}>$, $\mathcal{H}_2 = <H_2, \cdot, e, ^{-1}>$ be two canonical hypergroups whose elements have been renamed so that $H_1 \cap H_2 = \{e\}$, where e is the identity of canonical hypergroups $\mathcal{H}_1$ and $\mathcal{H}_2$.

Definition 1.6. A system $\mathcal{H}_1[\mathcal{H}_2]$ is called an *extension* of the canonical hypergroup $\mathcal{H}_1$ by the canonical hypergroup $\mathcal{H}_2$ if

$$\mathcal{H}_1[\mathcal{H}_2] = <M, *, e, ^I>,$$

where

$$M = H_1 \cup H_2, \quad e^I = e, \quad x^I = x^{-1},$$

$$e * x = x * e = x, \quad \text{for any } x \in M.$$

For all $x, y \in M \backslash \{e\}$, we have

$$x * y = \begin{cases} x \cdot y & x, y \in H_1 \\ x & x \in H_2, \quad y \in H_1 \\ y & x \in H_1, \quad y \in H_2 \\ x \cdot y & x, y \in H_2, \quad y \neq x^{-1} \\ x \cdot y \cup H_1 & x, y \in H_2, \quad y = x^{-1} \end{cases} \qquad (1.2)$$

If we consider $H_1 = \{e, a_1, a_2, \ldots, a_{n-1}\}$ and $H_2 = \{e, b_1, b_2, \ldots, b_{m-1}\}$, where $n, m \in \mathbb{N}^*$, we obtain the following table:

$*$	e	a_1	$\cdots$	a_{n-1}	b_1	$\cdots$	b_i	$\cdots$	b_{m-1}
e	e	a_1	$\cdots$	a_{n-1}	b_1	$\cdots$	b_i	$\cdots$	b_{m-1}
a_1	a_1	$a_1 a_1$	$\cdots$	$a_1 a_{n-1}$	b_1	$\cdots$	b_i	$\cdots$	b_{m-1}
$\vdots$	$\vdots$	$\vdots$	$\cdots$	$\vdots$	$\vdots$	$\cdots$	$\vdots$	$\cdots$	$\vdots$
a_{n-1}	a_{n-1}	$a_{n-1} a_1$	$\cdots$	$a_{n-1} a_{n-1}$	b_1	$\cdots$	b_i	$\cdots$	b_{m-1}
b_1	b_1	b_1	$\cdots$	b_1	$b_1 b_1$	$\cdots$	$b_1 b_i \cup H_1$	$\cdots$	$b_1 b_{m-1}$
$\vdots$	$\vdots$	$\vdots$	$\cdots$	$\vdots$	$\vdots$	$\cdots$	$\vdots$	$\cdots$	$\vdots$
b_i	b_i	b_i	$\cdots$	b_i	$b_i b_1 \cup H_1$	$\cdots$	$b_i b_i$	$\cdots$	$b_i b_{m-1}$
$\vdots$	$\vdots$	$\vdots$	$\cdots$	$\vdots$	$\vdots$	$\cdots$	$\vdots$	$\cdots$	$\vdots$
b_{m-1}	b_{m-1}	b_{m-1}	$\cdots$	b_{m-1}	$b_{m-1} b_1$	$\cdots$	$b_{m-1} b_i$	$\cdots$	$b_{m-1} b_{m-1}$

Without loss of generality, we suppose that $b_i = b_1^{-1}$, and for each element b_j, there is a unique element b_k such that $b_j = b_k^{-1}$ with i, j, $k \in \overline{1, m-1}$.

Theorem 1.9. *If $\mathcal{H}_1 = <H_1, \cdot, e, ^{-1}>$ and $\mathcal{H}_2 = <H_2, \cdot, e, ^{-1}>$ are two canonical hypergroups, then $\mathcal{H}_1[\mathcal{H}_2]$ is a canonical hypergroup.*

Proof. First of all, we prove that $\mathcal{H}_1[\mathcal{H}_2]$ is a hypergroup, i.e., (1) $(a * b) * c = a * (b * c)$, for all a, b, $c \in \mathcal{H}_1[\mathcal{H}_2]$, and (2) $a * \mathcal{H}_1[\mathcal{H}_2] = \mathcal{H}_1[\mathcal{H}_2] * a$, for all $a \in \mathcal{H}_1[\mathcal{H}_2]$. We remark that associativity can be proved in the same way as for polygroups. Therefore, we analyze relation (2) and commutativity.

For the commutativity of relation "$*$", we have $x * y = y * x$, for all $x, y \in M$. As $\mathcal{H}_1$ and $\mathcal{H}_2$ are canonical hypergroups, it follows that $x \cdot y = y \cdot x$, for all $x, y \in H_1$, and $x \cdot y = y \cdot x$, for all $x, y \in H_2$. If $x \in H_1$ and $y \in H_2$, then $x * y = y$ and $y * x = y$, so we have equality. Now, if $x \in H_2$, $y \in H_1$, we obtain $x * y = y * x = x$. Also, if $x, y \in H_2$, where $x \neq y^{-1}$ and $y \neq x^{-1}$, the equality is obvious. It remains to be shown that if $y = x^{-1}$, then $x = y^{-1}$. It is known that the elements of a canonical hypergroup have only one inverse; therefore,

$$y = x^{-1}, \text{ implies } e \in y \cdot x, \text{ i.e., } x \in y^{-1} \cdot e, \text{ so } x = y^{-1}.$$

Therefore, for $x, y \in H_2$, where $x = y^{-1}$, we have $x * y = y * x$. Hence, $\mathcal{H}_1[\mathcal{H}_2]$ is commutative.

The second condition is the reproducibility law:

$$a * \mathcal{H}_1[\mathcal{H}_2] = \mathcal{H}_1[\mathcal{H}_2] * a = \mathcal{H}_1[\mathcal{H}_2], \quad \text{for any } a \in \mathcal{H}_1[\mathcal{H}_2]. \quad (1.3)$$

So, we can write

$$a * M = M * a = M, \quad \text{for any } a \in M.$$

We have

$$a * M = \bigcup_{b \in M} (a * b) = \left(\bigcup_{b \in H_1} (a * b) \right) \cup \left(\bigcup_{b \in H_2} (a * b) \right).$$

If $a \in H_1$, then $\bigcup_{b \in H_1} (a * b) = H_1$ because $\mathcal{H}_1$ is a hypergroup and

$$\bigcup_{b \in H_2} (a * b) = \bigcup_{b \in H_2} (a \cdot b) = \bigcup_{b \in H_2} b = H_2.$$

So, in this case, we obtain $a * M = H_1 \cup H_2 = M$. The condition $M * a = M$ follows from the commutativity relation "$*$". If $a \in H_2$, then $\bigcup_{b \in H_2} (a \cdot b) = H_2$ because $\mathcal{H}_2$ is a hypergroup, and

$$\bigcup_{b \in H_2} (a * b) = \bigcup_{b \neq a^{-1}} (a \cdot b) \bigcup (a * a^{-1}) = [H_2 \backslash (a \cdot a^{-1})] \bigcup \left[(a \cdot a^{-1}) \bigcup H_1 \right]$$

$$= H_2 \cup H_1 = M.$$

Consequently, we have $a * M = M * a = M$, for all $a \in M$. This completes the proof. $\square$

Let us establish now a connection between the hearts of canonical hypergroups $\mathcal{H}_1$, $\mathcal{H}_2$, and $\mathcal{H}_1[\mathcal{H}_2]$.

Theorem 1.10. *If* $\mathcal{H}_1 = \ <H_1, \cdot, e,^{-1}>$ *and* $\mathcal{H}_2 = \ <H_2, \cdot, e,^{-1}>$ *are canonical hypergroups, then* $\omega_{\mathcal{H}_1[\mathcal{H}_2]} = H_1 \cup \omega_{\mathcal{H}_2}$.

Proof. First of all, we want to check that

$$H_1 \cup \omega_{\mathcal{H}_2} \subseteq \omega_{\mathcal{H}_1[\mathcal{H}_2]}. \tag{1.4}$$

Let $x \in H_1 \cup \omega_{\mathcal{H}_2}$. It follows that $x \in H_1$ or $x \in \omega_{\mathcal{H}_2}$. If $x \in H_1$, we note that there are z_1, $z_2 \in H_2$ with $z_1 = z_2^{-1}$ such that

$$\{e, x\} \subseteq z_1 * z_2^{-1} = z_1 \cdot z_1^{-1} \cup H_1 \Rightarrow x \in \omega_{\mathcal{H}_1[\mathcal{H}_2]}.$$

If $x \in \omega_{\mathcal{H}_2}$, then by using the definition of the heart of a hypergroup, there are $b_1, b_2, \ldots, b_p \in H_2$, $p \geq 1$ as

$$\{e, x\} \subseteq b_1 b_2 \cdots b_p.$$

Because $e \in b_1 b_2 \cdots b_p$, it means that there exist i, $j \in \{1, 2, \ldots, p\}$ such that $b_i = b_j^{-1}$. We consider $z_k = b_k$, $k \in \{1, 2, \ldots, p\}$,

$$S = z_1 * z_2 * \cdots * z_{i-1} * z_{i+1} * \cdots * z_{j-1} * z_{j+1} * \cdots * z_p,$$

and

$$S * (z_i * z_i^{-1}) = (z_1 z_2 \cdots z_{i-1} z_i \cdots z_p) \cup (S * H_1)$$

$$= (z_1 z_2 \cdots z_{i-1} z_i \cdots z_p) \cup S$$

$$= z_1 z_2 \cdots z_{i-1} z_i \cdots z_p.$$

So, there is $z_k \in M$, $z_k = b_k$, $k \in \{1, 2, \ldots, p\}$ with the following property:

$$\{e, x\} \subseteq z_1 * z_2 * \cdots * z_p \Rightarrow x \in \omega_{\mathcal{H}_1[\mathcal{H}_2]}.$$

Therefore, $H_1 \cup \omega_{\mathcal{H}_2} \subseteq \omega_{\mathcal{H}_1[\mathcal{H}_2]}$. In what follows, we prove the second inclusion:

$$\omega_{\mathcal{H}_1[\mathcal{H}_2]} \subseteq H_1 \cup \omega_{\mathcal{H}_2}. \tag{1.5}$$

Let $x \in \omega_{\mathcal{H}_1[\mathcal{H}_2]}$, which means that there are $z_1, z_2, \ldots, z_p \in M$ such that

$$\{e, x\} \subseteq z_1 * z_2 * \cdots * z_p. \tag{1.6}$$

We analyze several cases. If for all $i \in \{1, 2, \ldots, p\}$, we have $z_i \in H_1$, relation (1.6) becomes

$$\{e, x\} \subseteq z_1 z_2 \cdots z_p \Rightarrow x \in H_1.$$

If there is $j \in \{1, 2, \ldots, s\}$, $s < p$ such that $z_j \in H_2$ and $z_i \in H_1$, $i \neq j$. The indexes could be denoted as

$$z_1 * z_2 * \cdots * z_p = \underbrace{(z_1 * \cdots * z_j)}_{\in H_2} * \underbrace{(z_{j+1} \cdot z_{j+2} \cdots \cdots z_p)}_{\in H_1} = z_1 * \cdots * z_j$$

because if we consider $A = z_1 * \cdots * z_j$, $B = z_{j+1} \cdot z_{j+2} \cdots \cdots z_p$, and using the relation's "$*$" definition, we have

$$A * B = \bigcup_{a \in A, b \in B} (a * b) = \bigcup_{a \in A} a = A.$$

Relation (1.6) becomes

$$\{e, x\} \subseteq z_1 * \cdots * z_j.$$

$e \in z_1 * \cdots * z_j$, so there are $k, l \in \{1, 2, \ldots, j\}$ such that $z_k = z_l^{-1}$, whence

$$z_1 * \cdots * z_j = (z_1 \cdots z_{k-1} \cdot z_{k+1} \cdot z_{l-1} \cdot z_{l+1} \cdots z_j) * (z_k * z_k^{-1}).$$

In the same manner as in the previous case, $x \in \omega_{\mathcal{H}_2}$. Therefore, we get that there exist $z_j \in H_2$, $j \in \{1, 2, \ldots, s\}$, $s < p$ such that

$$\{e, x\} \subseteq z_1 \cdot z_2 \cdots \cdots z_j \quad \Rightarrow \quad x \in \omega_{\mathcal{H}_2}.$$

If $z_i \in H_2$, for all $i \in \{1, 2, \ldots, p\}$, then proceeding analogously, we have $x \in \omega_{\mathcal{H}_2}$. So, $\omega_{\mathcal{H}_1[\mathcal{H}_2]} \subseteq H_1 \cup \omega_{\mathcal{H}_2}$. Consequently, we have

$$\omega_{\mathcal{H}_1[\mathcal{H}_2]} = H_1 \cup \omega_{\mathcal{H}_2}.$$

This completes the proof. □

It is important to have some information about the periodicity of elements and to find the connection between the periodicity of elements from $\mathcal{H}_1[\mathcal{H}_2]$ and $\mathcal{H}_2$ in order to analyze Euler's function for the extension.

Proposition 1.1. *The periodicity of the elements of the canonical hypergroup $\mathcal{H}_1[\mathcal{H}_2]$ is the same as the periodicity of the elements of the canonical hypergroup $\mathcal{H}_2$:*

$$p(x)_{\mathcal{H}_1[\mathcal{H}_2]} = p(x)_{\mathcal{H}_2}, \quad \text{for all } x \in \mathcal{H}_2.$$

Proof. We recall the definition of periodicity: $p(x) = \min\{k \in \mathbb{N}^* | x^k \subseteq \omega_H\}$. So, for the canonical hypergroup $\mathcal{H}_1[\mathcal{H}_2]$, we have

$$p(x) = \min\{k \in \mathbb{N}^* | x^k \subseteq \omega_{\mathcal{H}_1[\mathcal{H}_2]}\} \tag{1.7}$$

$$= \min\{k \in \mathbb{N}^* | x^k \subseteq H_1 \cup \omega_{\mathcal{H}_2}\}.$$

Denote $p(x)_{\mathcal{H}_1[\mathcal{H}_2]} = m$. For $x \in H_1$, we have $p(x) = 1$; if $x \in H_2 \backslash H_1$, we obtain $x^m \subseteq H_1 \cup \omega_{\mathcal{H}_2}$. If $x^m \subseteq \omega_{\mathcal{H}_2}$, it follows that $p(x)_{\mathcal{H}_2} = m$, which implies that $p(x)_{\mathcal{H}_1[\mathcal{H}_2]} = p(x)_{\mathcal{H}_2}$. In the case of $x^m \subseteq H_1$, by using the definition of "$*$", we consider only the situation when

$$x * x^{-1} = x \cdot x^{-1} \cup H_1 = \{0\} \cup H_1, \quad \text{i.e.,} \quad p(x)_{\mathcal{H}_1[\mathcal{H}_2]} = 2.$$

However,

$$x \cdot x^{-1} = \{0\} \subseteq \omega_{\mathcal{H}_2}, \quad \text{i.e.,} \quad p(x)_{\mathcal{H}_2} = 2.$$

So, $p(x)_{\mathcal{H}_1[\mathcal{H}_2]} = p(x)_{\mathcal{H}_2}$, for all $x \in \mathcal{H}_2$. □

In what follows, we present a connection between $\varphi(\mathcal{H}_1[\mathcal{H}_2])$ and $\varphi(\mathcal{H}_2)$.

Theorem 1.11. *Let $\mathcal{H}_1 = \langle H_1, \cdot, e, ^{-1} \rangle$ and $\mathcal{H}_2 = \langle H_2, \cdot, e, ^{-1} \rangle$ be canonical hypergroups. Then, $\varphi(\mathcal{H}_1[\mathcal{H}_2]) = \varphi(\mathcal{H}_2)$.*

Proof. We remark that the elements of H_1 have periodicity of one; therefore, $exp\left(\mathcal{H}_1[\mathcal{H}_2]\right) = \exp\left(\mathcal{H}_2\right)$, and we obtain

$$\varphi(\mathcal{H}_1[\mathcal{H}_2]) = |\{x \in M : p(x) = \exp\left(\mathcal{H}_1[\mathcal{H}_2]\right)\}|$$
$$= |\{x \in H_2 : p(x) = \exp\left(\mathcal{H}_2\right)\}|$$
$$= \varphi\left(\mathcal{H}_2\right).$$

This completes the proof. □

Corollary 1.2. *If all the elements have periodicity equal to one, then it follows that* $\omega_{\mathcal{H}_1[\mathcal{H}_2]} = H_1 \cup H_2$, *which implies that* $\varphi\left(\mathcal{H}_1[\mathcal{H}_2]\right) = |H_1| + |H_2| - 1$.

Chapter 2

Introduction to Krasner Hyperrings

The more general structure that satisfies the ring-like axioms is the hyperring in the general sense: $(R, +, \cdot)$ is a hyperring if $+$ and $\cdot$ are two hyperoperations such that $(R, +)$ is a hypergroup and $\cdot$ is an associative hyperoperation, which is distributive with respect to $+$. There are different types of hyperrings. If only the addition $+$ is a hyperoperation and the multiplication $\cdot$ is a usual operation, then we say that R is an additive hyperring. A special case of this type is the hyperring introduced by Krasner [50]. Also, Krasner introduced a new class of hyperrings and hyperfields: the quotient hyperrings and hyperfields. In a long list of papers, Mittas studied first the hyperfield's additive part, i.e., the canonical hypergroup, as he named it, and then the hyperfield itself.

Throughout this chapter, by a "hyperring," we mean a Krasner hyperring.

2.1 Definition and Constructions of Krasner Hyperrings

We give first the definitions of a Krasner hyperring and a Krasner hyperfield.

Definition 2.1. A *Krasner hyperring* is an algebraic structure $(R, +, \cdot)$ which satisfies the following axioms:

(1) $(R, +)$ is a canonical hypergroup, i.e.:

    (i) for every $x, y, z \in R$, $x + (y + z) = (x + y) + z$;

(ii) for every $x, y \in R$, $x + y = y + x$;

(iii) there exists $0 \in R$ such that $0 + x = \{x\}$, for every $x \in R$;

(iv) for every $x \in R$, there exists a unique element $x' \in R$ such that $0 \in x + x'$

(we shall write $-x$ for x', and we call it the *opposite* of x);

(v) $z \in x + y$ implies $y \in -x + z$ and $x \in z - y$.

(2) $(R, \cdot)$ is a semigroup having zero as a bilaterally absorbing element, i.e., $x \cdot 0 = 0 \cdot x = 0$.

(3) Multiplication is distributive with respect to the hyperoperation $+$.

The following elementary facts follow easily from the axioms: $-(-x) = x$ and $-(x+y) = -x - y$, where $-A = \{-a \mid a \in A\}$. Also, for all $a, b, c, d \in R$, we have $(a + b) \cdot (c + d) \subseteq a \cdot c + b \cdot c + a \cdot d + b \cdot d$. In Definition 2.1, for simplicity of notations, we write sometimes xy instead of $x \cdot y$ and, in (iii), $0 + x = x$ instead of $0 + x = \{x\}$.

A Krasner hyperring $(R, +, \cdot)$ is called *commutative (with unit element)* if $(R, \cdot)$ is a commutative semigroup (with unit element).

Some students of Krasner, namely Mittas and Stratigopoulos have studied hyperrings and hyperfields. Others who made interesting contributions in relation to this topic include Corsini, Davvaz, Ch. Massouros, Nakassis, Vougiouklis, Konguetsof, Dramalidis, Spartalis, Pinotsis, Kemprasit, Stefanescu, Leoreanu, and Ameri. We pick up from their papers some constructions of hyperrings.

Example 2.1. Let $R = \{0, 1, 2\}$ be a set with the hyperoperation $+$ and the binary operation $\cdot$ defined as follows:

+	0	1	2
0	0	1	2
1	1	1	R
2	2	R	2

$\cdot$	0	1	2
0	0	0	0
1	0	1	2
2	0	2	2

Then, $(R, +, \cdot)$ is a Krasner hyperring.

Example 2.2. The first construction of a hyperring appeared in Krasner's paper [50], and it is the following one: Consider $(F, +, \cdot)$ as a field and G as a subgroup of $(F^*, \cdot)$, and take $F/G = \{aG \mid a \in F\}$

with the hyperaddition and the multiplication given by

$$aG \oplus bG = \{cG \mid c \in aG + bG\},$$
$$aG \odot bG = abG.$$

Then, $(F/G, \oplus, \odot)$ is a hyperring. If $(F, +, \cdot)$ is a unitary ring and G is a subgroup of the monoid $(F^*, \cdot)$ such that $xG = Gx$, for all $x \in F$, then $(F/G, \oplus, \odot)$ is a Krasner hyperring with identity.

Example 2.3. Let $(A, +, \cdot)$ be a ring and N a normal subgroup of its multiplicative semigroup. Then, the multiplicative classes $\overline{x} = xN$ $(x \in A)$ form a partition of R, and let $\overline{A} = A/N$ be the set of these classes. If for all $\overline{x}, \overline{y} \in \overline{A}$, we define

$$\overline{x} \oplus \overline{y} = \{\overline{z} \mid z \in \overline{x} + \overline{y}\}, \quad \text{and} \quad \overline{x} * \overline{y} = \overline{x \cdot y},$$

then the obtained structure is a Krasner hyperring.

Example 2.4. Let R be a commutative ring with identity. We set $\overline{R} = \{\overline{x} = \{x, -x\} \mid x \in R\}$. Then, $\overline{R}$ becomes a hyperring with respect to the hyperoperation $\overline{x} \oplus \overline{y} = \{\overline{x + y}, \ \overline{x - y}\}$ and the multiplication $\overline{x} \otimes \overline{y} = \overline{x \cdot y}$.

Example 2.5 (inspired by Lyndon [52, 53]). Let $(G, \cdot)$ be a group, $H = G \cup \{0, u, v\}$, where 0 is an absorbing element under multiplication and u, v are distinct orthogonal idempotents, with

$$a \cdot 0 = 0 \cdot a = 0,$$
$$u^2 = u,$$
$$v^2 = v,$$
$$uv = vu = 0,$$
$$ug = gu = u, \quad gv = vg = v \quad \text{for all } g \in G.$$

If we define the hypersum on H by

$$\begin{aligned} a + 0 &= \{a\} & &\text{for all } a \neq 0, \\ a + a &= \{a, 0\} & &\text{for all } a \in H, \\ a + b &= H \backslash \{a, b, 0\} & &\text{for all } a, b \in H \setminus \{0\} \quad \text{and} \quad a \neq b, \end{aligned}$$

then $(H, +, \cdot)$ is a Krasner hyperring.

Example 2.6 ([2]). If $(L, \wedge, \vee)$ is a relatively complemented distributive lattice, then we define a structure of Krasner hyperring on L by taking

$$a \oplus b := \{c \in L \mid a \wedge b = a \wedge c = b \wedge c\},$$
$$a \cdot b := a \vee b, \quad \text{for all } a, b \in L.$$

Example 2.7 ([70, 96]). Let $(G, \cdot)$ be a finite group with m elements, $m > 3$, and define a hyperaddition and a multiplication on $H = G \cup \{0\}$ by

$$a + 0 = 0 + a = \{a\} \quad \text{for all } a \in H,$$
$$a + a = \{a, 0\} \quad \text{for all } a \in G,$$
$$a + b = b + a = H \backslash \{a, b\} \quad \text{for all } a, b \in G, a \neq b,$$
$$a \odot 0 = 0 \quad \text{for all } a \in H,$$
$$a \odot b = a \cdot b \quad \text{for all } a, b \in G.$$

Then, $(H, +, \odot)$ is a Krasner hyperring.

Example 2.8 ([96]). If $(H, \leq, +)$ is a totally ordered group, then

$$x \oplus x = \{t \in H \mid t \leq x\} \quad \text{for all } x \in H,$$
$$x \oplus y = \{\max\{x, y\}\} \quad \text{for all } x, y \in H, \ x \neq y$$

defines a structure of canonical hypergroup on H. If $(H, +, \cdot)$ is a totally ordered ring (for example, $\mathbb{R}$), then $(H, \oplus, \cdot)$ is a Krasner hyperring.

Definition 2.2.

(1) A Krasner hyperring is called a *Krasner hyperfield* if $(R \setminus \{0\}, \cdot)$ is a group.
(2) A Krasner hyperring R is called a *hyperdomain* if R is a commutative Krasner hyperring with unit element, and $ab = 0$ implies that $a = 0$ or $b = 0$, for all $a, b \in R$.

Example 2.9. Let $F = \{0, 1\}$ be a set with the hyperoperation $+$ and the binary operation $\cdot$ defined as follows:

+	0	1
0	0	1
1	1	F

$\cdot$	0	1
0	0	0
1	0	1

Then, $(F, +, \cdot)$ is a Krasner hyperfield.

Example 2.10. Let $R = \{-1, 0, 1\}$ be a set with the hyperoperation $+$ and the binary operation $\cdot$ given as follows:

$+$	-1	0	1
-1	-1	-1	F
0	-1	0	1
1	F	1	1

$\cdot$	-1	0	1
-1	1	0	-1
0	0	0	0
1	-1	0	1

Then, $(F, +, \cdot)$ is a Krasner hyperfield.

Example 2.11 ([59]). Let $(H, +, \cdot)$ be a hyperfield. If we define a new hyperoperation on H as follows:

$$a \oplus b = (a + b) \cup \{a, b\}, \quad \text{if } a \neq -b, \ a, b \in H,$$
$$a \oplus (-a) = H \quad \text{for all } a \in H \setminus \{0\},$$
$$a \oplus 0 = 0 \oplus a = a \quad \text{for all } a \in H,$$

then $(H, \oplus, \cdot)$ is a new hyperfield. If $(H, +, \cdot)$ is a field, then $a \oplus b = \{a, b, a + b\}$, for $a \neq b, \ a, b \in H^*$.

2.2 Hyperideals

Definition 2.3. Let $(R, +, \cdot)$ be a Krasner hyperring and S be a non-empty subset of R. Then, S is said to be a *subhyperring* of R if $(S, +, \cdot)$ is itself a Krasner hyperring.

The subhyperring S of R is *normal* in R if and only if $x + S - x \subseteq S$, for all $x \in R$.

Definition 2.4. A *multiplicatively closed subset* (or *multiplicative subset*) is a subset S of a Krasner hyperring R such that the following two conditions hold: $xy \in S$, for all $x, y \in S$, and if $1 \in R$, then $1 \in S$.

Definition 2.5. A subhyperring I of a Krasner hyperring R is a *left* (*right*) *hyperideal* of R if $r \cdot a \in I$ ($a \cdot r \in I$), for all $r \in R$, $a \in I$. In particular, I is called a *hyperideal* if I is both a left and a right hyperideal.

It would be useful to have some criteria for deciding whether a given subset of a Krasner hyperring is a left (right) hyperideal or not. This is the purpose of the following lemma.

Lemma 2.1. *A non-empty subset I of a Krasner hyperring R is a left (right) hyperideal if and only if:*

(1) $a, b \in I$ *implies* $a - b \subseteq I$;
(2) $a \in I$, $r \in R$ *imply* $r \cdot a \in I$ $(a \cdot r \in I)$.

Proof.   It is straightforward.   □

Example 2.12. Let $(R, \cdot)$ be a semigroup with zero 0 such that $(R \setminus \{0\}, \cdot)$ is a group. Define the hyperoperation $+$ on R by

$$
x + y = \begin{cases}
y + x = \{x\} & \text{if } y = 0 \\
R \setminus \{x\} & \text{if } x = y \neq 0 \\
\{x, y\} & \text{if } x, y \in R \setminus \{0\} \quad \text{and} \quad x \neq y.
\end{cases}
$$

Then, $(R, +, \cdot)$ is a Krasner hyperring. Note that 0 is the zero of $(R, +, \cdot)$, and for each $x \in R$, x is the opposite of x in $(R, +)$. Since $Rx = R$, for all $x \in R \setminus \{0\}$, it follows that $\{0\}$ and R are the only hyperideals of $(R, +, \cdot)$.

Example 2.13. Define the hyperoperation $\oplus$ on the unit interval $[0, 1]$ by

$$
x \oplus y = \begin{cases}
\max\{x, y\} & \text{if } x \neq y \\
[0, x] & \text{if } x = y.
\end{cases}
$$

Then, $([0, 1], \oplus, \cdot)$ is a Krasner hyperring, where $\cdot$ is the usual multiplication. We can see that 0 is the zero of $([0, 1], \oplus, \cdot)$, and for every $x \in [0, 1]$, x is the opposite of x in $([0, 1], \oplus)$.

It is not difficult to check that for $a \in (0, 1]$, the intervals $[0, a]$ and $[0, a)$ are non-zero hyperideals of $([0, 1], \oplus, \cdot)$.

Now, suppose that I is a hyperideal of $([0, 1], \oplus, \cdot)$. Then, $0 \in I$. Let a be the supremum of I. If $a \in I$, then $[0, a] = a \oplus a \subseteq I \subseteq [0, a]$. This yields $I = [0, a]$. If $a \notin I$, then $a > 0$ and $I \subseteq [0, a)$. Assume that $m \in \mathbb{N}$ is such that $a - 1/m > 0$. Since a is the supremum of I, it follows that for every $n \in \mathbb{N}$, there exists an element $x_n \in I$ such that $a - 1/n \leq x_n < a$. This implies that

$$[0, a - 1/n] \subseteq [0, x_n] = x_n \oplus x_n \subseteq I,$$

for all $n \in \mathbb{N}$. Therefore, we can write

$$[0, a) = \bigcup_{\substack{n \in \mathbb{N} \\ n \geq m}} [0, a - 1/n] \subseteq I.$$

Hence, $I = [0, a)$. This shows that $\{[0, a] \,|\, a \in [0, 1]\} \cup \{[0, a) \,|\, a \in (0, 1]\}$ is the set of all hyperideals of the Krasner hyperring $([0, 1], \oplus, \cdot)$.

Definition 2.6. Let A and B be non-empty subsets of a hyperring R:

• The sum $A + B$ is defined by

$$A + B = \{x \mid x \in a + b, \text{ for some } a \in A, \ b \in B\}.$$

• The product AB is defined by

$$AB = \left\{ x \mid x \in \sum_{i=1}^{n} a_i b_i, a_i \in A, b_i \in B, n \in \mathbb{Z}^+ \right\}.$$

If A and B are hyperideals of R, then $A + B$ and AB are also hyperideals of R.

The following two corollaries are obtained directly from definitions.

Corollary 2.1. *Let A be a normal hyperideal of a Krasner hyperring R. Then:*

(1) $(A + x) + (A + y) = A + x + y$, *for all* $x, y \in R$;
(2) $A + x = A + y$, *for all* $y \in A + x$.

Proof. It is straightforward. □

Corollary 2.2. *Let A and B be hyperideals of a Krasner hyperring R with B normal in R. Then:*

(1) $A \cap B$ *is a normal hyperideal of A,*
(2) B *is a normal hyperideal of $A + B$.*

Proof. It is straightforward. □

Theorem 2.1. *Let R be a Krasner hyperring. Then, R is a ring if and only if $\langle 0 \rangle$ is a normal hyperideal of R.*

Proof. Suppose that $\langle 0 \rangle$ is a normal hyperideal of R. Then, for each $x \in R$, we have $x + \langle 0 \rangle - x \subseteq \langle 0 \rangle$. This implies that $x - x = 0$, for all $x \in R$.

Let x, y be two arbitrary elements of R. We consider the set $x + y$. If $a, b \in x + y$, then

$$a - b \subseteq (x + y) - (x + y) = (x - x) + (y - y) = 0 + 0 = \langle 0 \rangle.$$

Therefore, we get $a = b$. This means that $x + y$ is a singleton.

Conversely, if $x + y$ has only one element, for all $x, y \in R$, then $(R, +)$ is an abelian group. Moreover, for each $x \in R$, we have $x - x = 0$. This yields $x + \langle 0 \rangle - x \subseteq \langle 0 \rangle$. Therefore, $\langle 0 \rangle$ is normal in R. □

Definition 2.7. If I is a hyperideal of a Krasner hyperring R, then we define the relation $\equiv$ as follows:

$$x \equiv y \pmod{I} \quad \Leftrightarrow \quad x \in y + I.$$

Lemma 2.2. *The relation $\equiv$ defined in Definition 2.7 is an equivalence relation.*

Proof. (1) Suppose that $x \in R$. Since $x = x + 0 \in x + I$, it follows that $x \equiv x \pmod{I}$, i.e., $\equiv$ is reflexive.

(2) Let $x, y \in R$. If $x \equiv y \pmod{I}$, then $x \in y + I$, and so $x \in y + a$, for some $a \in I$. This yields $y \in x - a \subseteq x + I$, and hence $\equiv$ is symmetric.

(3) Now, suppose that $x, y, z \in R$ such that $x \equiv y \pmod{I}$ and $y \equiv z \pmod{I}$. Then, we have $x \in y + I$ and $y \in z + I$. Consequently, $x \in y + a$ and $y \in z + b$, for some $a, b \in I$. Thus, we can write $x \in z + a + b \subseteq z + I$. This means that $\equiv$ is transitive. □

We denote the equivalence class determined by the element $x \in R$ by $\overline{x}$.

Lemma 2.3. *If I is a hyperideal of a Krasner hyperring R, then $x + I = \overline{x}$.*

Proof. It is straightforward. □

Theorem 2.2. *Let R be a Krasner hyperring and I be a hyperideal of R. We define the hyperoperation $\oplus$ and the multiplication $\odot$ on the set $R/I = \{x + I \mid x \in R\}$ as follows:*

$$(x + I) \oplus (y + I) = \{z + I \mid z \in x + y\},$$
$$(x + I) \odot (y + I) = xy + I,$$

for all $x, y \in I$. Then, $(R/I, \oplus, \odot)$ is a Krasner hyperring.

Proof. The proof is left as an exercise to the reader. □

The Krasner hyperring R/I in Theorem 2.2 is called the *quotient hyperring* of R by I.

Theorem 2.3. *Let R be a Krasner hyperring and I be a hyperideal of R. Then, R/I is a ring if and only if I is a normal hyperideal of R.*

Proof. It follows by Theorems 2.1 and 2.2. □

Remark 2.1. Note that according to Theorem 2.2, a hyperideal I of a Krasner hyperring R gives us a Krasner hyperring, while according to Theorem 2.3, a normal hyperideal I gives us a ring (Figure 2.1).

2.3 Special Hyperideals

Definition 2.8. Let X be a subset of a Krasner hyperring R. Let $\{A_i \mid i \in J\}$ be the family of all hyperideals in R which contain X. Then, $\bigcap_{i \in J} A_i$ is called the *hyperideal generated by* X. This hyperideal is denoted by $\langle X \rangle$. If $X = \{x_1, x_2, \ldots, x_n\}$, then the hyperideal $\langle X \rangle$ is denoted $\langle x_1, x_2, \ldots, x_n \rangle$.

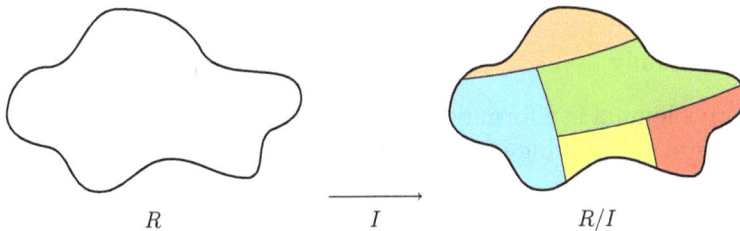

Fig. 2.1. If I is a normal hyperideal, then R/I is a ring.

Theorem 2.4. *Let R be a Krasner hyperring, $a \in R$ and $X \subset R$. Then:*

(1) *The principal hyperideal $\langle a \rangle$ is equal to the set*

$$\Big\{ t \mid t \in ra + as + na + k(a - a) + \sum_{i=1}^{m} r_i a s_i,$$

$$r, s, r_i, s_i \in R, m \in \mathbb{Z}^+ \ \text{and} \ n, k \in \mathbb{Z} \Big\}.$$

(2) *If R has a unit element, then*

$$\langle a \rangle = \Big\{ t \mid t \in k(a - a) + \sum_{i=1}^{m} r_i a s_i, \ r_i, s_i \in R, \ m, k \in \mathbb{Z}^+ \Big\}.$$

(3) *If a is at the center of R, then*

$$\langle a \rangle = \{ t \mid t \in ra + na + k(a - a), \ r \in R, n \in \mathbb{Z}^+ \},$$

where the center of R is the set $\{ x \in R \mid xy = yx \ \text{for all} \ y \in R \}$.

(4) *$Ra = \{ ra \mid r \in R \}$ is a left hyperideal in R, and $aR = \{ ar \mid r \in R \}$ is a right hyperideal in R. If R has a unit element, then $a \in aR \cap Ra$.*

(5) *If R has a unit element and a is at the center of R, then $Ra = \langle a \rangle = aR$.*

(6) *If R has a unit element and X is included at the center of R, then*

$$\langle a \rangle = \Big\{ t \mid t \in \sum_{i=1}^{m} r_i x_i, \ r_i \in R, x_i \in X, \ m \in \mathbb{Z}^+ \Big\}.$$

Proof. The proof is straightforward. □

Definition 2.9. A proper hyperideal M of R is a *maximal hyperideal* of R if the only hyperideals of R that contain M are M itself and R.

Zorn's lemma is a form of the axiom of choice, which is technically very useful for proving existence theorems.

Theorem 2.5. *Let R be a commutative Krasner hyperring with a unit element, and let I be a proper hyperideal of R. Then, there exists a maximal hyperideal of R containing I.*

Proof. From Zorn's lemma, it follows directly that every Krasner hyperring has a maximal ideal. The proof is left as an exercise to the reader. □

Definition 2.10. A proper hyperideal P of a Krasner hyperring R is called *prime* if for every pair of hyperideals A and B of R,

$$AB \subseteq P \implies A \subseteq P \quad \text{or} \quad B \subseteq P.$$

Lemma 2.4. *Let R be a commutative Krasner hyperring and P be a hyperideal of R such that $P \neq R$. Then, P is a prime hyperideal of R if and only if for every $a, b \in R$*

$$ab \in P \implies a \in P \quad \text{or} \quad b \in P. \tag{2.1}$$

Proof. Suppose that (2.1) holds. If A, B are hyperideals of R such that $AB \subseteq P$ and $A \nsubseteq P$, then there exists an element $a \in A \setminus P$. For every $b \in B$, $ab \in AB \subseteq P$, whence $a \in P$ or $b \in P$. Since $a \notin P$, we must have $b \in B$. This yields $B \subseteq P$, and consequently, P is a prime hyperideal.

Conversely, suppose that P is an arbitrary hyperideal of R and $ab \in P$. Then, the principal ideal $\langle ab \rangle$ is contained in P. Since R is commutative, it follows that $\langle ab \rangle \subseteq \langle a \rangle \langle b \rangle$, whence $\langle a \rangle \langle b \rangle \subseteq P$. If P is prime, then either $\langle a \rangle \subseteq P$ or $\langle b \rangle \subseteq P$, whence $a \in P$ or $b \in P$, respectively. □

Theorem 2.6. *If R is a commutative Krasner hyperring with a unit element, then each maximal hyperideal M of a hyperring R is a prime hyperideal.*

Proof. Let $x, y \in R$ be elements such that $xy \in M$ and $x \notin M$. The set $M + xR = \cup \{a + b \mid a \in M, b \in xR\}$ is a hyperideal of R containing both M and xR. Therefore, we have $x \in M + xR$ and $x \notin M$. Hence, we conclude that $1 \in M + xR$, and so $1 \in a + xr$, for some $a \in M$ and $r \in R$. Therefore, we obtain $y \in y(a + xr) = ya + yxr \subseteq M + M \subseteq M$. Hence, $y \in M$. □

Theorem 2.7. *Let R be a commutative Krasner hyperring with a unit element and $M \neq R$ be a hyperideal of R. Then, M is maximal if and only if R/M is a hyperfield.*

Proof. Suppose that M is a maximal hyperideal, and let $a \in R$ but $a \notin M$. It suffices to show that $a + M$ has a multiplicative inverse. Consider

$$U = \{u \in ar + b \mid r \in R, b \in M\}.$$

This is a hyperideal of R that contains M properly. Since M is maximal, we have $U = R$. Thus, $1 \in U$, so there exist $c \in R$ and $d \in M$ such that $1 \in ac + d$. Then, $1 + M \in (ac + d) + M = ac + (d + M) = ac + M = (a + M) \odot (c + M)$.

Now, suppose that R/M is a hyperfield and U is a hyperideal of R that contains M properly. Let $a \in U$, but $a \notin M$. Then, $a + M$ is a non-zero element of R/M, and so there there exists an element $b + M$ such that $(a + M) \odot (b + M) = 1 + M$. Since $a \in U$, we have $ab \in U$. Also, we have $1 - ab \subseteq M \subseteq U$. So, $1 \in (1 - ab) + ab \subseteq U$, which implies that $U = R$.   □

Theorem 2.8. *Let R be a commutative Krasner hyperring with a unit element and $P \neq R$ be a hyperideal of R. Then, P is prime if and only if R/P is a hyperdomain.*

Proof. Suppose that R/P is a hyperdomain and $ab \in P$. Then, $(a + P) \odot (b + P) = ab + P = P$. So, either $a + P = P$ or $b + P = P$, i.e., either $a \in P$ or $b \in P$. Hence, P is prime.

Now, suppose that P is prime and $(a + P) \odot (b + P) = 0 + P = P$. Then, $ab \in P$; therefore, $a \in P$ or $b \in P$. Thus, one of $a + P$ or $b + P$ is zero.   □

Definition 2.11. For a Krasner hyperring R, we define the *Jacobson radical* $J(R)$ of R as the intersection of all maximal hyperideals of R. If R does not have any maximal hyperideal, then we define $J(R) = R$.

Theorem 2.9. *Let R be a commutative Krasner hyperring with a unit element and I be a hyperideal of R. Then, $I \subseteq J(R)$ if and only if every element of $1 + I$ is invertible.*

Proof. Let $I \subseteq J(R)$, and suppose that there exists $x \in 1 + I$ such that x is not invertible. Clearly, $x \in 1 + a$, for some $a \in I$. Since x is not invertible, there exists a maximal hyperideal M such that $x \in M$. But $x \in 1 + a$ implies that $1 \in x - a \subseteq M$, which is a contradiction. Hence, every element of $1 + I$ is invertible.

Conversely, suppose that any element of $1 + I$ is invertible but $I \nsubseteq J(R)$. Thus, $I \nsubseteq M$, for some maximal hyperideal M of R. Then, there exists $a \in I \setminus M$. Therefore, $\langle M, a \rangle = R$. So, $1 \in m + ra$, for some $r \in R$ and $m \in M$; hence, $m \in 1 - ra \subseteq 1 + I$. Thus, m is invertible, which is a contradiction to the maximality of M. □

Corollary 2.3. *Let R be a commutative Krasner hyperring with a unit element and $U(R)$ be the set of all invertible elements in R. Then, an element $a \in R$ belongs to $J(R)$ if and only if $1 - ba \subseteq U(R)$ for all $b \in R$.*

Proof. It is straightforward. □

Definition 2.12. Let R be a Krasner hyperring. A finite chain of n normal hyperideals $A_0, A_1, \ldots, A_{n-1}$ of R,

$$R = A_0 \supset A_1 \supset \cdots \supset A_n = 0,$$

is called a *composition series of length n* for R, provided that A_{i-1}/A_i is *simple* $(i = 1, \ldots, n)$, i.e., if each term in the chain is maximal in its predecessor.

We shall consider a generalization of the Jordan–Holder theorem for Krasner hyperrings.

Theorem 2.10 (Jordan–Holder theorem). *If a Krasner hyperring R has a composition series, then any two composition series for R are equivalent.*

Proof. If R has a composition series, then denote by $c(R)$ the minimum length of such series for R. We shall prove by induction on $c(R)$. Clearly, if $c(R) = 1$, there is nothing to prove. So, assume that $c(R) = n > 1$ and that any hyperring with a composition series of length smaller than n has all of its composition series equivalent. Let

$$R = A_0 \supset A_1 \supset \cdots \supset A_n = 0 \qquad (2.2)$$

be a composition series of length n for R and

$$R = B_0 \supset B_1 \supset \cdots \supset B_m = 0 \qquad (2.3)$$

be a second composition series for R. If $A_1 = B_1$, then by inductive hypotheses and since $c(A_1) \leq n - 1$, if follows that the two series are equivalent. So, we may assume that $A_1 \neq B_1$. Then, since A_1 is a maximal hyperideal of R, it follows that $A_1 + B_1 = R$; hence,

$$R/A_1 = (A_1 + B_1)/A_1 \cong B_1/(A_1 \cap B_1)$$

and

$$R/B_1 = (A_1 + B_1)/B_1 \cong A_1/(A_1 \cap B_1).$$

Thus, $A_1 \cap B_1$ is maximal in both A_1 and B_1. Now, $A_1 \cap B_1$ has the composition series

$$A_1 \cap B_1 = C_0 \supset C_1 \supset \cdots \supset C_k = 0.$$

So,

$$A_1 \supset C_0 \supset C_1 \supset \cdots \supset C_k = 0 \quad \text{and} \quad B_1 \supset C_0 \supset C_1 \supset \cdots \supset C_k = 0$$

are the composition series for A_1 and B_1, respectively. Since $c(A_1) < n$, every two composition series for A_1 are equivalent, it follows that the following series are equivalent:

$$R = A_0 \supset A_1 \supset \cdots \supset A_n = 0 \quad \text{and}$$
$$R = A_0 \supset A_1 \supset C_0 \supset \cdots \supset C_k = 0.$$

In particular, $k < n-1$, so clearly $c(B_1) < n$. Thus, by our induction hypothesis, every two composition series for B_1 are equivalent. Thus, the two series,

$$R = B_0 \supset B_1 \supset \cdots \supset B_m = 0 \quad \text{and}$$
$$R = B_0 \supset B_1 \supset C_0 \supset \cdots \supset C_k = 0,$$

are equivalent. Since

$$R/A_1 \cong B_1/C_0 \quad \text{and} \quad R/B_1 \cong A_1/C_0,$$

it follows that the series (2.2) and (2.3) are equivalent.   □

2.4   The Existence of Non-quotient Krasner Hyperrings

The existence of non-quotient Krasner hyperrings and hyperfields plays a very important and quite determinative role in the independence, self-sufficiency, and further development of the theory of hyperrings and hyperfields. The existence of such hyperrings and hyperfields was proved in Refs. [58–60]. Moreover the construction methods used there endowed this theory with new interesting classes of hyperrings and hyperfields. The results in this paragraph were obtained by Massouros *et al.* [58–60, 70].

Let $(R, +, \cdot)$ be a ring and G a subset of R. G is called a *multiplicative subgroup* of R if and only if $(G, \cdot)$ is a group. Moreover, if G is such that $R = RG$ and $rG = Gr$, for all r in R, then G is called a *normal subgroup* of R. We remark that only rings with an identity element admit normal subgroups. As already mentioned, a normal subgroup G of R induces an equivalence relation P in R and a partition of R in equivalence classes which inherits a hyperring structure from R. Hyperrings obtained via this construction are called *quotient hyperrings* and are denoted by R/G or R/P. The following results answer the questions:

(1) Are all hyperrings embeddable in quotient hyperrings?
(2) Are all hyperrings generated by a set of orthogonal idempotents embeddable into quotient hyperrings?
(3) Are all primitive hyperrings embeddable into quotient hyperrings?
(4) Are all hyperfields embeddable into quotient hyperrings?

Actually, as already mentioned, one can generalize the notion of a quotient hyperring as follows: Assume that P is an equivalence relation in R, and for each r in R, $P(r)$ is the equivalence class to which r belongs. Assume that for all a and b in R, $P(a)P(b)$ is a subset of $P(ab)$. Let R/P be the set of all equivalence classes in R, and for each subset X of R, let the P-closure $P(X)$ of X be the set of all equivalence classes that intersect X. Clearly, multiplication in R induces an associative multiplication in R/P, provided that the product of any two classes $P(a)$ and $P(b)$ is defined to be the P-closure of the set in R. Similarly, Rs addition induces a commutative hyperoperation $\oplus$ in R/P, provided that one defines $P(a) \oplus P(b)$ to be the P-closure of the set $P(a) + P(b)$ in R. If P is such that $(R/P, \oplus, \cdot)$ is a hyperring (as a rule, it is not), then R/P is called a *partition hyperring*. One can verify that $(R/P, \oplus, \cdot)$ is a hyperring if and only if P satisfies the following conditions:

(1) $P(0)$ is a bilateral ideal of R such that for every a in R, $a + P(0)$ is a subset of $P(a)$;
(2) for every a in R, $P(-a) = -P(a)$;
(3) P is such that $\oplus$ is associative and the multiplication in R/P is left and right distributive over $\oplus$.

Clearly, condition (3) is a restatement of the problem, and it would be interesting to derive conditions on P that ensure that $(R/P, \oplus, \cdot)$ is a hyperring. In light of this, Krasner's original construction can be seen as a proof of the fact that if P is induced by a normal subgroup G, then R/P inherits a hyperring structure from R. We shall show that there are hyperrings that are not embeddable in partition hyperrings and that there are partition hyperrings that are not embeddable in quotient hyperrings. It turns out, though, that the class of partition hyperfields and the class of quotient hyperfields are one and the same.

Massouros introduced a hyperring which is not isomorphic to a quotient hyperring because it contains more than one right unit (a quotient hyperring R/G has a single unit, the G's image). Nevertheless, a quotient hyperring R/G can have subhyperrings that do not contain the G's image. Thus, such a subhyperring H could contain more than one unit, either on the left or on the right (evidently, not both). Indeed, Massouros' idea of construction was to consider a ring A with an identity element 1 and to define a ring $R = A x A$, in which the addition is defined component by component and the multiplication is defined via the following rule: $(a, b) \cdot (c, d) = (a(c+d), b(c+d))$. If we now set $G = \{(1, 0), (-1, 0)\}$, then G is not a normal subgroup of R (it fails to satisfy $rG = Gr$, for all r in R). Nevertheless, G induces an equivalence relation P in R such that R/P inherits a hyperring structure from R. We observe that $rG = -rG$ in R/P, and R/P has more than one right unit (all rG with $r = (a, b)$ and $a + b = 1$).

As already mentioned, the existence of multiple units from the right shows that the hyperring in question is not isomorphic either to quotient hyperrings or to quotient subhyperrings that contain the image of the normal group which induces the hyperring structure. However, there exist quotient subhyperrings that do not contain a unit element, and the above construction is embeddable in a quotient hyperring. Indeed, assume that for every semigroup S and every ring A, $A[S]$ is the semigroup ring from S over A, i.e., the set of all mappings from S to A that have finite support. This set can be endowed with a ring structure where the algebra of a semigroup over a field is defined. Indeed, for any two such functions f and g, it suffices to define:

$(f + g)(s) = f(s) + g(s)$, for every s in S; and
$(fg)(r) = \sum f(s)g(t)$, where $r \in S$ and (s, t) ranges over all pairs such that $st = r$.

We observe that for every subsemigroup T of S, the elements of $A[T]$ can be identified with the elements of $A[S]$ whose support is a subset of T, i.e., $A[T]$ can be isomorphically embedded in $A[S]$. Let X be a left zero semigroup (which means that $xy = x$, for all x and y in X) of at least two elements. Let X^e be the smallest semigroup with an identity element e that contains X, and assume that A has an identity element 1, such that $1 + 1$ is not zero. We observe that A is isomorphic to $A[\{e\}]$ and can be isomorphically mapped onto $A[X^e]$ (identifying each a in A with the function that maps X to $\{0\}$ and e to a); therefore, $F = \{-1, 1\}$ is a normal subgroup of $A[X^e]$. If Y is any two-element subset of X, then we have:

- $A[Y]$ is isomorphic to Massouros' ring and isomorphic to a subring of $A[X^e]$ (Y is a subsemigroup of X^e).
- F induces a partition of $A[Y]$ such that $A[Y]/F$ is isomorphic to a subhyperring of $A[X^e]/F$.
- $A[Y]/F$ is isomorphic to Massouros' hyperring.

We remark that F is not embeddable in $A[Y]$. However, if g maps Y onto $\{0, 1\}$, then F introduces in $A[Y]$ the same partition as $\{-g, g\}$ in $A[Y]$, and this group is isomorphic to the group that Massouros used in order to partition $A[Y]$. In what follows, we propose to use the following symbols and terminology:

(1) H^* will represent a hyperring whose elements are $0^*, a^*, b^*, \ldots$.
(2) R/P will represent a partition hyperring (it is assumed that P is such that R/P inherits a hyperring structure from R).

(3) If H' is a subhyperrring of R/P, we denote its elements by $0', a', b', \ldots$. If H^* and H' are isomorphic, then we assume that the images of $0^*, a^*, b^*, \ldots$ are $0', a', b', \ldots$, respectively. We note that a' can also be seen as a subset of R since it is a P equivalence class.

(4) An equivalence P is said to be *induced by a group G* if and only if the classes of P are of the form rG and G is a multiplicative subgroup of R.

The following two propositions link the cardinality of $a^* \oplus b^*$ to the cardinality of b' as a subset of R (clearly, it is assumed that H^* is embeddable in a partition hyperring, R/P). They are a blueprint for constructing counterexamples and for proving non-embeddability in partition hyperrings. The second proposition is, after all, a "counting lemma"; therefore, it provides a natural method for constructing counterexamples.

Theorem 2.11. *If P is an equivalence relation that induces a hyperring structure in R/P, then $I = P(0)$ is an ideal of R. Furthermore, $a + I$ is a subset of $P(a)$, for every a in R, and P induces a partition $P^* = P/I$ over $R^* = R/I$. Finally, R/P and R^*/P^* are isomorphic hyperrings.*

Corollary 2.4. *If a hyperring H^* is embeddable in a partition hyperring R/P, then we can assume without loss of generality that $P(0) = \{0\}$.*

Theorem 2.12. *Assume that a hyperring H^* is embeddable in a partition hyperring R/P for which $P(0) = \{0\}$, and assume that there are two elements a^* and b^* in H^* such that for every $c^* \in a^* \oplus b^*$, $c^* \oplus (-c^*)$ and $b^* \oplus (-b^*)$ have only 0^* in common. Then, the cardinality of b' cannot exceed the cardinality of $a^* \oplus b^*$. (Clearly, b' is the image of b^*, and when we speak about its cardinality, we view b' as a subset of R.)*

Proof. Indeed, if a is an element in a' and if b_1 and b_2 are distinct elements of b', then $a + b_1$ and $a + b_2$ belong to different equivalence classes in R/P. Therefore, there is an injection from b' into $a' \oplus b'$ in R/P, and as a result, there is an injection from b' into $a^* \oplus b^*$. This injection is not always a surjection because the element a of a' is arbitrary but fixed. $\qquad\square$

In the case when P is induced by a group G, we obtain that $aG + bG = (a + bG)G$. Therefore, the mapping we described in the above proof is onto.

Corollary 2.5. *Under the assumptions in Theorem 2.12, if P is induced by a group, then b' and $a^* \oplus b^*$ have the same cardinality.*

Now, we shall see how the above propositions can be used in the construction of counterexamples. Theorem 2.12 and its corollary can be used in the construction of hyperrings that are not embeddable in quotient hyperrings.

Theorem 2.13. *There are partition hyperrings that are not embeddable in quotient hyperrings.*

Proof. Clearly, commutative hypergroups can be seen as hyperrings where every product is zero. Let $\mathbb{Q}$ be the set of all rational numbers, and let L be the set of all irreducible fractions of the form k/m with $m = 1$ or $m = 2$. Clearly, L is an additive subgroup of $\mathbb{Q}$, and if it is equipped with the type of multiplication mentioned above (all products being zero), then L is a ring. Let P be defined as follows: i/k and j/m are equivalent if and only if either $k = m = 2$ or $i + j = 0$. Then, L/P is a partition hyperring H^* whose elements are $0^* = \{0\}$, $x^* = \{(2i + 1)/2 \mid i = 0, -1, 1, -2, 2, -3, 3, \ldots\}$, and $d^*(i) = \{-i, i\}$, for $i = 1, 2, \ldots$.

One can prove that L/P is not embeddable in a quotient hyperring R/G by way of contradiction. Indeed, Corollary 2.5 can be used then to prove that for each i, $d'(i)$ has two elements and their sum is zero. Furthermore, we can prove by induction that these two elements can be chosen in such a way that $d^*(i) = \{-id, id\}$ for $i = 1, 2, \ldots$. If L/P were embeddable into R/G, then x' would be of the form $x' = xG$, for every x in x'. If it were also true that $x + x = 0$, then for every y in x', $y + y = 0$. Furthermore, since for every i, $d^*(i)$ is in $x^* \oplus x^*$, we can prove that the above d is of the form $x_1 + y_1$, where x_1 and y_1 are in x'. However, $d + d$ is not zero, while, under our assumptions, $(x_1 + y_1) + (x_1 + y_1)$ is. Therefore, if x is in x', $x + x$ cannot be zero.

Furthermore, we observe that since $x' + x' = \{0', d'(1), d'(2), \ldots\}$, $x + x$ must belong to some $d'(i)$. This i cannot be even because if i were equal to $2m$, then both $x + md$ and $x - md$ would belong to x' $(x^* \oplus d^*(m) = x^*)$. But then, either $x + md$ or $x - md$ must satisfy $t + t = 0$, in contradiction to what we just proved. Thus, there is an m

such that $x + x$ is in $d'(2m+1)$. If $x + x = (2m+1)d$, then it suffices to take $z = x - md$ in order to obtain an element z in x' such that $z + z = d$. If, on the other hand, $x + x = -(2m+1)d$, one can achieve the same result by taking $z = (m+1)d - x$. Thus, we can always assume that the element x we chose satisfies $x + x = d$.

It ensues then that $x' = \{-x, x, -3x, 3x, -5x, 5x, \dots\}$. Indeed, if y is in x', then either $y + y = (2m+1)d$ or $y + y = -(2m+1)d$ for a well-chosen m. In the first instance, it suffices to consider t, $t = y - (2m+1)x$. In the second, $t = y + (2m+1)x$. In both cases, $t + t = 0$, and we can prove that t cannot belong either to x' or to any $d'(i)$, $i = 1, 2, \dots$. Hence, $t = 0$; therefore, $x' = \{-x, x, -3x, 3x, -5x, 5x, \dots\}$.

Finally, we observe that if $x' = xG = (3x)G$, then $x = (3x)g$, for some g in G. However, since $x' = xG$, xg must be of the form $(2k+1)x$. Thus, $x = 3(2k+1)x$, which implies that $(3k+1)d = (6k+2)x = 0$. But $3k+1$ is not zero, and if its absolute value is n, then $(3k+1)d$ belongs to $d'(n)$. Therefore, the above line of reasoning produces a contradiction, which implies that H^* is not embeddable in a quotient hyperring. □

While the above construction shows quite conclusively that there are hyperrings that are not embeddable in quotient hyperrings, it leaves open the possibility that all hyperrings generated by a set of multiplicative idempotents may be embeddable in quotient hyperrings.

Theorem 2.14. *Let T^* be a multiplicative group, and let H^* be the disjoint union of $\{0^*, u^*, v^*\}$ and T^*. Then, H^* can be endowed with a hyperring structure if one defines an hyperaddition $\oplus$ and a multiplication as follows:*

(1) *for every a^* in H^*, $a^* \oplus 0^* = 0^* \oplus a^* = \{a^*\}$;*
(2) *for every a^* other than zero, $a^* \oplus a^* = \{0^*, a^*\}$;*
(3) *for all distinct a^* and b^*, $a^* \oplus b^* = H^* \setminus \{a^*, b^*, 0^*\}$, provided that neither a^* nor b^* is 0^*;*
(4) *for every a^* in H^*, $a^* 0^* = 0^* a^* = 0^*$;*
(5) *$u^* u^* = u^*$, $v^* v^* = v^*$ and $u^* v^* = v^* u^* = 0^*$;*
(6) *for every t^* in T^*, $u^* t^* = t^* u^* = u^*$ and $v^* t^* = t^* v^* = v^*$;*
(7) *the multiplication of H^* and T^* are identical over T^*.*

Proof. The proof of this theorem is quite straightforward, albeit long. □

The important part is that we can prove the following theorem using Theorem 2.14.

Theorem 2.15. *If H^* is embeddable in a partition hyperring R/P, then the following assertions hold:*

(1) $u' \cup 0'$ *and* $v' \cup 0'$ *are finite fields, viewed as subsets of R.*
(2) u' *and* v' *are isomorphic to subgroups of* T^*.
(3) *The isomorphic images* $f(u')$ *and* $f(v')$ *of* u' *and* v' *are normal subgroups of* T^*.
(4) *There are homomorphisms from* u' *onto* $T^*/f(v')$ *and from* v' *onto* $T^*/f(u')$.

Before we delineate a proof of the theorem, let us observe that one, among many, way of constructing hyperrings not embeddable in quotient hyperrings is to let T^* have a prime number element q such that $q + 1$ is not a power of 2, e.g., $q = 5$. Indeed, if Theorem 2.15 holds, then either u' or v' would be isomorphic to T^*, and we would have a finite field of $q + 1$ elements. Since the latter is impossible, we have produced a class of hyperrings that are not embeddable in partition hyperrings. *A fortiori*, they cannot be embedded in quotient hyperrings.

Proof. If H^* is embeddable in a partition hyperring R/P, then we could use Theorem 2.12 to prove that the images u' and v' of u^* and v^*, respectively, are finite subsets of R having at most q elements. Moreover, since u^* and v^* are multiplicative idempotents, it follows that u' and v' are multiplicative semigroups of R. We can construct a semigroup homomorphism f that maps $u'xv'$ onto T^* by defining f as follows: $f(u, v) = t^*$ if and only if $u+v$ is a member of t' (as already indicated, t' is the isomorphic image of t^* when H^* is mapped onto a subhyperring of R/P). Since $u^*v^* = v^*u^* = 0^*$ and $0'$ has a single element, R's zero, for any two pairs (u_1, v_1) and (u_2, v_2) from $u'xv'$, $(u_1 + v_1)(u_2 + v_2)$ is equal to $u_1u_2 + v_1v_2$. Therefore, $f(u_1u_2, v_1v_2) = f(u_1, v_1)f(u_2, v_2)$, and f is a semigroup homomorphism from $u'xv'$ onto T^*.

Now, let (u_0, v_0) be a multiplicative idempotent in $u'xv'$ (such an idempotent exists because $u'xv'$ is finite). Then, if $u = u_0$, we obtain a homomorphism $f(u_0, v)$ that maps v' onto a subset $f(v')$ of T^*. Similarly, if $v = v_0$, we obtain a homomorphism $f(u, v_0)$, that maps u' into a subset $f(u')$ of T^*. It is then elementary to check that $f(u_0, v)$ and $f(u, v_0)$ are injections and all finite semigroups of a group are groups. It ensues that u', v', and $u'xv'$ are multiplicative groups and that f is a group homomorphism from $u'xv'$ onto T^*. It also follows that (u_0, v_0) is the unity of $u'xv'$ and that $\{u_0\}xv'$ and $u'x\{v_0\}$ are normal subgroups of $u'xv'$. Then, u' is isomorphic to $u'xv'/\{u_0\}xv'$ and u' is homomorphic to $T^*/f(v')$. Similarly, v' is homomorphic to $T^*/f(u')$. It is easy now to check that $u' \cup \{0\}$ and $v' \cup \{0\}$ are finite fields (u' and v' are finite groups, $u' + u' = u' \cup \{0\}$, and $v' + v' = v' \cup \{0\}$. We also note that if the cardinality of T^* is a prime number q, then either u' or v' is isomorphic to T^*. Indeed, if $f(u')$ is not T^*, it must be equal to $\{e^*\}$, where e^* is the identity of T^*. But then, the finite group v' is isomorphic to $f(v')$, which is a subgroup of T^*, and homomorphic to $T^*/\{e^*\}$, and the only way this can happen is if v' has q elements.   □

A similar construction can be used in order to show that there exist hyperfields that are not embeddable into partition hyperrings. Given that each hyperfield is an irreducible and faithful module over itself, it follows that there are primitive rings that are not embeddable into partition hyperfields. Indeed, let T^* be any finite group of $m > 3$ elements, and define a hyperfield structure over $H^* = T^* \cup \{0*\}$ as follows:

(1) $a^*0^* = 0^*a^* = 0^*$, for every a^* in H^*;
(2) $a^* \oplus 0^* = 0^* \oplus a^* = \{a^*\}$, for every a^* in H^*;
(3) $a^* \oplus a^* = \{a^*, 0^*\}$, for every a^* in T^*;
(4) $a^* \oplus b^* = b^* \oplus a^* = T^* \setminus \{a^*, b^*\}$, for every a^* and b^* in T^*, provided that a^* and b^* are distinct.

Structures that satisfy such properties are fairly common. Indeed, it suffices to consider the field of complex numbers $\mathbb{C}$ and the multiplicative group $\mathbb{R}^*$ of all nonzero reals in order to obtain a quotient hyperfield $\mathbb{C}/\mathbb{R}^*$ that has properties (3) and (4). We can prove the following result.

Theorem 2.16. *If the above constructed H^* is embeddable in a partition hyperring R/P, and if H' is the isomorphic image of H^*, then the following statements hold:*

(1) *The isomorphism maps the unit e^* of T^* onto a finite multiplicative subgroup of R that will be called e' in what follows.*

(2) *If H_1 is the subset of R that corresponds to H', then e' and P induce the same partition on H_1.*

(3) *$e' \cup \{0\}$ is a field of $m-1$ elements, while H_1 is a field of $m(m-2)+1 = (m-1)^2$ elements.*

Note that if (1)–(3) hold, then we can choose T^* in such a way that H^* cannot be embeddable in a partition hyperring. Indeed, all finite fields are commutative, their cardinality is a power of a prime, and the multiplicative group of their non-zero elements is cyclic. One can choose then either m or the structure of T^* in such a way that H^* is not embeddable into a partition hyperring.

Proof. By Theorem 2.12, all non-zero elements of H' correspond to finite subsets of R with at most $m-2$ elements. Let H_1 be the union of all these subsets. We start by observing that e' is a finite set multiplicatively closed, without divisors of zero ($e'e' = e'$). Furthermore, we have $e' \oplus e' = \{e', 0'\}$, and so if x and y are distinct elements of e', then $x - y$ is in e'. It follows that for every a in e', the mappings $x \longmapsto ax$ and $x \longmapsto xa$ are injections from e' into e'. Therefore, e' is a group. The same reasoning can be used for $H_1 \setminus \{0\}$. Since $H' \setminus \{0\}$ is isomorphic to T^*, $H_1 \setminus \{0\}$ has no divisors of zero and is multiplicatively closed. If x and y are distinct elements of H_1, $x - y$ also belongs to H_1, and we deduce that for every a in $H_1 \setminus \{0\}$, the mappings $x \longmapsto ax$ and $x \longmapsto xa$ are injections. Since H_1 is finite, it follows that $H_1 \setminus \{0\}$ is a group, e' is a subgroup of $H_1 \setminus \{0\}$, and these two groups share the same identity element e. Let x' be any non-zero element of H'. Since H' is a hyperfield, there is a y' such that $y'x' = x'y' = e'$ in H'. It ensues that if y is any element of y', then yx' and $x'y$ are subsets of e' in H_1. The inverse x of y in $H_1 \setminus \{0\}$ is clearly an element of x'. Multiplying by x, we obtain that x' is a subset of xe' and $e'x$ in H_1. On the other hand, since $x'e' = e'x' = x'$ in H', it follows that xe' and $e'x$ must be subsets of x' in H_1. Hence, $x' = xe = ex$, for some x in x'. However, if this property is true for one x in x', it is true for every x in x' (it suffices to remark that e' is a group). Since P is induced by e' over H_1, each class in H_1 has exactly $m-2$ elements (Theorem 2.12), except for $0'$, which contains only 0. Therefore, H_1 is a field of $m(m-2)+1$

elements, while $e' \cup \{0\}$ is a field ($e^* \oplus e^* = \{e^*, 0^*\}$) of $(m-2)+1$ elements.   □

We shall construct a class of hyperfields which contains hyperfields that are not isomorphic to quotient ones. The idea here is to take a group G and introduce a hyperfield structure over $H = G \cup \{0\}$ as follows:

(1) for every h in H, $h \oplus 0 = 0 \oplus h = \{h\}$;
(2) $g_1 \oplus g_2 = \{g_1, g_2\}$, for every two distinct elements of G;
(3) $g \oplus g = H \setminus \{g\}$, for every g in G.

One can prove that G can be chosen in such a way that H is not embeddable in a quotient hyperfield. Indeed, we have the following theorem.

Theorem 2.17. *If G is not trivial and $gg = e$ for every g in G, then H is not isomorphic to a quotient hyperfield.*

Proof. This can be proven by way of contraposition. If H was isomorphic to a quotient hyperfield F/Q, then:

(1) since $gg = e$ in G, Q contains all squares in F other than zero;
(2) since for each element g in G, $g \oplus g = H \setminus \{g\}$, it follows that $Q = -Q$ and $Q + Q = F \setminus Q$.

However, if all the squares of F and their opposites are in Q, then we obtain a contradiction. If the characteristic of F is not two, then each element of F is the difference of two squares. If the characteristic of F is two, then the sum of two squares is a square. Otherwise, $Q + Q$ cannot be equal to $F \setminus Q$.   □

Finally, one can prove that if a Cartesian product of hyperrings is embeddable in a quotient hyperring, then every term of the product which is a hyperfield must be isomorphic to a quotient hyperfield. Thus, one can produce hyperrings that are not embeddable in quotient hyperrings.

Therefore, the structures and counterexamples presented in this section proved that the theory of hyperrings is not a straightforward extension of ring theory.

2.5   Hypervaluations

In this section, all Krasner hyperrings are considered commutative, and we define a hypervaluation on a commutative Krasner hyperring. For this, as in the classical case, we need a mapping from R onto an ordered group G. If the element $\infty \notin G$, then we define $a \cdot \infty = \infty \cdot a = \infty \cdot \infty = \infty > a$, for all $a \in G$. Then, we prove some interesting results concerning this concept.

We say that an arbitrary group G is partially ordered by $\leq$ if $(G, \leq)$ is a poset in which $a < b$ implies $ga < gb$ and $ag < bg$, for all $g \in G$. We see that if $a < a'$ and $b < b'$, then $ab < a'b'$. Consequently, G has a submonoid $P = \{g \in G \mid 1 \leq g\}$ called the positive cone. The following properties are straightforward, writing P^{-1} for $\{a^{-1} \mid a \in P\}$:

(1) $P \cap P^{-1} = \{1\}$.
(2) If $\leq$ is a total order, then $P \cup P^{-1} = G$.

Theorem 2.18. *If G is a totally ordered group, then $G_\infty = G \cup \{\infty\}$ is a hyperring with the hyperoperation $\oplus$ having the following properties:*

(1) $a < b \implies a \oplus b = \{a\}$, *for all $a, b \in G_\infty$;*
(2) $a \oplus a = \{g \in G_\infty \mid a \leq g\}$,

and the multiplication $a \odot b = a \vee b$, for all $a, b \in G_\infty$.

Proof.   This theorem is due to Nakano (see [69]). Nakano showed that this structure is an m-ring. Note that in an m-ring, we don't have the conditions (iii) and (iv) in Definition 2.1, and instead of condition (v), we have $a \in b \oplus c \implies b \in a \oplus c$. Therefore, it is enough to show that the conditions (iii), (iv), and (v) hold. If we set $0 = \infty$, then $a \oplus 0 = a$. Since $0 \in a \oplus a = \{g \in G_\infty \mid a \leq g\}$, then $a = -a$. Hence, condition (v) of Definition 2.1 clearly holds. $\square$

The symbol ∞ is usually adjoined to an ordered group G in such a way that $a \cdot \infty = \infty \cdot a = \infty \cdot \infty = \infty > a$, for all $a \in G$. As above, we denote $G_\infty = G \cup \{\infty\}$.

Definition 2.13. Let R be a hyperring. By a *hypervaluation* on R, we mean a map μ from R onto G_∞, where G is a totally ordered abelian group G, such that the following conditions are satisfied:

(1) $\mu(0) = \infty$.
(2) $\mu(xy) = \mu(x) \cdot \mu(y)$, for all $x, y \in R$.
(3) $\mu(-x) = \mu(x)$, for all $x \in R$.
(4) $z \in x + y \implies \mu(z) \geq \min\{\mu(x), \mu(y)\}$, for all $x, y, z \in R$.

Lemma 2.5. *In* (4), *we have* $\mu(z) = \min\{\mu(x), \mu(y)\}$ *whenever* $\mu(x) \neq \mu(y)$.

Proof. Suppose that $z \in x + y$, $\mu(z) > \min\{\mu(x), \mu(y)\}$, and $\mu(x) < \mu(y)$, then $\mu(z) > \mu(x)$. Since $z \in x + y$, we have $x \in z - y$, and so

$$\mu(x) \geq \min\{\mu(z), \mu(-y)\} = \min\{\mu(z), \mu(y)\} > \mu(x),$$

which is a contradiction. $\square$

If $\mu : R \to G_\infty = G \cup \{\infty\}$ is a hypervaluation, we say that (R, μ, G) is a hypervaluated hyperring.

Theorem 2.19. *Let* (R, μ, G) *be a hypervalued hyperring. We set*

$$R^\mu = \{x \in R \mid \mu(x) \geq 1\} \quad \text{and} \quad P^\mu = \{x \in R \mid \mu(x) > 1\}.$$

Then:

(1) R^μ *is a subhyperring of* R.
(2) P^μ *is a prime hyperideal of* R^μ.
(3) *the set* $\mu^{-1}(\infty)$ *is a prime hyperideal of* R *contained in* P^μ.

Proof. (1) Let $x, y \in R^\mu$. We must show that $x - y \subseteq R^\mu$ and $xy \in R^\mu$. For every $z \in x - y$, we have $\mu(z) \geq \min\{\mu(x), \mu(-y)\} = \min\{\mu(x), \mu(y)\} \geq 1$, and so $z \in R^\mu$. Hence, $x - y \subseteq R^\mu$. Also, we have $\mu(xy) = \mu(x)\mu(y) \geq 1$, thus $xy \in R^\mu$.

(2) Clearly, P^μ is a hyperideal of R^μ. Therefore, we show that P^μ is a prime hyperideal of R^μ. Let $x, y \in R^\mu$ be elements such that $xy \in P^\mu$ and $x \notin P^\mu$. Since $x \in R^\mu$ and $x \notin P^\mu$, then $\mu(x) = 1$. From $xy \in P^\mu$, we have $\mu(xy) > 1$, and so $1 < \mu(xy) = \mu(x)\mu(y) = \mu(y)$, which implies that $y \in P^\mu$.

(3) Suppose that $x, y \in \mu^{-1}(\infty)$ and $r \in R$. Then, $\mu(x) = \mu(y) = \infty$. For every $z \in x - y$, we have $\mu(z) \geq \min\{\mu(x), \mu(y)\} = \infty$, and so $\mu(z) = \infty$ or $z \in \mu^{-1}(\infty)$, which

implies that $x - y \subseteq \mu^{-1}(\infty)$. Also, we have $\mu(rx) = \mu(r)\mu(x) = \mu(r)\infty = \infty$, which implies that $rx \in \mu^{-1}(\infty)$. Therefore, $\mu^{-1}(\infty)$ is a hyperideal of R. Now, let $x, y \in R$ be elements such that $xy \in \mu^{-1}(\infty)$ and $x \notin \mu^{-1}(\infty)$. Then, $\mu(xy) = \infty$ and $\mu(x) \neq \infty$. Since $\mu(xy) = \mu(x)\mu(y) = \infty$, we get $\mu(y) = \infty$ or $y \in \mu^{-1}(\infty)$. Thus, $\mu^{-1}(\infty)$ is a prime hyperideal of R. Clearly, we have $\mu^{-1}(\infty) \subseteq P^\mu$.

$\square$

We say that μ is a nontrivial hypervaluation if $\mu(R) \neq \{\infty\}$.

The class of hypervalued hyperrings has many properties which are similar to the corresponding properties of the class of Manis valuation rings. In what follows, we present some of these common properties.

Theorem 2.20. *Let R be a hyperring, let A be a subhyperring of R, and let P be a proper prime hyperideal of A. Then, the following statements are equivalent:*

(1) $(\forall x \in R \setminus A)(\exists y \in P)\ xy \in A \setminus P$.
(2) *There exists a hypervaluation $\mu : R \to G_\infty$ such that $R^\mu = A$ and $P^\mu = P$.*

Proof. $[2 \Rightarrow 1]$ If an element x belongs to the set $R \setminus A$, then $\mu(x) < 1$; therefore, $\mu(x) \neq \infty$, i.e., $\mu(x) \in G$. Thus, we have $\mu(x)^{-1} = \mu(y)$, for some $y \in R$. Further, we get $1 < \mu(x)^{-1} = \mu(y)$, i.e., $y \in P^\mu = P$. Finally, $\mu(xy) = \mu(x)\mu(y) = 1$; therefore, $xy \in R^\mu \setminus P^\mu$.

$(1 \Rightarrow 2)$ Note that (1) implies the following property:

$$(\forall x, y \in R)\ xy \in P \Rightarrow x \in P \quad \text{or} \quad y \in P.$$

Now, for each $x \in R$, we set $(P : x)_R = \{z \in R \mid xz \in P\}$, and then we define the equivalence relation $\sim$ by

$$x \sim y \quad \text{if and only if } (P : x)_R = (P : y)_R.$$

We denote the equivalence class of an element $x \in R$ by $\mu(x)$. We define the multiplication on the set $R / \sim$ by $\mu(x) \cdot \mu(y) = \mu(xy)$. One can prove that the set $G = \{\mu(x) \mid x \in R\} \setminus \{\mu(0)\}$ is a totally ordered group with respect to the above-defined multiplication, where the ordering is given by

$$\mu(x) \leq \mu(y) \quad \text{if and only if } (P : x)_R \subseteq (P : y)_R.$$

Moreover, we take $\mu(0) = \infty$. It can be proved that μ is a hypervaluation. Indeed, let $z \in x + y$ and $\mu(x) < \mu(y)$. First, suppose that $\mu(z) < \mu(x)$, i.e., $zu \notin P$ and $ux \in P$, for some $u \in R$. We show that $yu \notin P$. Suppose that $yu \in P$. Then,

$$zu \in (x + y)u = xu + yu \in P + P \subseteq P,$$

i.e., $zu \in P$, which is a contradiction. From the assumption $\mu(x) < \mu(y)$, we conclude that $yt \in P$ and $xt \notin P$, for some $t \in R$. Thus, we obtain $xt \cdot yu = xu \cdot yt \in P$, and so we have $xt \in P$ or $yu \in P$, which is a contradiction. In the same way, we obtain a contradiction in the case $\mu(x) < \mu(z)$. Thus, we have proved that $\mu(z) = \mu(x)$, for all $z \in x + y$, if $\mu(x) \neq \mu(y)$.

If $\mu(x) = \mu(y)$, we must show that $\mu(z) \geq \mu(x)$ for any $z \in x + y$, i.e., $\{u \in R \mid ux \in P\} \subseteq \{u \in R \mid uz \in P\}$. Let $u \in R$ and $ux \in P$. Since $\mu(x) = \mu(y)$, we have $uy \in P$ and $uz \in ux + uy \subseteq P + P \subseteq P$. Hence, $(P : x)_R \subseteq (P : z)_R$, i.e., $\mu(x) \leq \mu(z)$. The rest of the proof may be done easily, and it is left to the reader. □

In Theorem 2.20, the pair (A, P) is called a *hypervaluation pair* of R.

Theorem 2.21. *Let R be a hyperring, and let $\mu : R \to G_\infty$ be a nontrivial hypervaluation on R. Then, we have:*

(1) $\mu^{-1}(\infty) = \{x \in R \mid (\forall y \in R \backslash R^\mu)\ xy \in R^\mu\}$.
(2) *If P is a prime hyperideal of R^μ such that $P \subseteq P^\mu$ and $P \not\subseteq \mu^{-1}(\infty)$, then $\mu^{-1}(\infty) \subseteq P$.*

Proof. (1) Let $x \in R$ be an element such that $\mu(x) = \infty$ and $y \in R \setminus R^\mu$. Then, we have $\mu(xy) = \mu(x)\mu(y) = \infty$, and hence $xy \in R^\mu$. Now, let us suppose that for an element $x \in R$ and for every $y \in R \setminus R^\mu$, we have $xy \in R^\mu$. Then, $\mu(x) = \infty$. In fact, if $\mu(x) < 1$, we can take $y = x \in R \setminus R^\mu$ and deduce that $\mu(xy) < 1$, i.e., $xy \notin R^\mu$. So, we consider the case of $1 \leq \mu(x) < \infty$. If $\mu(x) > 1$, we take $y \in R$ such that $\mu(y) = \mu(x)^{-1} < 1$ and $\mu(y^2) < \mu(y)$. Therefore, it follows that $xy^2 \notin R^\mu$ and $y^2 \in R \setminus R^\mu$, which is a contradiction. Finally, if $\mu(x) = 1$, take any $y \in R \setminus R^\mu$, and hence $\mu(xy) < \mu(x) = 1$. Then, it follows that $xy \notin R^\mu$ and $y \in R \setminus R^\mu$, which is a contradiction. Thus, we can conclude that $\mu(x) = \infty$.

(2) Let P be a prime hyperideal of R^μ, and let $P \subseteq P^\mu$ be such that $P \not\subseteq \mu^{-1}(\infty)$. Take $p \in P$ with $\mu(p) < \infty$, and let $n \in \mu^{-1}(\infty)$. Then, $\mu(p)^{-1} = \mu(x)$, for some $x \in R \setminus R^\mu$. It is immediate that $nx \in P^\mu$. Further, we have

$$xp \in R^\mu \setminus P^\mu \subseteq R^\mu \setminus P \;\Rightarrow\; xp \cdot n = xn \cdot p \in P^\mu P \subseteq P \;\Rightarrow\; n \in P,$$

since $xp \notin P$. Therefore, we obtain that $\mu^{-1}(\infty) \subseteq P$. $\square$

Theorem 2.22. *Let R be a hyperring, and let $\mu : R \to G_\infty$ and $\lambda : R \to H_\infty$ be nontrivial hypervaluations on R. Then, $R^\mu = R^\lambda$ and $P^\mu = P^\lambda$ if and only if $\lambda = f \circ \mu$, for some order preserving isomorphism $f : G_\infty \to H_\infty$.*

Proof. Let us suppose that $R^\mu = R^\lambda$ and $P^\mu = P^\lambda$. We set $f(\mu(x)) = \lambda(x)$, for every $\mu(x) \neq \infty$ and $f(\infty_G) = \infty_H$. This definition is correct since $\infty_G \neq \mu(x) = \mu(y)$ implies $\lambda(x) = \lambda(y) \neq \infty_H$. In fact, $\mu(y)^{-1} \in G$, and hence $\mu(y)^{-1} = \mu(z)$, for some $z \in R$. Thus, we have

$$1 = \mu(x)\mu(y)^{-1} = \mu(x)\mu(z) \quad \Rightarrow \quad xz \in R^\mu \setminus P^\mu$$
$$= R^\lambda \setminus P^\lambda \quad \Rightarrow \quad \lambda(xz) = 1.$$

Analogously, from $1 = \mu(x)\mu(y)^{-1}$ and $\mu(y)\mu(z) = \mu(yz)$, it follows that $\lambda(x)\lambda(y) \neq \infty_H$. Finally, from Theorem 2.21, it follows that $\mu^{-1}(\infty_G) = \lambda^{-1}(\infty_H)$. It is easy to see that f is a homomorphism. In fact, if $\mu(x) < \mu(y)$, then $\lambda(x) < \lambda(y)$. Otherwise, we obtain $\lambda(y) \leq \lambda(x) < \infty$, and so $\lambda(y)^{-1} = \lambda(z)$, for some $z \in R$. Furthermore, $1 \leq \lambda(x)\lambda(y)^{-1} = \mu(x)\mu(z)$ implies that $xz \in R^\lambda = R^\mu$, and hence $1 \leq \mu(xz)$. But $\lambda(z) = \lambda(y)^{-1} \neq \infty$ implies that $\mu(z) \neq \infty$. Therefore, $yz \in R^\lambda \setminus P^\lambda$, and so $yz \in R^\mu \setminus P^\mu$, i.e., $1 = \mu(yz)$. From $1 \leq \mu(xz)$, it follows that $\mu(y) = \mu(z)^{-1} \leq \mu(x)$. Thus, we obtain $\mu(y) \leq \mu(x)$, which is a contradiction.

It remains to be proven that the following property holds:

$$(\forall a, b \in G_\infty) \quad f(a + b) = f(a) \oplus f(b).$$

Since $a + \infty_G = a$ in G and $a_1 + \infty_H = a_1$ in H, it suffices to consider the case of $a, b \in G$. Let $a, b \in G$ and $x, y \in R$ be such that $a = \mu(x)$

and $b = \mu(y)$. Then, $f(a) + f(b) = f(\mu(x)) + f(\mu(y))$. Therefore, if $\mu(x) < \mu(y)$, we have $\lambda(x) < \lambda(y)$, and hence

$$f(\mu(x) + \mu(y)) = f(\mu(x)) = \lambda(x) = f(\mu(x)) + f(\mu(y)).$$

If $\mu(x) = \mu(y)$, then $\lambda(x) = \lambda(y)$, and so the set $f(\mu(x) + \mu(y))$ is equal to

$$\{f(\mu(z)) \mid \mu(x) \le \mu(z)\} = \{\lambda(z) \mid \lambda(x) \le \lambda(z)\} = \lambda(x) + \lambda(y),$$

as required.   □

2.6   Semigroups Admitting Krasner Hyperring Structure

We say that a semigroup $(S, \cdot)$ *admits a Krasner hyperring (ring) structure* if there is a hyperoperation (operation) $+$ on $S^0 = S \cup \{0\}$ such that $(S^0, +, \cdot)$ is a hyperring (ring). In Ref. [47], Kemprasit and Punkla gave a necessary and sufficient condition for a set X so that some transformation semigroups on X admit a hyperring structure.

Example 2.14. Let G be a group and define a hyperoperation $+$ on G^0 by

$$x + 0 = 0 + x = \{x\} \quad \text{for all } x \in G^0,$$
$$x + x = G^0 \backslash \{x\} \quad \text{for all } x \in G^0 \backslash \{0\},$$
$$x + y = \{x, y\} \quad \text{for all } x, y \in G^0 \backslash \{0\} \text{ with } x \ne y.$$

We can easily verify that $(G^0, +, \cdot)$ is a Krasner hyperring. From this fact, we conclude that every group admits a hyperring structure.

Let $\mathcal{SR}$ and $\mathcal{SHR}$ denote the class of all semigroups admitting a ring structure and the class of all semigroups admitting a hyperring structure, respectively. Then, $\mathcal{SHR}$ contains $\mathcal{SR}$ as a subclass. Semigroups belonging to the class $\mathcal{SR}$ have long been studied, for example see Refs. [45,46]. Kemprasit and Punkla characterized some standard transformation semigroups belonging to $\mathcal{SHR}$.

Let X be an arbitrary set, and let

$P_X = $ the partial transformation semigroup on X,

$T_X = $ the full transformation semigroup on X,

I_X = the one to one partial transformation semigroup on X,

G_X = the symmetric group on X,

M_X = the semigroup of all one to one transformations of X,

E_X = the semigroup of all onto transformations of X.

Observe that

$$G_X \subseteq M_X \subseteq I_X \subseteq P_X,$$
$$G_X \subseteq M_X \subseteq T_X \subseteq P_X,$$
$$G_X \subseteq E_X \subseteq T_X \subseteq P_X.$$

The following facts are known. If X contains more than two elements, then the center of G_X is $\{1_X\}$, where 1_X is the identity map on X. For $\alpha \in P_X$, $\alpha^2 = \alpha$ if and only if $Im\alpha \subseteq Dom\alpha$ and $x\alpha = x$, for all $x \in Im\alpha$, where $Dom\alpha$ and $Im\alpha$ denote the domain and image of α, respectively. For convenience, the following notation will be used.

For distinct $a_1, a_2, \ldots, a_n \in X$, let $(a_1, a_2, \ldots, a_n)$ be the element of G_X defined by

$$(a_1, a_2, \ldots, a_n)(x) = \begin{cases} a_{i+1} & \text{if } x = a_i \text{ for some } i \in \{1, 2, \ldots, n-1\}, \\ a_1 & \text{if } x = a_n, \\ x & \text{otherwise.} \end{cases}$$

For $A \subseteq X$, $A \neq \emptyset$ and $x \in X$, let $A_x \in P_X$ be defined by $DomA_x = A$ and $ImA_x = \{x\}$.

Let X be an arbitrary set and $|X|$ denote the cardinality of X. First, we note that if $|X| = 0$, then all of the previous transformation semigroups contain exactly one element. If $|X| = 1$, $P_X = I_X \cong (\mathbb{Z}_2, \cdot)$ and $T_X = M_X = E_X = G_X$, which contains exactly one element. Hence, if $|X| \leq 1$, all of these transformation semigroups belong to $\mathcal{SR}$ ($\subseteq \mathcal{SHR}$).

Theorem 2.23. *Let S be P_X or I_X. Then, $S \in \mathcal{SHR}$ if and only if $|X| \leq 1$.*

Proof. Assume that $S \in \mathcal{SHR}$. Then, there exists a hyperoperation $+$ on S^0 such that $(S^0, +, \cdot)$ is a hyperring. To prove that $|X| \leq 1$, suppose to the contrary. Let $a, b \in X$ be such that $a \neq b$.

The element $\{a\}_a$ of S maps a in b. Similarly, we can define other elements of S such that $\{a\}_b$ and $\{b\}_a$. Since $\emptyset \neq \{a\}_a + \{a\}_b \subseteq S$, there exists an element $\alpha \in S$ such that $\alpha \in \{a\}_a + \{a\}_b$. Thus, $\{a\}_a \alpha \in \{a\}_a(\{a\}_a + \{a\}_b)$. However, $\{a\}_a(\{a\}_a + \{a\}_b) = \{a\}_a + 0 = \{a\}_a$, so $\{a\}_a \alpha = \{a\}_a$. This implies that $a \in Dom\alpha$ and $\alpha a = a$. Consequently, we have $\alpha \{a\}_a = \{a\}_a$. Hence, $\{a\}_a \in (\{a\}_a + \{a\}_b)\{a\}_a$. Since $(\{a\}_a + \{a\}_b)\{a\}_a = \{a\}_a + \{a\}_b$, it follows that

$$0 = \{b\}_a\{a\}_a \in \{b\}_a(\{a\}_a + \{a\}_b) = 0 + \{a\}_a = \{a\}_a.$$

This is a contradiction. Hence, $|X| \leq 1$. The converse follows from what we have mentioned above.   □

Theorem 2.24. $T_X \in \mathcal{SHR}$ *if and only if* $|X| \leq 1$.

Proof.   Assume that $T_X \in \mathcal{SHR}$, and suppose that $|X| > 1$. Then, there exists a hyperoperation $+$ on T_X^0 such that $(T_X^0, +, \cdot)$ is a hyperring. Let us note that for $\alpha, \beta \in T_X^0$, $\beta\alpha = 0$ implies that $\alpha = 0$ or $\beta = 0$. Let a and b be distinct elements of X, and define $\alpha, \beta : X \to X$ by

$$\alpha(x) = \begin{cases} a & \text{if } x \in \{a, b\}, \\ x & \text{otherwise;} \end{cases}$$

$$\beta(x) = \begin{cases} b & \text{if } x \in \{a, b\}, \\ x & \text{otherwise.} \end{cases}$$

Then, $\alpha, \beta \in T_X$, $\alpha^2 = \alpha$ and $\beta^2 = \beta$. It is easy to see that $\alpha\beta = \alpha$. Thus,

$$\alpha - \alpha = \alpha^2 - \alpha\beta = \alpha(\alpha - \beta).$$

Since $0 \in \alpha - \alpha$ and $\alpha \neq 0$, we have $0 \in \alpha - \beta$. Consequently, we have $\alpha = \beta$, which is a contradiction. As mentioned previously, the converse holds.   □

Theorem 2.25. *Let* S *be* M_X *or* E_X. *Then,* $S \in \mathcal{SHR}$ *if and only if* X *is finite.*

Proof.   If X is finite, then $S = G_X$, so $S \in \mathcal{SHR}$ since every group is in $\mathcal{SHR}$.

Conversely, assume that $S \in \mathcal{SHR}$. In order to show that X is finite, suppose to the contrary that X is infinite. Let $+$ be a hyperoperation on S^0 such that $(S^0, +, \cdot)$ is a hyperring. Again, let us note that for $\alpha, \beta \in S^0$, $\alpha\beta = 0$ implies $\alpha = 0$ or $\beta = 0$. Since for every $\alpha \in G_X$,

$$(-1_X)\alpha = -(1_X\alpha) = -(\alpha 1_X) = \alpha(-1_X),$$

it follows that -1_X is at the center of G_X. Since X is infinite, the center of G_X is $\{1_X\}$. Therefore, $-1_X = 1_X$. Consequently, we have $-\alpha = \alpha$, for all $\alpha \in S$, so $0 \in \alpha + \alpha$, for all $\alpha \in S$. Next, let a and b be distinct elements of X. Since X is infinite, $|X| = |X \setminus \{a,b\}|$, so there exists a one-to-one map γ from X onto $X \setminus \{a,b\}$ and a map λ from $X \setminus \{a,b\}$ onto X. Define $\mu : X \to X$ by $\mu(a) = \mu(b)$ and $\mu(x) = \lambda(x)$, for all $x \in X \setminus \{a,b\}$. Then, $\gamma \in M_X$ and $\mu \in E_X$. Denote the transposition map of a and b by (a,b). Then, we have $(a,b)\gamma = \gamma$ and $\mu(a,b) = \mu$.

Case 1: $S = M_X$. Since $0 \in \gamma + \gamma$ and $\gamma + \gamma = (a,b)\gamma + 1_X\gamma = ((a,b) + 1_X)\gamma$, we have $0 \in ((a,b) + 1_X)\gamma$. However, $\gamma \neq 0$, so $0 \in (a,b) + 1_X$. Then, (a,b) is an inverse of 1_X in $(S^0, +)$. This is a contradiction since 1_X is the unique inverse of 1_X in $(S^0, +)$.

Case 2: $S = E_X$. Since $0 \in \mu + \mu$ and $\mu + \mu = \mu(a,b) + \mu 1_X = \mu((a,b)+1_X)$, we have $0 \in \mu((a,b) + 1_X)$. However, $\mu \neq 0$, so $0 \in (a,b) + 1_X$. This implies that (a,b) is an inverse of 1_X in $(S^0, +)$, which is a contradiction because 1_X is the unique inverse of 1_X in $(S^0, +)$.

Therefore, the theorem is completely proved. $\square$

The following two corollaries are obtained from the fact that $S\mathcal{R} \subseteq S\mathcal{HR}$, Theorems 2.23 and 2.24, respectively, and the paragraph before Theorem 2.23.

Corollary 2.6. *Let S be P_X or I_X. Then, $S \in S\mathcal{R}$ if and only if $|X| \leq 1$.*

Corollary 2.7. *$T_X \in S\mathcal{R}$ if and only if $|X| \leq 1$.*

We know that $G_X \in S\mathcal{HR}$ for any cardinality of X, and if X is finite and $|X| \geq 3$, $G_X \notin S\mathcal{R}$. The following theorem shows that the condition $n \leq 2$ is necessary and sufficient for G_X to belong to $S\mathcal{R}$.

Theorem 2.26. $G_X \in \mathcal{SR}$ *if and only if* $|X| \leq 2$.

Proof. Assume that $G_X \in \mathcal{SR}$. Let $+$ be an operation on G_X^0 such that $(G_X^0, +, \cdot)$ is a ring. In order to show that $|X| \leq 2$, suppose to the contrary that $|X| > 2$. Let a, b, and c be distinct elements of X. Then, $(a, b, c) + (a, c) = \alpha$, for some $\alpha \in G_X^0$, where (a, b, c) is a cycle of length 3.

Case 1: $\alpha = 0$. Then, $(a, b, c) + (a, c) = 0$, so

$$0 = ((a, b, c) + (a, c))(a, c) = (a, b, c)(a, c) + 1_X = (b, c) + 1_X,$$
$$0 = (a, c)((a, b, c) + (a, c)) = (a, c)(a, b, c) + 1_X = (a, b) + 1_X.$$

These imply that $(a, b) + 1_X = 0 = (b, c) + 1_X$, which is a contradiction since $(a, b) \neq (b, c)$.

Case 2: $\alpha \neq 0$. Then, $(b, c) + 1_X = ((a, b, c) + (a, c))(a, c) = \alpha(a, c)$, and so

$$1_X + (b, c) = (b, c)((b, c) + 1_X) = (b, c)\alpha(a, c).$$

Then, we have $\alpha(a, c) = (b, c)\alpha(a, c)$. Since α and (a, c) are in the group G_X, it follows that $(b, c) = 1_X$, which is a contradiction.

Conversely, assume that $|X| \leq 2$. If $|X| \leq 1$, then $G_X \in \mathcal{SR}$. If $|X| = 2$, it is clear that $G_X^0 \cong (\mathbb{Z}_3, \cdot)$, so $G_X \in \mathcal{SR}$. $\qquad \square$

We use Theorems 2.25 and 2.26 to obtain the following theorem.

Theorem 2.27. *Let* S *be* M_X *or* E_X. *Then,* $S \in \mathcal{SR}$ *if and only if* $|X| \leq 2$.

Proof. Let $S \in \mathcal{SR}$. Then, $S \in \mathcal{SHR}$, so by Theorem 2.25, X is finite. Thus, $S = G_X$. Hence, $|X| \leq 2$ by Theorem 2.26.

Conversely, if $|X| \leq 2$, then $S = G_X$, so $S \in \mathcal{SR}$ by Theorem 2.26. $\qquad \square$

Now, we characterize multiplicative interval semigroups on $\mathbb{R}$, admitting a Krasner hyperring structure.

Theorem 2.28 ([46]). *A subset* S *of* $\mathbb{R}$ *is a multiplicative interval semigroup on* $\mathbb{R}$ *if and only if* S *is one of the following types:*

(1) $\mathbb{R}$,
(2) $\{0\}$,
(3) $\{1\}$,
(4) $(0, \infty)$,

(5) $[0, \infty)$,
(6) (a, ∞) *where* $a \geq 1$,
(7) $[a, \infty)$ *where* $a \geq 1$,
(8) $(0, b)$ *where* $0 < b \leq 1$,
(9) $[0, b]$ *where* $0 < b \leq 1$,
(10) $(0, b]$ *where* $0 < b \leq 1$,
(11) $[0, b)$ *where* $0 < b \leq 1$,
(12) (a, b) *where* $-1 \leq a < 0 < a^2 \leq b \leq 1$,
(13) $[a, b]$ *where* $-1 \leq a < 0 < a^2 \leq b \leq 1$,
(14) $(a, b]$ *where* $-1 \leq a < 0 < a^2 \leq b \leq 1$,
(15) $[a, b)$ *where* $-1 < a < 0 < a^2 < b \leq 1$.

Lemma 2.6. *If S is a multiplicative interval semigroup on $\mathbb{R}$ such that $S \subseteq [0, \infty)$, then $S \in \mathcal{SHR}$.*

Proof. By assumption, S is one of types (2)–(11) in Theorem 2.28. Clearly, $S \in \mathcal{SHR}$ if S is of type (2) or (3).

Case 1: S is one of types (4), (5), and (8)–(11). Then, S^0 can be considered $S^0 = [0, \infty)$, $S^0 = [0, b)$, $[0, b]$, for some $b > 0$, with $b \leq 1$. Define the hyperoperation $\oplus$ on S^0 by

$$x \oplus x = [0, x] \qquad \text{for all} \quad x \in S^0 \quad \text{and}$$
$$x \oplus y = \{\max\{x, y\}\} \quad \text{for all distinct} \quad x, y \in S^0.$$

Then, $(S^0, \oplus)$ is a canonical hypergroup. In order to show that $(S^0, \oplus, \cdot)$ is a hyperring where $\cdot$ is the multiplication on S^0, set $x, y, z \in S^0$. We have

$$y \oplus z = \begin{cases} \{z\} & \text{if } y < z, \\ \{y\} & \text{if } z < y, \\ [0, y] & \text{if } y = z. \end{cases}$$

But $x \geq 0$, so

$$x(y \oplus z) = \begin{cases} \{xz\} & \text{if } y < z, \\ \{xy\} & \text{if } z < y, \\ [0, xy] & \text{if } y = z. \end{cases}$$

Since $x \geq 0$, we obtain $x(y \oplus z) = xy \oplus xz$.

Case 2: S is of types (6) and (7). Then, $S = (a, \infty)$ or $[a, \infty)$, for some $a \geq 1$, so S^0 can be considered $S \cup \{0\}$, where 0 is the zero real number. Define the hyperoperation $\oplus$ on S^0 by

$$x \oplus x = 0 \oplus x = \{x\} \quad \text{for all } x \in S^0,$$
$$x \oplus x = [x, \infty) \cup \{0\} \quad \text{for all } x \in S, \quad \text{and}$$
$$x \oplus y = \{\min\{x, y\}\} \quad \text{for all distinct } x, y \in S.$$

Then, $(S^0, \oplus)$ is a canonical hypergroup. We claim that $(S^0, \oplus, \cdot)$ is a Krasner hyperring, where $\cdot$ is a multiplication on S^0. Set $x, y, z \in S^0$. If $x = 0$, then $x(y \oplus z) = \{0\} = xy \oplus xz$. Assume that $x \neq 0$. Then, $x \geq a \geq 1$. We have

$$y \oplus z = \begin{cases} \{z\} & \text{if } y = 0 \text{ or } 0 < z < y, \\ \{y\} & \text{if } z = 0 \text{ or } 0 < y < z, \\ [y, \infty) \cup \{0\} & \text{if } 0 < y = z. \end{cases}$$

The fact that $x > 0$ implies

$$x(y \oplus z) = \begin{cases} \{xz\} & \text{if } y = 0 \text{ or } 0 < z < y, \\ \{xy\} & \text{if } z = 0 \text{ or } 0 < y < z, \\ [xy, \infty) \cup \{0\} & \text{if } 0 < y = z. \end{cases}$$

Since $x > 0$, we have that $x(y \oplus z) = xy \oplus xz$.

Hence, the lemma is completely proved. $\qquad\square$

Lemma 2.7. *Let S be a multiplicative interval semigroup on $\mathbb{R}$ such that $S \not\subseteq [0, \infty)$. Then, $S \in \mathcal{SHR}$ if and only if for every $x \in S$, $-x \in S$.*

Proof. By assumption, S is one of types (1) and (12)–(15). Then, $S^0 = S$. First, assume that for every $x \in S$, $-x \in S$. Define the hyperoperation $\oplus$ on S by

$$x \oplus x = \{x\} \quad \text{for all } x \in S,$$
$$x \oplus y = y \oplus x = \{x\} \quad \text{for all } x, y \in S \text{ with } |y| < |x|, \quad \text{and}$$
$$x \oplus (-x) = [-|x|, |x|] \quad \text{for all } x \in S.$$

Then, $(S, \oplus)$ is a canonical hypergroup. Let $x, y, z \in S$. Then,

$$
y \oplus z =
\begin{cases}
\{y\} & \text{if } y = z, \\
\{z\} & \text{if } |y| < |z|, \\
\{y\} & \text{if } |z| < |y|, \\
[-|y|, |y|] & \text{if } z = -y.
\end{cases}
$$

If $x \geq 0$, then $x[-|y|, |y|] = [-x|y|, x|y|] = [-|xy|, |xy|]$, and if $x < 0$, then $x[-|y|, |y|] = [x|y|, -x|y|] = [-|xy|, |xy|]$. Therefore, we have

$$
x(y \oplus z) =
\begin{cases}
\{xy\} & \text{if } y = z, \\
\{xz\} & \text{if } |y| < |z|, \\
\{xy\} & \text{if } |z| < |y|, \\
[-|xy|, |xy|] & \text{if } z = -y.
\end{cases}
$$

So, we have $x(y \oplus z) = xy \oplus xz$. Hence, $(S, \oplus, \cdot)$ is a hyperring where $\cdot$ is the multiplication on S. Therefore, $S \in \mathcal{SHR}$.

For the converse, assume that there exists $c \in S$ such that $-c \notin S$. Then, S is one of types (12)–(15). In order to show that $S \notin \mathcal{SHR}$, suppose to the contrary that $S \in \mathcal{SHR}$. Then, there exists a hyperoperation $\oplus$ on S such that $(S, \oplus, \cdot)$ is a hyperring where $\cdot$ is the multiplication on S. Set $K = \{x \in S \mid -x \in S\}$. Then, $c \in S \setminus K$, and there exists $e > 0$ such that $K = [-e, e]$ or $K = (-e, e)$. If $x \in S \setminus K$, then $-x \notin S$ and $0 \in x \oplus y$, for some $y \in S$, so

$$
0 \in x(x \oplus y) = x^2 \oplus xy \quad \text{and} \quad 0 \in (x \oplus y)y = xy \oplus y^2,
$$

which implies that $x^2 = y^2$, and hence $y = x$. This proves that

$$
0 \in x \oplus x \quad \text{for all } x \in S \setminus K.
$$

Hence, $0 \in xy \oplus xy$ for all $x \in S \setminus K$ and $y \in S$.

We claim that

$$
\text{for every} \quad x \in K \setminus \{0\}, \quad 0 \notin x \oplus (-x) \quad \text{and}
$$
$$
\text{for every} \quad y \in x \oplus (-x), \quad -y \in x \oplus (-x).
$$

In order to prove this, let $x \in K \setminus \{0\}$. We have $0 \in cx \oplus cx$. If $0 \in x \oplus (-x)$, then $0 \in cx \oplus (-cx)$, which implies that $cx = -cx$, which is a contradiction. Hence, $0 \notin x \oplus (-x)$. Next, let $y \in x \oplus (-x)$. Then,

$$-xy \in (-x^2) \oplus x^2 = x(x \oplus (-x)).$$

It follows that $-xy = xt$, for some $t \in x \oplus (-x)$, so $-y = t \in x \oplus (-x)$.

Case 1: $K = [-e, e]$. Hence, there exists $y \in K$ such that $y > 0$ and $y \in e \oplus (-e)$. Then,

$$0 < \frac{y}{e} \leq 1 \quad \text{and} \quad cy \subset ce \oplus (-ce).$$

Since $(S, \oplus)$ is reversible, then $-ce \in ce \oplus cy$. Since $0 < y/e \leq 1$,

$$0 < \frac{cy}{e} \leq c \quad \text{if } c > 0 \quad \text{and}$$

$$0 > \frac{cy}{e} \geq c \quad \text{if } c < 0.$$

It follows that $cy/e \in S$ since S is an interval on $\mathbb{R}$ and $0, c \in S$. Hence,

$$-ce \in ce \oplus \left(\frac{cy}{e}\right) e = \left(c \oplus \frac{cy}{e}\right) e,$$

which implies that $-ce = te$, for some $t \in c \oplus cy/e$. Consequently, $-c = t \in S$, which is a contradiction.

Case 2: $K = (-e, e)$. Then, $e \notin K$. Let $d \in K$ be such that $d > 0$. Then, $0 < d < e$. So, there exists $z \in K$ such that $z > 0$ and $z \in d \oplus (-d)$. Thus, $0 < z < e$.

Subcase (i): $c \geq e$. Since $c \in S$, $e \in S \setminus K$, so $-e \notin S$. Since $z \in d \oplus (-d)$, $ez \in ed \oplus (-ed)$. Since $(S, \oplus)$ is reversible, $-ed \in ed \oplus ez$. If $z \leq d$, then $0 < ez/d \leq e$, so $ez/d \in S$ and hence

$$-ed \in ed \oplus \left(\frac{ez}{d}\right) d = \left(e \oplus \frac{ez}{d}\right) d,$$

whence $-e \in e \oplus ez/d \subseteq S$, which is a contradiction.

Next, assume that $z > d$. Then, $0 < ed/z < e$. Thus, $ed/z \in K$. From the fact that $ez \in ed \oplus (-ed)$,

$$e^2 d \in \frac{ed}{z}(ed \oplus (-ed)) = \frac{e^2 d^2}{z} \oplus \left(-\frac{e^2 d^2}{z}\right) = ed\left(\frac{ed}{z} \oplus \left(-\frac{ed}{z}\right)\right),$$

which implies that $e \in (ed/z) \oplus (-ed/z)$. So, $-e \in (ed/z) \oplus (-ed/z) \subseteq S$, which is a contradiction.

Subcase (ii): $c \leq -e$. Then, $-e \in S \setminus K$, and so $e \notin S$. Since $z \in d \oplus (-d)$, $(-e)z \in (-e)d \oplus (-e)(-d)$. Since $(S, \oplus)$ is reversible,

$$ed = (-e)(-d) \in (-e)d \oplus (-e)z = (-ed) \oplus (-ez).$$

First, assume that $z \leq d$. Then, $0 > -ez/d \geq -e$, so $-ez/d \in S$. Hence,

$$ed \in (-ed) \oplus \left(-\frac{ez}{d}\right)d = \left((-e) \oplus \left(-\frac{ez}{d}\right)\right)d,$$

whence $e \in (-e) \oplus (-ez/d) \subseteq S$, which is a contradiction.

Next, assume that $z > d$. Then, $0 > -ed/z > -e$. Thus, $-ed/z \in K$. Since $(-e)z \in (-e)d \oplus (-e)(-d)$, we have

$$e^2 d \in -\frac{ed}{z}((-ed) \oplus ed)$$
$$= \frac{e^2 d^2}{z} \oplus \left(-\frac{e^2 d^2}{z}\right) = (-e)d\left(\left(-\frac{ed}{z}\right) \oplus \frac{ed}{z}\right).$$

Consequently, we have that $-e \in (ed/z) \oplus (-ed/z)$.

So, $e \in (ed/z) \oplus (-ed/z) \subseteq S$, which is a contradiction.

Hence, the lemma is completely proved. $\qquad\square$

We note that $(\mathbb{R}, \cdot) \in \mathcal{SR} \subseteq \mathcal{SHR}$. The hyperoperation $\oplus$ defined for the case $S = \mathbb{R}$ makes $(\mathbb{R}, \oplus, \cdot)$ a hyperring, which is not a ring.

Now, we are ready to state our main result.

Theorem 2.29. *Let S be a multiplicative interval semigroup on $\mathbb{R}$. Then, $S \in \mathcal{SHR}$ if and only if either $S \subseteq [0, \infty)$ or for every $x \in S$, $-x \in S$.*

Proof. It follows directly from Lemmas 2.6 and 2.7. $\qquad\square$

By Theorem 2.28, Theorem 2.29 is equivalent to the following.

Theorem 2.30. *A multiplicative interval semigroup S on $\mathbb{R}$ belongs to $\mathcal{SHR}$ if and only if S is one of the following types:*

(1) $\mathbb{R}$,

(2) $\{0\}$,

(3) $\{1\}$,

(4) $(0, \infty)$,

(5) $[0, \infty)$,

(6) (a, ∞) *where* $a \geq 1$,

(7) $[a, \infty)$ *where* $a \geq 1$,

(8) $(0, b)$ *where* $0 < b \leq 1$,

(9) $[0, b]$ *where* $0 < b \leq 1$,

(10) $(0, b]$ *where* $0 < b \leq 1$,

(11) $[0, b)$ *where* $0 < b \leq 1$,

(12) $(-c, c)$ *where* $0 < c^2 \leq c \leq 1$,

(13) $[-c, c]$ *where* $0 < c^2 \leq c \leq 1$.

Remark 2.2. It is easy to show that there are exactly six types of additive interval semigroups on $\mathbb{R}$, as follows: $\mathbb{R}$; $\{0\}$; (a, ∞), where $a \geq 0$; $[a, \infty)$, where $a \geq 0$; $(-\infty, b)$, where $b \leq 0$; $(-\infty, b]$, where $b \leq 0$.

Let S be an additive interval semigroup on $\mathbb{R}$ and $S \neq \{0\}$. Then, S has no zero. Thus, S^0 can be considered $S \cup \{-\infty\}$, where $x + (-\infty) = -\infty + x = -\infty$, for all $x \in S \cup \{-\infty\}$. Hence, under the usual order on $\mathbb{R}$ and defining $x \geq -\infty$ for all $x \in S \cup \{-\infty\}$, we have that $S \cup \{-\infty\}$ is a totally ordered set having $-\infty$ as its minimum element. If S is of type (1), (5), or (6), by following the proof of Case 1 of Lemma 2.6, we have $S \in \mathcal{SHR}$. If S is type (3) or (4), we have $S \in \mathcal{SHR}$ by following the proof of Case 2 of Lemma 2.6.

Therefore, every additive interval semigroup on $\mathbb{R}$ belongs to the class $\mathcal{SHR}$.

In continuation of this section, we let x' be the opposite of x in any hyperring $(A, +, \cdot)$.

Let V be a vector space over a division ring R and n a positive integer. We denote by $L_R(V)$ and $G_R(V)$ the semigroup of all linear transformations $\alpha : V \to V$ with respect to the composition and the unit group of $L_R(V)$, respectively. Then,

$$G_R(V) = \{\alpha : V \to V \mid \alpha \text{ is an isomorphism}\}.$$

Let $M_n(R)$ and $G_n(R)$ denote, respectively, the full $n \times n$ matrix semigroup over R and the matrix group of all invertible $n \times n$ matrices over R, i.e., $G_n(R)$ is the unit group of $M_n(R)$. It is known that if $\dim_R V = n$, then

$$L_R(V) \cong M_n(R) \quad \text{and} \quad G_R(V) \cong G_n(R).$$

Since every group is in $\mathcal{SHR}$, we have that $G_R^0(V) \in \mathcal{SHR}$.

We know that $L_R(V) \in \mathcal{SR} \subseteq \mathcal{SHR}$ with the usual addition $+$. Then, in the ring $(L_R(V), +, \cdot)$, $\alpha' = -\alpha$, for all $\alpha \in L_R(V)$.

If k is any cardinal number such that $0 \leq k \leq \dim_R V$, then

$$S = G_R(V) \cup \{\alpha \in I_R(V) \mid \dim_R \operatorname{Im} \alpha \leq k\}$$

is clearly a subsemigroup of $L_R(V)$ containing $G_R(V)$.

We denote the center of R by $Z(R)$. It is clear that for $a \in Z(R)$, $a1_V \in L_R(V)$, where 1_V is the identity map on V and $a1_V$ is defined in the usual sense, i.e., $(a1_V)(v) = a(1_V)(v) = av$, for all $v \in V$.

Lemma 2.8. *Assume that B is a basis of V. Let $\alpha \in L_R(V)$ be such that $\alpha\beta = \beta\alpha$, for all $\beta \in G_R(V)$, and for every $v \in B$, $\alpha(v) = a_v v$, for some $a_v \in R$. Then, there exists $a \in Z(R)$ such that $\alpha = a1_V$.*

Proof. It is trivial if $B = \emptyset$. Assume that $B \neq \emptyset$. In order to show that $a_v \in Z(R)$ for every $v \in B$, let $u \in B$ and $b \in R \setminus \{0\}$. Define $\beta \in L_R(V)$ by

$$\beta(v) = \begin{cases} bu & \text{if } v = u, \\ v & \text{if } v \in B \setminus \{u\}. \end{cases}$$

Since $(B \setminus \{u\}) \cup \{bu\}$ is a basis of V, $\beta \in G_R(V)$. Then, $\alpha\beta = \beta\alpha$ by hypothesis. Thus, $(\beta\alpha)(u) = \beta a_u u = (a_u b)u$ and $(\alpha\beta)(u) = \alpha(bu) = (ba_u)u$, which implies that $a_u b = ba_u$. Hence, $a_u \in Z(R)$.

Next, we shall show that $a_v = a_{v'}$ for all $v, v' \in B$. If $|B| = 1$ or $a_v = 0$, for all $v \in B$, there is nothing to prove. Assume that $|B| > 1$ and $a_w \neq 0$, for some $w \in B$. Let $z \in B \setminus \{w\}$. Define $\gamma \in L_R(V)$ by

$$\gamma(v) = \begin{cases} z & \text{if } v = w, \\ a_w w & \text{if } v = z, \\ v & \text{if } v \in B \setminus \{w, z\}. \end{cases}$$

Since $a_w \neq 0$, $(B \setminus \{w\}) \cup \{a_w w\}$ is a basis of V, so $\gamma \in G_R(V)$. By hypothesis, $\alpha\gamma = \gamma\alpha$. Then, $(\gamma\alpha)(w) = \gamma(a_w w) = a_w z$ and $(\alpha\gamma)(w) = \alpha(z) = a_z z$. These imply that $a_z = a_w$. Hence, the lemma is proved.

□

Lemma 2.9. *Let $\alpha \in L_R(V)$, and assume that $\alpha\beta = \beta\alpha$, for all $\beta \in G_R(V)$. Then, there exists $a \in Z(R)$ such that $\alpha = a1_V$.*

Proof.   Let B be a basis of V. In order to show that for every $v \in B$

$$\alpha(v) = a_v v \quad \text{for some } a_v \in R,$$

suppose to the contrary that it is not true. Then, there exists $u \in B$ such that $\alpha(u) \neq bu$, for all $b \in R$. Then, $\alpha(u) \neq u$ and $\{u, \alpha(u)\}$ is linearly independent. Let B' be a basis of V containing $\{u, \alpha(u)\}$. Consequently, $\{u + \alpha(u)\} \cup (B' \setminus \{\alpha(u)\})$ is a basis of V. Define $\beta \in L_R(V)$ by

$$\beta(v) = \begin{cases} u + \alpha(u) & \text{if } v = \alpha(u), \\ v & \text{if } v \in B' \setminus \{\alpha(u)\}. \end{cases}$$

Then, $\beta \in G_R(V)$ since $\{u + \alpha(u)\} \cup (B' \setminus \{\alpha(u)\})$ is a basis of V. By hypothesis, $\alpha\beta = \beta\alpha$. However, $(\beta\alpha)(u) = u + \alpha(u)$ and $(\alpha\beta)(u) = \alpha(u)$, so we have $u = 0$, which is a contradiction. Now, by Lemma 2.8, $\alpha = a1_V$, for some $a \in Z(R)$.

□

Theorem 2.31. *Let S be a subsemigroup of $L_R(V)$ containing $G_R(V)$. Assume that $\oplus$ is a hyperoperation on S^0 such that $(S^0, \oplus, \cdot)$ is a hyperring. Then,*

$$\alpha' = \alpha, \quad \text{for all } \alpha \in S^0, \quad \text{or} \quad \alpha' = -\alpha, \quad \text{for all } \alpha \in S^0,$$

where α' represents the opposite of α, with respect to $\oplus$.

Proof.   The result is trivially true if $V = \{0\}$. Assume that $V \neq \{0\}$. Since $G_R(V) \subseteq S$, $1_V \in S$. Then, for every $\alpha \in G_R(V)$, we have

$$1'_V \alpha = (1_V \alpha)' = \alpha' = (\alpha 1_V)' = \alpha 1'_V.$$

By Lemma 2.9, $1'_V = a1_V$, for some $a \in Z(R)$. Since $(1'_V)^2 = 1_V$, we obtain $(a1_V)^2 = 1_V$. If $u \in V \setminus \{0\}$, then $a^2 u = (a1_V)^2(u) = 1_V(u) = u$. We deduce that $a^2 = 1$, so $a = \pm 1$ since R is a division ring. Consequently, we have $0 \in 1_V \oplus 1_V$ or $0 \in 1_V \oplus (-1_V)$, which implies that

$$0 \in (\alpha \oplus \alpha) \quad \text{for all } \alpha \in S^0 \quad \text{or} \quad 0 \in (\alpha \oplus -\alpha) \quad \text{for all } \alpha \in S^0.$$

Hence, the theorem is proved. $\qquad\qquad\qquad\qquad\qquad\qquad\qquad\square$

Corollary 2.8. *Let S be a subsemigroup of $M_n(R)$ containing all nonsingular matrices in $M_n(R)$. Assume that $\oplus$ is a hyperoperation on S^0 such that $(S^0, \oplus, \cdot)$ is a hyperring. Then,*

$$A' = A, \quad \text{for all } A \in S^0, \quad \text{or} \quad A' = -A, \quad \text{for all } A \in S^0,$$

where A' is the additive inverse of A in $(S^0, \oplus, \cdot)$.

We know that $G_R(V) \in \mathcal{SHR}$ for any dimension of V. As a consequence, we have that $G_R(V) \in \mathcal{SR}$ only for the case of $\dim_R V \leq 1$.

Corollary 2.9. $G_R(V) \in \mathcal{SR}$ *if and only if* $\dim_R V \leq 1$.

Proof. Since $G_R(V) = \{0\}$ if $\dim_R V = 0$ and $G_R(V) \cong R \setminus \{0\}$ if $\dim_R V = 1$, we have that $G_R(V) \in \mathcal{SR}$ if $\dim_R V \leq 1$.

Conversely, assume that $G_R(V) \in \mathcal{SR}$, and suppose that $\dim_R V > 1$. Let $\oplus$ be an operation on $G_R^0(V)$ such that $(G_R^0(V), \oplus, \cdot)$ is a ring. By Theorem 2.31,

$$\alpha \oplus \alpha = 0 \quad \text{for all } \alpha \in G_R^0(V) \text{ or}$$
$$\alpha \oplus (-\alpha) = 0 \quad \text{for all } \alpha \in G_R^0(V).$$

Let B be a basis of V. By assumption, $|B| > 1$. Let u and w be distinct elements of B. Define $\beta \in L_R(V)$ by

$$\beta(v) = \begin{cases} w & \text{if } v = u, \\ u & \text{if } v = w, \\ v & \text{if } v \in B \setminus \{u, w\}. \end{cases}$$

Then, $\beta \in G_R(V)$, $\beta \neq 1_V$ and $\beta^2 = 1_V$. Since u and w are linearly independent over R, $w \neq -u$, so $\beta \neq -1_V$. We have that $1_V \oplus \beta \neq 0$

and $1_V \oplus (-\beta) \neq 0$. Then, we have

$$0 \neq (1_V \oplus \beta)^2 = 1_V \oplus \beta \oplus \beta \oplus \beta^2$$
$$= 1_V \oplus \beta \oplus \beta \oplus 1_V$$

and

$$0 \neq (1_V \oplus \beta)(1_V \oplus (-\beta)) = 1_V \oplus (-\beta) \oplus \beta \oplus (-\beta^2)$$
$$= 1_V \oplus (-\beta) \oplus \beta \oplus (-1_V).$$

Therefore, we obtain a contradiction. Hence, the corollary holds. $\square$

Corollary 2.10. $G_n(R) \in \mathcal{SR}$ *if and only if* $n = 1$.

Proof. It is an immediate consequences of Corollary 2.9. $\square$

We note here that if $\dim_R V = 0$, then $L_R(V) = \{0\}$ and $L_R(V) \setminus G_R(V) = \emptyset$, and if $\dim_R V = 1$, then $L_R(V) \cong R$ and $L_R(V) \setminus G_R(V) = \{0\}$.

Theorem 2.32. *Assume that* $\dim_R V > 1$ *and let* S *be a subsemigroup of* $L_R(V)$ *containing* $L_R(V) \setminus G_R(V)$. *If* $S \in \mathcal{SHR}$, *then* $S = L_R(V)$.

Proof. Let $\oplus$ be a hyperoperation on S such that $(S, \oplus, \cdot)$ is a hyperring, where $\cdot$ is the operation on S. In order to show that $S = L_R(V)$, let $\alpha \in L_R(V)$. If $\alpha \notin G_R(V)$, then $\alpha \in L_R(V) \setminus G_R(V) \subseteq S$. Suppose that $\alpha \in G_R(V)$. Let B be a basis of V. Then, $|B| \geq 2$. Let $u \in B$ be fixed. Then, $\{\alpha(u)\}$ is not a basis of V. Let $\beta, \gamma \in L_R(V)$ be defined by

$$\beta(v) = \begin{cases} \alpha(v) & \text{if } v \in B \setminus \{u\}, \\ 0 & \text{if } v = u; \end{cases}$$

and

$$\gamma(v) = \begin{cases} \alpha(u) & \text{if } v = u, \\ 0 & \text{if } v \in B \setminus \{u\}. \end{cases}$$

Then, $\beta, \gamma \in L_R(V) \setminus G_R(V) \subseteq S$, so $\beta \oplus \gamma \subseteq S$. For each $w \in B$, let $\lambda_w \in L_R(V)$ be defined by

$$\lambda_w(v) = \begin{cases} w & \text{if } v = w, \\ 0 & \text{if } v \in B \setminus \{w\}. \end{cases}$$

Then, $\lambda_w \in S$ for all $w \in B$, so $(\beta \oplus \gamma)\lambda_w = \beta\lambda_w \oplus \gamma\lambda_w$ for all $w \in B$. We clearly have

$$\beta\lambda_u = 0, \quad (\gamma\lambda_u)(u) = \alpha(u),$$
$$\beta\lambda_v(v) = \alpha(v) \text{ for all } v \in B \setminus \{u\}, \text{ and}$$
$$\gamma\lambda_v = 0 \text{ for all } v \in B \setminus \{u\}.$$

Consequently, we have

$$(\beta \oplus \gamma)\lambda_u = \{\gamma\lambda_u\}$$

and

$$(\beta \oplus \gamma)\lambda_v = \{\beta\lambda_v\} \quad \text{for all } v \in B \setminus \{u\}.$$

Let $\eta \in \beta \oplus \gamma$. Then, we have

$$\eta\lambda_u = \gamma\lambda_u \quad \text{and} \quad \eta\lambda_v = \beta\lambda_v \quad \text{for all } v \in B \setminus \{u\}.$$

These imply that

$$\eta(u) = (\eta\lambda_u)(u) = (\gamma\lambda_u)(u) = \alpha(u)$$

and

$$\eta(v) = (\eta\lambda_v)(v) = (\beta\lambda_v)(v) = \alpha(v) \quad \text{for all } v \in B \setminus \{u\}.$$

It follows that $\alpha = \eta \in \beta \oplus \gamma \subseteq S$. Hence, we prove that $S = L_R(V)$, as required. $\square$

Corollary 2.11. *Assume that $n > 1$, and let S be a subsemigroup of $M_n(R)$ containing all singular matrices in $M_n(R)$. If $S \in \mathcal{SHR}$, then $S = M_n(R)$.*

Proof. It is a consequence of Theorem 2.32. $\square$

Chapter 3

Homomorphisms and Isomorphisms

3.1 Homomorphisms and Good Homomorphisms

Definition 3.1. A *homomorphism* from a Krasner hyperring $(R, +, \cdot)$ onto a Krasner hyperring $(R', +, \cdot)$ is a function $f : R \to R'$ such that

$$f(x + y) \subseteq f(x) + f(y) \quad \text{and} \quad f(x \cdot y) = f(x) \cdot f(y),$$

for all $x, y \in R$.

Definition 3.2. Let $f : R \to R'$ be a homomorphism of Krasner hyperrings. The *kernel* of f (denoted by $Kerf$ is $\{x \in R \mid f(x) = 0\}$. If A is a subset of R, then $f(A) = \{y \in R' \mid y = f(x), \text{ for some } a \in A\}$ is the *image* of f and denoted as Imf. If B is a subset of R', then $f^{-1}(B) = \{x \in R \mid f(x) \in R'\}$ is the *inverse image* of B.

Note that $Kerf$ may be empty.

Example 3.1. Consider the Krasner hyperring $(R, \oplus, \cdot)$ defined in Example 2.13. Define $f[0, 1] \to [0, 1]$ by $f(x) = 1$, for all $x \in [0, 1]$. Since $1 \oplus 1 = [0, 1]$ and $1 \cdot 1 = 1$, it follows that f is a homomorphism from the Krasner hyperring $([0, 1], \oplus, \cdot)$ onto itself. Now, we observe that $Kerf = \emptyset$.

Now, we present the general properties of a Krasner hyperring homomorphism with a non-empty kernel.

Theorem 3.1. *Let f be a homomorphism from a Krasner hyperring R onto a Krasner hyperring R' such that the kernel of f is non-empty. Then, the following statements hold:*

(1) $f(0) = 0$;
(2) $f(-x) = -f(x)$, *for all* $x \in R$;
(3) $Kerf$ *is a hyperideal of* R.

Proof. (1) Suppose that $a \in Kerf$. Then, $f(a) = 0$, and so we obtain

$$\{f(0)\} = f(0)+0 = f(0)+f(a) \supseteq f(0+a) = f(\{a\}) = \{f(a)\} = \{0\}.$$

This yields $f(0) = 0$.

(2) Let x be an arbitrary element of R. Since $0 \in x - x$, it follows that $f(0) \in f(x - x) \subseteq f(x) + f(-x)$, and so $0 \in f(x) + f(-x)$. This implies that $f(-x) \in -f(x) + 0$. Therefore, we get $f(-x) = -f(x)$, for all $x \in R$.

(3) By Part (1), we have $0 \in Kerf$. Suppose that $x, y \in Kerf$ and $r \in R$. Then, $f(x) = 0 = f(y)$. Since $f(x + y) \subseteq f(x) + f(y) = 0 + 0 = \{0\}$, it follows that $x + y \subseteq Kerf$. On the other hand, since $f(-x) = -f(x)$, we get $f(-x) = 0$, and so $-x \in Kerf$. Moreover, $f(rx) = f(r)f(x) = f(r)0 = 0$ and $f(xr) = f(x)f(r) = 0f(r) = 0$. Consequently, $rx \in Kerf$ and $xr \in Kerf$. This shows that $Kerf$ is a hyperideal of R. □

Definition 3.3. A function f from $(R, +, \cdot)$ onto $(R', +, \cdot)$ is said to be a *good (or strong) homomorphism* if

$$f(x + y) = f(x) + f(y), \quad f(xy) = f(x)f(y) \quad \text{and} \quad f(0) = 0,$$

for all $x, y \in R$.

An *isomorphism* from $(R, +, \cdot)$ onto $(R', +, \cdot)$ is a bijective good homomorphism. The Krasner hyperrings $(R, +, \cdot)$ and $(R', +, \cdot)$ are said to be *isomorphic*, and we write $R \cong R'$ if there is an isomorphism from R onto R'.

Theorem 3.2. *If f is a good homomorphism from R onto R', then Imf is a subhyperring of R'.*

Proof. Let x, y be two arbitrary elements of R. Then, we have

$$f(x) + f(y) = f(x + y) \subseteq Imf,$$
$$-f(x) = f(-x) \in Imf,$$
$$f(x)f(y) = f(xy) \in Imf.$$

Therefore, Imf is a subhyperring of R'. $\qquad\square$

Theorem 3.3. *Let f be a good homomorphism from R onto R'. Then, f is one to one if and only if $Kerf = \{0\}$.*

Proof. Let $y, z \in R$ be such that $f(y) = f(z)$. Then, $f(y) - f(y) = f(z) - f(y)$. It follows that $f(0) \in f(z - y)$, and so there exists $x \in z - y$ such that $0 = f(0) = f(x)$. Thus, if $Kerf = \{0\}$, $x = 0$, whence $y = z$.

Now, let $x \in Kerf$. Then, $f(x) = 0 = f(0)$. Thus, if f is one to one, we conclude that $x = 0$. $\qquad\square$

Example 3.2. Suppose that $([0, 1], \oplus, \cdot)$ is a Krasner hyperring defined as in Example 2.13. Define $f : [0; 1] \to [0; 1]$ by $f(0) = 0$ and $f(x) = 1$, for all $x \in (0, 1]$. Since

$$f(x \oplus y) = \begin{cases} f(\{0\}) = \{0\} = 0 \oplus 0 = f(x) \oplus f(y) & \text{if } x = y = 0 \\ f(\{x\}) = \{1\} = 1 \oplus 0 = f(x) \oplus f(y) & \text{if } x > y = 0 \\ f(\{x\}) = \{1\} \subset [0, 1] = 1 \oplus 1 = f(x) \oplus f(y) & \text{if } x > y > 0 \\ f([0; x]) = \{0, 1\} \subset [0, 1] = 1 \oplus 1 = f(x) \oplus f(y) & \text{if } x = y > 0 \end{cases}$$

$$f(xy) = f(x)f(y) = \begin{cases} 0 & \text{if } x = 0 \quad \text{or} \quad y = 0 \\ 1 & \text{if } x \neq 0 \quad \text{and} \quad y \neq 0, \end{cases}$$

it follows that f is a homomorphism from $([0, 1], \oplus, \cdot)$ onto itself, which is not a good homomorphism. Also, $Kerf = \{0\}$, but f is not one to one. In addition, $f([0, 1]) = \{0, 1\}$, which is not a subhyperring of the Krasner hyperring $([0, 1], \oplus, \cdot)$.

3.2 Isomorphism Theorems of Krasner Hyperrings

In what follows, we present the isomorphism theorems of Krasner hyperrings.

Theorem 3.4 (First isomorphism theorem). *If f is a good homomorphism from R onto R', then $R/Kerf \cong Imf$.*

Proof. Taking $K = Kerf$, we define $\rho : R/K \to Imf$ by setting $\rho(x + K) = f(x)$, for all $x \in R$. We prove first that ρ is well defined. Let $x, y \in R$ such that $x + K = y + K$. Then, $x \in y + K$, and so $x \in y + a$, for some $a \in K$. Consequently, we get

$$f(x) \in f(y + a) = f(y) + f(a) = f(y) + 0 = \{f(y)\},$$

which implies that $f(x) = f(y)$. Hence, $\rho(x + K) = \rho(y + K)$. This proves that the map ρ is well defined. Clearly, ρ is onto. To show that ρ is one to one, suppose that $\rho(x + K) = \rho(y + K)$, or, equivalently, $f(x) = f(y)$. This means that $0 \in f(x) - f(y) = f(x - y)$. Thus, there is $z \in x - y$ such that $f(z) = 0$. This yields $z \in K$. Now, we have

$$z \in x - y \quad \Rightarrow \quad x \in z + y \quad \Rightarrow \quad x \in y + K.$$

Thus, $x + K = y + K$, and so ρ is one to one. To show that ρ is a good homomorphism, suppose that $x, y \in R$. Then,

$$\begin{aligned}
\rho\big((x + K) \oplus (y + K)\big) &= \rho\{z + K \mid z \in x + y\}) = \{\rho(z + K) \mid z \in x + y\} \\
&= \{f(z) \mid z \in x + y\} = f(x + y) = f(x) + f(y) \\
&= \rho(x + K) + \rho(y + K),
\end{aligned}$$

and

$$\begin{aligned}
\rho\big((x + K) \odot (y + K)\big) &= \rho(xy + K) = f(xy) = f(x)f(y) \\
&= \rho(x + K)\,\rho(y + K).
\end{aligned}$$

In addition, we have $\rho(I) = \rho(0 + I) = f(0) = 0$. Therefore, ρ is an isomorphism of Krasner hyperrings. $\square$

We are now in a position to state and prove the second and third isomorphism theorems of Krasner hyperrings.

Theorem 3.5 (Second isomorphism theorem). *If I and J are hyperideals of a Krasner hyperring R, then $J/(I \cap J) \cong (I + J)/I$.*

Proof. We define $\rho : J \to (I+J)/I$ by $\rho(x) = x + I$, for all $x \in J$. Then, for every $a, b \in J$, we have

$$\rho(a+b) = \rho(\{x \,|\, x \in a+b\}) = \{\rho(x) \,|\, x \in a+b\}$$
$$= \{x + I \,|\, x \in a + b\} = (a+I) \oplus (b+I)$$
$$= \rho(a) \oplus \rho(b)$$

and

$$\rho(ab) = ab + I = (a+I) \odot (b+I) = \rho(a) \odot \rho(b).$$

Moreover, $\rho(0) = I$. This shows that ρ is a good homomorphism.

Now, let $x + I \in (I+J)/J$ be arbitrary. Then, $x \in y + I$, for some $y \in I + J$. That is, $y \in a + b$, for some $a \in I$, and $b \in J$. Since $y \in b + I$, it follows that $y + I = b + I$. So, we have

$$\rho(b) = b + I = y + I = x + I.$$

This shows that ρ is onto. Therefore, by the first isomorphism theorem of Krasner hyperrings, we get $J/Ker\rho \cong (I+J)/I$. Finally, we obtain

$$Ker\rho = \{a \in J \,|\, \rho(a) = 0\} = \{a \in J \,|\, a + I = I\}$$
$$= \{a \in J \,|\, a \in I\} = I \cap J.$$

This completes the proof. □

Theorem 3.6 (Third isomorphism theorem). *If I and J are hyperideals of a Krasner hyperring R such that $I \subseteq J$, then $(R/I)/(J/I) \cong R/J$.*

Proof. We define $\rho : R/I \to R/J$ by $\rho(x + I) = x + J$, for all $x \in R$. Then, ρ is a good homomorphism of Krasner hyperrings with kernel J/I. Moreover, ρ is onto. Therefore, by the first isomorphism theorem of Krasner hyperrings, we obtain $(R/I)/(J/I) \cong R/J$. □

3.3   The Hyperring of Fractions

Let R be a commutative Krasner hyperring with identity, and let S be a multiplicatively closed subset of R such that $0 \notin S$. The relation $\sim$ is defined on $R \times S$ as follows: $(a, s) \sim (b, t)$ if and only if there exists $u \in S$ such that $uta = usb$. This is an equivalence relation on the set $R \times S$. The equivalence class of (a, s) is denoted by a/s, and we let $S^{-1}R$ be the quotient set. On $S^{-1}R$, the hyperoperation $\oplus$ is defined by

$$\frac{a}{s} \oplus \frac{b}{t} = \left\{ \frac{c}{st} \mid c \in ta + sb \right\} = \frac{ta + sb}{st},$$

and multiplication is defined in the standard way. One can easily verify all conditions of the definition. We prove only condition (v). Suppose that $\frac{z}{r} \in \frac{x}{s} \oplus \frac{y}{t} = \frac{xt+ys}{st}$. Then, there exists $v \in xt + ys$ such that $\frac{z}{r} = \frac{v}{st}$. Hence, there exists $u \in S$ such that $uzst = urv$. Thus, $uzst \in ur(xt + ys) = urxt + urys$, and so $urxt \in uzst - urys$ and $urys \in -urxt + uzst$. Therefore,

$$\frac{x}{s} = \frac{urtx}{urts} \in \frac{uzst - urys}{urts} = \frac{uzst}{urts} - \frac{urys}{urts} = \frac{z}{r} - \frac{y}{t}$$

and

$$\frac{y}{t} \in \frac{-x}{s} \oplus \frac{z}{r}.$$

Therefore, $S^{-1}R$ is a Krasner hyperring.

The hyperring $S^{-1}R$ is usually called the *hyperring of fractions* of R.

Theorem 3.7. *Let S be a multiplicatively closed subset of a commutative Krasner hyperring R with identity. If I and J are hyperideals in R, then:*

(1) $S^{-1}(I + J) = S^{-1}I + S^{-1}J,$
(2) $S^{-1}(IJ) = (S^{-1}I)(S^{-1}J),$
(3) $S^{-1}(I \cap J) = S^{-1}I \cap S^{-1}J.$

Proof. The proof is left as an exercise to the reader. □

Lemma 3.1. *The natural mapping $\varphi : R \to S^{-1}R$, where $\varphi(r) = r/1$, is a good homomorphism.*

Proof. It is straightforward. □

Theorem 3.8. *Let I be a hyperideal of a commutative Krasner hyperring R with identity. Then, $S \cap I \neq \emptyset$ if and only if $S^{-1}I = S^{-1}R$.*

Proof. If $t \in S \cap I$, then $t/t = 1/1 \in S^{-1}I$. Thus, for every $r/s \in S^{-1}R$, we have $1/1 \odot r/s \in S^{-1}I$. Since $r/s = 1/1 \odot r/s$, we get $r/s \in S^{-1}I$, and this proves $S^{-1}R \subseteq S^{-1}I$.

Conversely, assume that $S^{-1}I = S^{-1}R$. If we consider the natural homomorphism $\varphi : R \to S^{-1}R$, then $\varphi(1) = 1/1$. On the other hand, $\varphi(1) \in S^{-1}R$; consequently, $\varphi(1) \in S^{-1}I$, and so $\varphi(1) = a/s$, for some $a \in I$, $s \in S$. Now, we have $1/1 = a/s$; therefore, there exists $t \in S$ such that $ts = ta$. Since $ts \in S$ and $ta \in I$, we get $I \cap S \neq \emptyset$. $\qquad\square$

Theorem 3.9. *Let* $\varphi : R \to S^{-1}R$ *be the natural homomorphism and* I *be a hyperideal in* R. *Then*:

(1) $I \subseteq \varphi^{-1}(S^{-1}I)$.
(2) *If* $I = \varphi^{-1}(J)$ *for some hyperideal* J *in* $S^{-1}R$, *then* $S^{-1}I = J$. *In other words, every hyperideal in* $S^{-1}R$ *is of the form* $S^{-1}I$, *for some hyperideal* I *in* R.
(3) *If* P *is a prime hyperideal in* R *and* $S \cap P = \emptyset$, *then* $S^{-1}P$ *is a prime hyperideal in* $S^{-1}R$ *and* $\varphi^{-1}(S^{-1}P) = P$.

Proof. The proof of (1) is obvious. To prove (2), suppose that $I = \varphi^{-1}(J)$. If $r/s \in S^{-1}I$, then $r \in I$, and so $\varphi(r) \in J$ or $r/1 \in J$. Therefore, $1/s \odot r/1 \in J$. Since $r/s = 1/s \odot r/1$, it follows that $r/s \in J$. This implies $S^{-1}I \subseteq J$. Now, let $r/s \in J$. Then, $\varphi(r) = r/1 = r/1 \odot s/s = r/s \odot s/1 \in J$. This yields $\varphi(r) \in J$, which implies that $r \in \varphi^{-1}(J)$ or $r \in I$. Therefore, $r/s \in S^{-1}I$, and this shows that $J \subseteq S^{-1}I$.

(3) We know that $S^{-1}P$ is a hyperideal such that $S^{-1}P \neq S^{-1}R$. If $r/s \odot r'/s' \in S^{-1}P$, then $rr'/ss' \in S^{-1}P$. Then, there exist $a \in P$ and $t \in S$ such that $rr'/ss' = a/t$. Consequently, there exists $u \in S$ such that $utrr' = uss'a$. Since $ut \in S$ and $S \cap P = \emptyset$, it follows that $rr' \in P$, whence $r \in P$ or $r' \in P$. So, $r/s \in S^{-1}P$ or $r'/s' \in S^{-1}P$. Therefore, $S^{-1}P$ is prime. Moreover, $P \subseteq \varphi^{-1}(S^{-1}P)$ by (1).

Conversely, assume that $r \in \varphi^{-1}(S^{-1}P)$. Then, $\varphi(r) \in S^{-1}P$. Since $\varphi(r) = r/1$, it follows that there exist $a \in P$ and $t \in S$ such that $r/1 = a/t$, and so there is $u \in S$ such that $urt = ua$. Since $ua \in P$, it follows that $urt \in P$. Now, we have $utr \in P$ and $ut \notin P$; therefore, $r \in P$, and the proof is complete. $\qquad\square$

Theorem 3.10. *Let* S *be a multiplicatively closed subset of a commutative Krasner hyperring* R *with identity. Then, there is a* *one-to-one correspondence between the set $\mathcal{U}$ of prime hyperideals of R, which are disjoint with S, and the set $\mathcal{V}$ of prime hyperideals of $S^{-1}R$, given by $P \longmapsto S^{-1}P$.*

Proof. We define $\psi : \mathcal{U} \to \mathcal{V}$ by $P \longmapsto S^{-1}P$. By Theorem 3.9(3), this assignment defines a one-to-one function. So, it is enough to show that ψ is onto. Assume that J is a prime hyperideal of $S^{-1}R$, and let $P = \varphi^{-1}(J)$. Since $S^{-1}P = J$, by Theorem 3.9(3), we need only show that P is prime. If $ab \in P$, then $\varphi(a)\varphi(b) = \varphi(ab) \in J$. Since J is prime in $S^{-1}R$, it follows that $\varphi(a) \in J$ or $\varphi(b) \in J$. This yields either $a \in \varphi^{-1}(J) = P$ or $b \in \varphi^{-1}(J) = P$. Therefore, P is prime. □

Let R be a commutative Krasner hyperring with identity and P be a prime hyperideal of R. Then, $S = R\backslash P$ is a multiplicatively closed subset of R. The hyperring of fractions $S^{-1}R$ is called the *localization* of R at P and is denoted by R_P. If I is a hyperideal of R, then the hyperideal $S^{-1}I$ in R_P is denoted by I_P.

Theorem 3.11. *Let P be a prime hyperideal in a hyperring R.*

(1) *There is a one-to-one correspondence between the set of prime hyperideals of R, which are contained in P, and the set of prime hyperideals of R_P, given by $Q \longmapsto Q_P$.*
(2) *The hyperideal P_P in R_P is the unique maximal hyperideal of R_P.*

Proof. The proof is left as an exercise to the reader. □

Let R be a commutative Krasner hyperring with identity and A be a subhyperring of R. For a multiplicatively closed subset S of A, we can form a large hyperring of quotients $A_{[S]} = \{x \in R \mid (\exists s \in S)$ $xs \in A\}$. In fact, $A_{[S]}$ is a subhyperring of R and $A \subseteq A_{[S]}$. For a hyperideal I of A, its large extension is defined as the set $[I]A_{[S]} = \{x \in R \mid (\exists s \in S) \, xs \in I\}$, which is a hyperideal of $A_{[S]}$. If $S = A\backslash P$, for a given prime hyperideal P of A, then we shall write $A_{[P]}$ instead of $A_{[A\backslash P]}$. It is evident that the following equality holds: $([P]A_{[P]}) \cap A = P$.

Let R be a commutative Krasner hyperring with identity and S be a multiplicatively closed set in R. The natural map $\varphi_S : R \to S^{-1}R$

given by $\varphi_S(a) = a/1$, for all $a \in R$, is a good homomorphism. Note that φ_S has the following properties:

- For every $s \in S$, the element $\varphi_S(s) = s/1$ is a unit of $S^{-1}R$, having an inverse of $1/s$.
- If $a \in Ker\varphi_S$, then there exists $s \in S$ such that $sa = 0$.
- Each element a/s of $S^{-1}R$ can be written as $a/s = \varphi_S(a)\big(\varphi_S(s)\big)^{-1}$ because

$$\frac{a}{s} = \frac{a}{1}\frac{1}{s} = \frac{a}{s}\left(\frac{s}{1}\right)^{-1} = \varphi_S(a)\big(\varphi_S(s)\big)^{-1}.$$

Theorem 3.12 (Universal property). *Let S be a multiplicatively closed set of the commutative Krasner hyperring R with identity and $\varphi_S : R \to S^{-1}R$ denote the natural homomorphism. Suppose that R' is another commutative Krasner hyperring with identity and $\theta : R \to R'$ is a good homomorphism with the property that $\theta(s)$ is a unit of R', for all $s \in S$. Then, there is a unique good homomorphism $\psi : S^{-1}R \to R'$ such that $\psi \circ \varphi_S = \theta$, i.e., the diagram in Figure 3.1 is commutative. In particular, we have*

$$\psi(a/s) = \theta(a)\big(\theta(s)\big)^{-1}, \qquad (3.1)$$

for all $a \in R$ and $s \in S$.

Proof. First, we show that the formula given (3.1) is unambiguous. Let $a, b \in R$ and $s, t \in S$ such that $a/s = b/t$ in $S^{-1}R$. Then, there is $u \in S$ such that $uta = usb$. Applying the good homomorphism θ

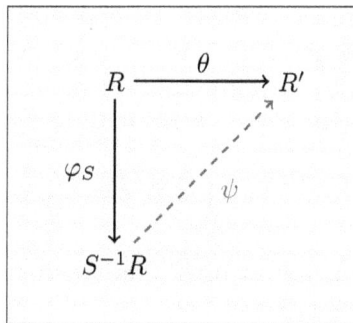

Fig. 3.1.   Universal property of rings of fractions.

to this equation, we get

$$\theta(u)\theta(t)\theta(a) = \theta(u)\theta(s)\theta(b). \tag{3.2}$$

On the other hand, by hypothesis, $\theta(u)$, $\theta(s)$, and $\theta(t)$ are invertible in R'. Multiplying (3.2) by the product of their inverses, we obtain

$$\theta(a)\big(\theta(s)\big)^{-1} = \theta(b)\big(\theta(t)\big)^{-1}.$$

This means that we can define a map $\psi : S^{-1}R \to R'$ by the formula given in (3.1), i.e., this map is well defined. Now, we show that ψ is a good homomorphism. For any a/s, b/t in $S^{-1}R$, we have

$$\psi\left(\frac{a}{s} + \frac{b}{t}\right) = \psi\left(\frac{ta + sb}{st}\right) = \psi(ta + sb)\big(\psi(st)\big)^{-1}$$

$$= \psi(t)\psi(a)\big(\psi(s)\big)^{-1}\big(\psi(t)\big)^{-1} + \psi(s)\psi(b)\big(\psi(s)\big)^{-1}\big(\psi(t)\big)^{-1}$$

$$= \psi(a)\big(\psi(s)\big)^{-1} + \psi(b)\big(\psi(t)\big)^{-1} = \psi\left(\frac{a}{s}\right) + \psi\left(\frac{b}{t}\right)$$

and

$$\psi\left(\frac{a}{s}\frac{b}{t}\right) = \psi\left(\frac{ab}{st}\right) = \psi(ab)\big(\psi(st)\big)^{-1} = \psi(a)\big(\psi(s)\big)^{-1}\psi(b)\big(\psi(t)\big)^{-1}$$

$$= \psi\left(\frac{a}{s}\right)\psi\left(\frac{b}{t}\right).$$

Moreover, we have $\psi(1/1) = \psi(1)\big(\psi(1)\big)^{-1} = 1 \cdot 1 = 1$. Also, we observe that $\psi \circ \varphi_S = \theta$ because, for all $a \in R$, we have $(\psi \circ \varphi_S)(a) = \psi(a/1) = \theta(a)\big(\theta(1)\big)^{-1} = \theta(a)$. It remains to be shown that ψ is the only good homomorphism with the stated properties. Assume that $\psi' : S^{-1}R \to R'$ is a good homomorphism such that $\psi' \circ \varphi_S = \theta$. Then, for every $a \in R$, we have $\psi'(a/1) = (\psi' \circ \varphi_S)(a) = \theta(a)$. In particular, for $s \in S$, we have $\psi'(s/1) = \theta(s)$. Now, we can write

$$\psi'\left(\frac{1}{s}\right)\theta(s) = \psi'\left(\frac{1}{s}\right)\psi'\left(\frac{s}{1}\right) = \psi'\left(\frac{1}{s}\frac{s}{1}\right) = \psi'(1_{S^{-1}R}) = 1_{R'}.$$

This yields that $\theta(s)$ is a unit in R', and we have

$$\psi'\left(\frac{1}{s}\right) = \big(\theta(s)\big)^{-1}.$$

Therefore, for all $a \in R$ and $s \in S$, we have

$$\psi'\left(\frac{a}{s}\right) = \psi'\left(\frac{1}{s}\frac{1}{s}\right) = \psi'\left(\frac{a}{1}\right)\psi'\left(\frac{1}{s}\right) = \theta(a)\big(\theta(s)\big)^{-1}.$$

Consequently, we have $\psi = \psi'$, and so there exists precisely one good homomorphism with the desired properties. □

Corollary 3.1. *Let R and R' be two commutative Krasner hyper-rings with identity and S be a multiplicatively closed set in R. Let $\theta : R \to R'$ be a good homomorphism with the following properties:*

(1) *If $s \in S$, then $\theta(s)$ is a unit in R';*
(2) *For every $a \in R$, if $\theta(a) = 0$, then there is $s \in S$ such that $sa = 0$;*
(3) *Every element of R' has the form $\theta(a)\big(\theta(s)\big)^{-1}$, where $a \in R$ and $s \in S$.*

Then, there exists a unique isomorphism $\psi : S^{-1}R \to R'$ such that $\psi \circ \varphi_S = \theta$.

Proof. By Theorem 3.12, there is a unique good homomorphism θ such that $\psi \circ \varphi_S = \theta$. Moreover, in the proof of Theorem 3.12, we observe that for any $a/s \in S^{-1}R$,

$$\psi\left(\frac{a}{s}\right) = \theta(a)\big(\theta(s)\big)^{-1}.$$

Hence, by property (3), we deduce that θ is onto. On the other hand, if $\theta(a)\big(\theta(s)\big)^{-1} = 0$, then $\theta(a) = 0$. By property (2), there exists $t \in S$ such that $ta = 0$. This implies that $a/s = 0$ in $S^{-1}R$. Therefore, θ is an isomorphism, as desired. □

Chapter 4

Generalizations of Hyperideals

4.1   2-Absorbing Hyperideals

In Ref. [41], Kamali Ardekani and Davvaz gave a characterization of a new generalization of prime hyperideals in Krasner hyperrings by introducing 2-absorbing hyperideals. In this section, we review their results.

Definition 4.1. A proper hyperideal I of a Krasner hyperring $(R, +, \cdot)$ is called a 2-*absorbing hyperideal* if $a \cdot b \cdot c \in I$ implies that $a \cdot b \in I$ or $a \cdot c \in I$ or $b \cdot c \in I$, for all $a, b, c \in R$.

Example 4.1. Let $(G, \odot)$ be a group and $H = G \cup \{0, u, v\}$, where 0 is an absorbing element under multiplication and u, v are distinct orthogonal idempotents with

$$u \odot v = v \odot u = 0; \quad u \odot u = u; \quad v \odot v = v;$$

$$a \odot 0 = 0 \odot a = 0, \quad \text{for all } a \in H;$$

$$u \odot g = g \odot u = u; \quad v \odot g = g \odot v = v, \quad \text{for all } g \in G.$$

Let us define hyperoperation $\oplus$ on H as follows:

$$a \oplus 0 = 0 \oplus a = \{a\}; \quad a \oplus a = \{0, a\}, \quad \text{for all } a \in H;$$

$$a \oplus b = b \oplus a = H \setminus \{0, a, b\}, \quad \text{for all } a, b \in H \setminus \{0\} \quad \text{and} \quad a \neq b.$$

Then, $(H, \oplus, \odot)$ is a Krasner hyperring. Put $I = \{0, u\}$ and $J = \{0\}$. Obviously, I and J are 2-absorbing hyperideals. The hyperideal I is prime but J is not because $u \odot v = 0 \in J$ while $u, v \notin J$.

Example 4.2. Let $(R, +, \cdot)$ be a hyperdomain and

$$M = \left\{ \begin{pmatrix} x_1 & x_2 & x_3 \\ 0 & 0 & x_4 \\ 0 & 0 & 0 \end{pmatrix} \middle| x_1, x_2, x_3 \in R \right\}.$$

Put $I = \left\{ \begin{pmatrix} 0 & 0 & a \\ 0 & 0 & b \\ 0 & 0 & 0 \end{pmatrix} \middle| a, b \in R \right\}$, and define the hyperoperation $\oplus$ on M as

$$\begin{pmatrix} x_1 & x_2 & x_3 \\ 0 & 0 & x_4 \\ 0 & 0 & 0 \end{pmatrix} \oplus \begin{pmatrix} x_1' & x_2' & x_3' \\ 0 & 0 & x_4' \\ 0 & 0 & 0 \end{pmatrix} = \left\{ \begin{pmatrix} y_1 & y_2 & y_3 \\ 0 & 0 & y_4 \\ 0 & 0 & 0 \end{pmatrix} \middle| y_i \in x_i + x_i', 1 \leq i \leq 4 \right\},$$

where for $1 \leq i \leq 4$, we have $x_i, x_i', y_i \in R$. Then, M, with the hyperoperation $\oplus$ and usual multiplication of matrices, is a Krasner hyperring. Furthermore, it is not difficult to see that I is a 2-absorbing hyperideal of M.

One can show that a non-zero proper hyperideal I of a Krasner hyperring R is a 2-absorbing hyperideal if and only if whenever $I_1 \cdot I_2 \cdot I_3 \subseteq I$, for some hyperideals I_1, I_2, I_3 of R, then $I_1 \cdot I_2 \subseteq I$ or $I_2 \cdot I_3 \subseteq I$ or $I_1 \cdot I_3 \subseteq I$.

Hereinafter, the Krasner hyperring $(R, +, \cdot)$ is commutative with unit element.

Theorem 4.1. *Let I be a 2-absorbing hyperideal of R. Then, one of the following statements is valid:*

(1) $\sqrt{I} = P$ *is a prime hyperideal of R and $P^2 \subseteq I$.*
(2) $\sqrt{I} = P_1 \cap P_2$. $P_1 P_2 \subseteq I$ *and* $(\sqrt{I})^2 \subseteq I$, *where P_1 and P_2 are the only distinct prime hyperideals of R that are minimal over I.*

Proof. We prove this statement in three steps:
Step 1. $\sqrt{I}$ is a 2-absorbing hyperideal of R.
Suppose that $x, y, z \in R$ such that $xyz \in \sqrt{I}$. By assumption, $(xyz)^2 \in I$. Thus, $x^2 y^2 z^2 \in I$, and this implies that $(xy)^2 = x^2 y^2 \in I$, $(xz)^2 = x^2 z^2 \in I$, or $(yz)^2 = y^2 z^2 \in I$. Therefore, at least one of xy, xz, and yz belongs to $\sqrt{I}$.

Step 2. There are at most two prime hyperideals of R that are minimal over I.

Suppose that P_1 and P_2 are prime hyperideals of R that are minimal over I. Then, there are $x_1 \in P_1 \setminus P_2$ and $x_2 \in P_2 \setminus P_1$. Also, there exist $c_2 \in R \setminus P_1$, $c_1 \in R \setminus P_2$, and $m, n \in \mathbb{N}$ such that $c_2 x_1^n, c_1 x_2^m \in I$. This implies that $c_2 x_1, c_1 x_2 \in I \subseteq P_1 \cap P_2$ because I is a 2-absorbing hyperideal. Consequently, $c_1 \in P_1 \setminus P_2$ and $c_2 \in P_2 \setminus P_1$. Hence, $(c_1 + c_2) \cap P_1 = \emptyset$ since if $t \in (c_1 + c_2) \cap P_1$, then $c_2 \in -c_1 + t \subseteq P_1$, which contradicts $c_2 \notin P_1$. In the same way, $(c_1 + c_2) \cap P_2 = \emptyset$. Therefore, for all $t \in c_1 + c_2$, we have $t x_2 \notin P_1$ and $t x_1 \notin P_2$, which lead to $t x_1, t x_2 \notin I$. On the other hand, $(c_1 + c_2) x_1 x_2 \subseteq I$. Thus, for all $t \in c_1 + c_2$, we get $t x_1 x_2 \in I$, which implies that $x_1 x_2 \in I$.

Now, suppose that P_3 is a prime hyperideal of R that is minimal over I and $P_3 \neq P_1, P_2$. Consequently, there are $y_1 \in P_1 \setminus (P_2 \cup P_3)$ and $y_2 \in P_2 \setminus (P_1 \cup P_3)$. Then, based on the previous argument $y_1 y_2 \in I \subseteq P_3$, which leads to $y_1 \in P_3$ or $y_2 \in P_3$, which is a contradiction.

Step 3. In this step, we prove the principal assertion of the theorem. Suppose that $x, y \in \sqrt{I}$. Then, $x^2, y^2 \in I$, and so $x(x + y)y \subseteq I$. Therefore, for all $t \in x + y$, we have $xt \in I$, $xy \in I$, or $ty \in I$ because I is a 2-absorbing hyperideal. If $xt \in I$, then $xt \in x(x+y) = x^2 + xy$ and, consequently, $xy \in -x^2 + xt \subseteq I$. Similarly, $ty \in I$ yields $xy \in I$. Therefore, we have $(\sqrt{I})^2 \subseteq I$. By Steps (1) and (2), $\sqrt{I} = P$ is a prime hyperideal of R or $\sqrt{I} = P_1 \cap P_2$, where P_1 and P_2 are the only distinct prime hyperideals of R that are minimal over I. If $\sqrt{I} = P$, then $P^2 = (\sqrt{I})^2 \subseteq I$. If $\sqrt{I} = P_1 \cap P_2$, then for all $y \in \sqrt{I}$, $z_1 \in P_1 \setminus P_2$, and $z_2 \in P_2 \setminus P_1$, we have $y + z_1 \subseteq P_1 \setminus P_2$. By the same argument in Step 2, we get $z_1 z_2 \in I$ and $(y + z_1) z_2 \subseteq I$. Thus, for all $s \in y z_2 + z_1 z_2$, we have $y z_2 \in s - z_1 z_2 \subseteq I$. Similarly, $y z_1 \in I$, and this follows that $P_1 P_2 \subseteq I$. □

Theorem 4.2. *Let I be a hyperideal of R. Then, I is a 2-absorbing hyperideal of R if and only if $(I :_R x)$ is a prime hyperideal of R containing $\sqrt{I}$, for all $x \in \sqrt{I} \setminus I$.*

Proof. By Theorem 4.1, either $\sqrt{I} = P$ or $\sqrt{I} = P_1 \cap P_2$, where P is a prime hyperideal and P_1, P_2 are non-zero distinct prime hyperideals of R that are minimal over I. We prove the statement for the case of $\sqrt{I} = P_1 \cap P_2$ since by putting $P_1 = P_2 = P$ in $\sqrt{I} = P_1 \cap P_2$, we get $\sqrt{I} = P$.

Suppose that I is a 2-absorbing hyperideal of R. According to Theorem 4.1, we conclude that $xP_1, xP_2 \subseteq I$, for all $x \in \sqrt{I} \setminus I$. This means that $P_1, P_2 \subseteq (I :_R x)$ and, consequently, $\sqrt{I} \subseteq (I :_R x)$. Assume that $yz \in (I :_R x)$, where $y, z \in R$ and $x \in \sqrt{I} \setminus I$. Clearly, the statement is valid when $y \in P_1 \cup P_2$ or $z \in P_1 \cup P_2$. Then, we prove it for $y, z \notin P_1 \cup P_2$. In this case, we have $yz \notin P_1 \cap P_2 = \sqrt{I}$, which leads to $yz \notin I$. Hence, by assumption, we get $y \in (I :_R x)$ or $z \in (I :_R x)$, which implies that $(I :_R x)$ is a prime hyperideal.

Conversely, suppose that for all $x \in \sqrt{I} \setminus I$, we have $(I :_R x)$ is a prime hyperideal. First, we prove that $(I :_R x)$ is a prime hyperideal of R, for all $x \in (P_1 \cup P_2) \setminus \sqrt{I}$. For this purpose, we show that for all $x \in P_1 \setminus P_2$, $(I :_R x) = P_2$, and with the same argument, $(I :_R x) = P_1$, for all $x \in P_2 \setminus P_1$. Suppose that $x \in P_1 \setminus P_2$ and $y \in P_2$. Then, $xy \in P_1 P_2$, and so $xy \in I$ by Theorem 4.1. This yields $y \in (I :_R x)$. Now, assume that $y \in (I :_R x)$. Then, $yx \in I \subseteq \sqrt{I} = P_1 \cap P_2$, which implies that $y \in P_2$. Consequently, $(I :_R x) = P_2$ is a prime hyperideal, as desired. By hypothesis, this means that $(I :_R x)$ is a prime hyperideal of R, for all $x \in (P_1 \cup P_2) \setminus I$.

Now, we show that I is a 2-absorbing hyperideal. To do this, suppose that $xyz \in I$, where $x, y, z \in R$. Then, $yz \in (I :_R x)$. Obviously, at least one of x, y, z belongs to $(P_1 \cup P_2) \setminus I$. In order to prove the assertion, without loss of generality, assume that $x \in (P_1 \cup P_2) \setminus I$. In this case, the previous argument leads to $y \in (I :_R x)$ or $z \in (I :_R x)$. Consequently, $yx \in I$ or $zx \in I$, which implies that I is a 2-absorbing hyperideal. $\qquad\square$

Theorem 4.3. *Let I be a 2-absorbing hyperideal and $P = P_1$ and P_2 be prime hyperideals of R. Then:*

(1) *if $\sqrt{I} = P$, then $(I :_R x)$ is a 2-absorbing hyperideal of R, for all $x \in R \setminus P$ with $\sqrt{(I :_R x)} = P$, and $\Omega = \{(I :_R x) \mid x \in R\}$ is a totally ordered set;*

(2) *if $\sqrt{I} = P_1 \cap P_2$, then $(I :_R x)$ is a 2-absorbing hyperideal of R, for all $x \in R \setminus (P_1 \cup P_2)$ with $\sqrt{(I :_R x)} = P_1 \cap P_2$, and $\Omega = \{(I :_R x) \mid x \in R \setminus (P_1 \triangle P_2)\}$ is a totally ordered set;*

(3) *if* $\sqrt{I} = P_1 \cap P_2$, *then* $(I :_R x) = P_2$, *for all* $x \in P_1 \setminus P_2$, *and* $(I :_R x) = P_1$, *for all* $x \in P_2 \setminus P_1$.

Proof. (1) Suppose that $abc \in (I :_R x)$, where $a, b, c \in R$ and $x \in R \setminus P$. Then, $ax \in I$, $bcx \in I$, or $abc \in I$ by hypothesis. If $ax \in I$, then $a \in (I :_R x)$, which implies that $at \in (I :_R x)$, for all $t \in R$. For $bcx \in I$, we have $bc \in (I :_R x)$. If $abc \in I$, then $ab \in I$, $ac \in I$, or $bc \in I$. Therefore, $ab \in (I :_R x)$, $ac \in (I :_R x)$, or $bc \in (I :_R x)$. Based on the above results, we find that $(I :_R x)$ is a 2-absorbing hyperideal.

It is straightforward that $I \subseteq (I :_R x) \subseteq P$, and so $\sqrt{(I :_R x)} = P$. To prove the final assertion, suppose that $x, y \in R \setminus P$, which leads to $xy \in R \setminus P$. We have $(I :_R x) \subseteq (I :_R xy)$ and $(I :_R y) \subseteq (I :_R xy)$, which means that $(I :_R x) \cup (I :_R y) \subseteq (I :_R xy)$. Moreover, if $z \in (I :_R xy)$, then by hypothesis, we get $xz \in I$ or $yz \in I$ since $xy \notin I$. In other words, $z \in (I :_R x)$ or $z \in (I :_R y)$, which means that $(I :_R xy) \subseteq (I :_R x) \cup (I :_R y)$. Therefore, $(I :_R xy) = (I :_R x) \cup (I :_R y)$ and, consequently, $(I :_R x) = (I :_R xy)$ or $(I :_R y) = (I :_R xy)$. Then, $\Omega_1 = \{I :_R x \mid x \in R \setminus P\}$ is a totally ordered set. Now, we prove that $\Omega_2 = \{(I :_R x) \mid x \in P \setminus I\}$ is a totally ordered. For this purpose, it is enough to show that $(I :_R y) \subseteq (I :_R x)$ or $(I :_R x) \subseteq (I :_R y)$, for all $x, y \in P \setminus I$. If $w \in (I :_R y) \cap P$, then $w \in (I :_R x)$, which follows that $(I :_R y) \subseteq (I :_R x)$, as desired. If $w \in (I :_R y) \setminus P$, then for all $z \in (I :_R x) \setminus (I :_R y)$, we get $zw \notin P$. On the other hand, $z(x + y)w \subseteq I$ and $(z(x + y)) \cap I = \emptyset$, which implies that $(x + y)w \subseteq I$. Thus, $xw \in sw - yw \subseteq I$, for all $s \in x + y$. Therefore, $w \in (I :_R x)$ and, consequently, $(I :_R y) \subseteq (I :_R x)$. Therefore, Ω_1 is totally ordered. Based on the above results, we conclude that $\Omega = \Omega_1 \cup \Omega_2 = \{(I :_R x) \mid x \in R\}$ is a totally ordered set.

(2) In view of $I \subseteq (I :_R x) \subseteq P_1 \cap P_2$, we find that $P_1 \cap P_2 = \sqrt{I} \subseteq \sqrt{I :_R x} \subseteq \sqrt{P_1 \cap P_2} = P_1 \cap P_2$. Consequently, $\sqrt{(I :_R x)} = P_1 \cap P_2$. The proofs of the remaining assertions are the same as those in Part (1).

(3) Suppose that $\sqrt{I} = P_1 \cap P_2$ and $x \in P_1 \setminus P_2$. Then, for all $t \in (I :_R x)$, we have $tx \in I \subseteq \sqrt{I} = P_1 \cap P_2 \subseteq P_2$. Therefore, $t \in P_2$, which means that $(I :_R x) \subseteq P_2$. On the other hand, if $t \in P_2$, then $xt \in P_1 P_2$. Hence, $t \in (I :_R x)$ and, consequently, $P_2 \subseteq (I :_R x)$. The above results follow that $P_2 = (I :_R x)$. Similarly, $P_1 = (I :_R x)$ can be deduced from $x \in P_2 \setminus P_1$. $\square$

Theorem 4.4. *Let I be a 2-absorbing hyperideal of R such that $I \neq \sqrt{I}$. Then:*

(1) *If $x \in \sqrt{I} \setminus I$ and $y \in R$ such that $yx \notin I$, then $(I :_R yx) = (I :_R x)$.*

(2) *If $x, y \in \sqrt{I} \setminus I$, then $(I :_R fx + dy) = (I :_R x)$, for all $f, d \in R$ such that $fd \notin (I :_R x)$. In particular, $(I :_R x + y) = (I :_R x)$.*

Proof.   (1) Suppose that $c \in (I :_R yx)$, where $x \in \sqrt{I} \setminus I$ and $y \in R$. Then, $cy \in (I :_R x)$ which means that $c \in (I :_R x)$ by Theorem 4.2. Therefore, $(I :_R yx) \subseteq (I :_R x)$. It is clear that $(I :_R x) \subseteq (I :_R yx)$, and consequently, the statement is valid.

(2) Suppose that $x, y \in \sqrt{I} \setminus I$. Then, $(I :_R x) \subset (I :_R y)$ or $(I :_R y) \subset (I :_R x)$ by Theorem 4.3. In order to establish the assertion, without loss of generality, assume that $(I :_R x) \subset (I :_R y)$, which leads to $(I :_R x) \subset (I :_R y) \subseteq (I :_R dy)$ and $(I :_R x) \subseteq (I :_R fx)$. Therefore, for all $t \in (I :_R x)$, we get $t(dy + fx) \subseteq I$, and so $(I :_R x) \subseteq (I :_R dy + fx)$. To prove the equality, suppose that there exists $s \in dy + fx$ such that $(I :_R x) \neq (I :_R s)$. By applying Theorem 4.3, there exists $z \in (I :_R y) \cap (I :_R s)$ such that $z \notin (I :_R x)$ because $(I :_R x) \subseteq (I :_R y)$ and $(I :_R x) \subseteq (I :_R dy+fx)$. Since $zs \in z(dy + fx)$, $zfx \in -zdy + zs \subseteq I$, which means that $zf \in (I :_R x)$. Therefore, $z \in (I :_R x)$ or $f \in (I :_R x)$, which is a contradiction.   □

Definition 4.2. Let I be a non-zero properly hyperideal of R and $Z_R(R/I) = \{r + I \in R/I \mid \exists s \in R \setminus I \text{ such that } rs \in I\}$. Then, I is called *Q-primal* if $Z_R(R/I) = Q/I$, for some prime hyperideal Q of R containing I.

Theorem 4.5. *Let I be a 2-absorbing hyperideal of R such that $I \neq \sqrt{I}$. Then, I is a Q-primal hyperideal of R, where $Q = \bigcup_{x \in (\sqrt{I} \setminus I)} (I :_R x)$.*

Proof.   Suppose that $a + I \in Q/I$. Then, there exists $x \in \sqrt{I} \setminus I$ such that $a \in (I :_R x)$. Therefore, $ax \in I$, which follows that $a + I \in Z_R(R/I)$. To prove $Z_R(R/I) \subseteq Q/I$, assume that $a + I \in Z_R(R/I)$, where $a \notin I$. Then, there is $b \in R \setminus I$ such that $ab \in I$. By Theorem 4.1, we can distinguish two cases:

Case 1: $\sqrt{I} = P$ is a hyperideal of R. Then, we have $ab \in P$ and, consequently, $a \in P \setminus I$ or $b \in P \setminus I$. Therefore, $a \in (I :_R a)$ or $a \in (I :_R b)$, which implies that $a + I \in Q/I$.

Case 2: $\sqrt{I} = P_1 \cap P_2$, where P_1 and P_2 are the only distinct prime hyperideals of R that are minimal over I. If $a \in \sqrt{I} \setminus I$ or $b \in \sqrt{I} \setminus I$, then by applying the same argument as for Case (1), we find $a + I \in Q/I$. Now, suppose that $a, b \notin \sqrt{I} \setminus I$. Therefore, a belongs to $P_1 \setminus P_2$ or $P_2 \setminus P_1$, which leads to $a \in (I :_R b)$, by Theorem 4.2. Hence, $a + I \in Q/I$ and, consequently, $Z_R(R/I) \subseteq Q/I$.

Thus, in both cases, we have $Z_R(R/I) = Q/I$. Moreover, since, $I \neq \sqrt{I}$, then Theorem 4.3 implies that $\Omega = \{(I :_R x) \mid x \in \sqrt{I} \setminus I\}$ is a set of linear ordered (prime) hyperideals of R. Therefore,

$$Z_R(R/I) = \bigcup_{(I :_R x) \in \Omega} ((I :_R x)/I)$$

is a hyperideal of R/I. □

Theorem 4.6. *Let R' be a commutative Krasner hyperring with unit element and $\varphi : R \longrightarrow R'$ be a good homomorphism.*

(1) *If I' is a 2-absorbing hyperideal of R', then $\varphi^{-1}(I')$ is a 2-absorbing hyperideal of R.*

(2) *If φ is an epimorphism and I is a 2-absorbing hyperideal of R containing $\ker \varphi$, then $\varphi(I)$ is a 2-absorbing hyperideal of R'.*

Proof. (1) Suppose that $abc \in \varphi^{-1}(I')$, then $\varphi(a)\varphi(b)\varphi(c) \in I'$. Therefore, at least one of $\varphi(ab)$, $\varphi(bc)$, and $\varphi(ac)$ belongs to I', which implies that $ab \in \varphi^{-1}(I')$ or $bc \in \varphi^{-1}(I')$ or $ac \in \varphi^{-1}(I')$.

(2) Assume that $a', b', c' \in R'$ such that $a'b'c' \in \varphi(I)$. Then, there are $a, b, c \in R$ such that $\varphi(a) = a'$, $\varphi(b) = b'$ and $\varphi(c) = c'$. Therefore, $\varphi(abc) = a'b'c' \in \varphi(I)$, which implies that there is $i \in I$ such that $(abc - i) \cap \ker \varphi \neq \emptyset$. Consider $t \in (abc - i) \cap \ker \varphi$. We conclude that $abc \in t + i \subseteq \ker \varphi + I \subseteq I$. This implies that $ab \in I$ or $ac \in I$ or $bc \in I$. Consequently, at least one of $a'b'$, $a'c'$, and $b'c'$ belongs to $\varphi(I)$. □

The following corollary is deduced directly from Theorem 4.6.

Corollary 4.1. *Let I and J be distinct proper hyperideals of R. If $J \subseteq I$ and I is a 2-absorbing hyperideal of R, then I/J is a 2-absorbing hyperideal of R/J.*

Theorem 4.7. *Let R_1, R_2 be Krasner hyperrings and $R = R_1 \times R_2$.*

(1) *If I_1 (I_2, resp.) is a 2-absorbing hyperideal of R_1 (R_2, resp.), then $I_1 \times R_2$ ($R_1 \times I_2$, resp.) is a 2-absorbing hyperideal of R.*

(2) *If J is a 2-absorbing hyperideal of R, then either $J = I_1 \times R_2$ ($J = R_1 \times I_2$, resp.), where I_1 (I_2, resp.) is a 2-absorbing hyperideal of R_1 (R_2, resp.) or $I = I_1 \times I_2$, where I_1 (I_2, resp.) is a prime hyperideal of R_1 (R_2, resp.).*

Proof. (1) It is straightforward.

(2) Suppose that J is a proper 2-absorbing hyperideal of R. Then, $J = I_1 \times I_2$, where for $i = 1, 2$, we have that I_i is a hyperideal of R_i. Assume that $I_2 = R_2$ and $R' = R/(\{0\} \times R_2)$. Therefore, $I_1 \neq R_1$ and $J' = J/(\{0\} \times R_2)$ is a 2-absorbing hyperideal of R' by Corollary 4.1. This follows that I_1 is a 2-absorbing hyperideal of R_1 since $R' \cong R_1$ and $I_1 \cong J'$. In the same way, $I_1 = R_1$ implies that I_2 is a 2-absorbing hyperideal of R_2.

To complete the proof, it is enough to show that if $I_1 \neq R_1$ and $I_2 \neq R_2$, then I_i is a prime hyperideal of R_i, where $i = 1, 2$. Assume that at least one of I_i is not prime, e.g., I_1. Therefore, there are $a, b \in R_1$ such that $ab \in I_1$, but $a, b \notin I_1$. Putting $x = (a, 1)$, $y = (1, 0)$, and $z = (b, 1)$, we get $xyz = (ab, 0) \in J$, while $xy = (a, 0), xz = (ab, 1), yz = (b, 0)$ do not belong to J, and this contradicts the assumption. □

Theorem 4.8. *Let I be a hyperideal of R and S be a multiplicatively closed subset of R. In addition, let $S^{-1}R$ be the Krasner hyperring of quotients of R.*

(1) *If I is a 2-absorbing hyperideal of R and $S \cap I = \emptyset$, then $S^{-1}I$ is a 2-absorbing hyperideal of $S^{-1}R$.*

(2) *If $S^{-1}I$ is a 2-absorbing hyperideal of $S^{-1}R$ and $S \cap Z_R$ $(R/I) = \emptyset$, then I is 2-absorbing hyperideal of R.*

Proof. (1) Suppose that $a, b, c \in R$ and $s, t, k \in S$ such that $(a/s)(b/t)(c/k) \in S^{-1}I$. Then, there exists $u \in S$ such that $uabc \in I$. Hence, $uab \in I$ or $uac \in I$ or $bc \in I$ by hypothesis. If $uab \in I$, then $(a/s)(b/t) = (uab)/(ust) \in S^{-1}I$. Also, $uac \in I$ implies that $(a/s)(c/k) = (uac)/(usk) \in S^{-1}I$, and $bc \in I$ leads to $(b/t)(c/k) \in S^{-1}I$. By the above result, $S^{-1}I$ is a 2-absorbing hyperideal.

(2) Suppose that $a, b, c \in I$ such that $abc \in I$. In this case, we have $(abc)/1 = (a/1)(b/1)(c/1) \in S^{-1}I$. Hence, $(a/1)(b/1) \in S^{-1}I$ or $(b/1)(c/1) \in S^{-1}I$ or $(a/1)(c/1) \in S^{-1}I$ since $S^{-1}I$ is a 2-absorbing hyperideal. If $(a/1)(b/1) \in S^{-1}I$, then there exists $u \in S$ such that $uab \in I$. This implies that $ab \in I$ since $S \cap Z_R(R/I) = \emptyset$.

Similarly, $(b/1)(c/1) \in S^{-1}I$ $((a/1)(c/1) \in S^{-1}I$, resp.) which leads to $bc \in I$ ($ac \in I$, resp.) Consequently, I a 2-absorbing hyperideal. $\square$

Definition 4.3. A proper hyperideal I of R is called an *irreducible precisely* if I cannot be expressed as the intersection of two strictly larger hyperideals of R.

The following theorem shows the relationship between irreducible and 2-absorbing hyperideals.

Theorem 4.9. *Let I be an irreducible hyperideal of R and $P = P_1$, P_2 be distinct prime hyperideals of R.*

(1) *If $\sqrt{I} = P$, then I is a 2-absorbing hyperideal if and only if $P^2 \subseteq I$ and $(I :_R x) = (I :_R x^2)$, for all $x \in R \setminus P$.*
(2) *If $\sqrt{I} = P_1 \cap P_2$, then I is 2-absorbing hyperideal if and only if $P_1 P_2 \subseteq I$ and $(I :_R x) = (I :_R x^2)$, for all $x \in R \setminus P_1 \cap P_2$.*

Proof. (1) To prove the necessity part, it is only necessary to check $(I :_R x^2) \subseteq (I :_R x)$, for all $x \in R \setminus P$ because it is clear that $(I :_R x) \subseteq (I :_R x^2)$ and $P^2 \subseteq I$ by Theorem 4.1.

Suppose that $y \in (I :_R x^2)$. Then, $yx \in I$ or $x^2 \in I$. If $x^2 \in I$, then $x \in P$, which is a contradiction. Then, $yx \in I$, which implies that $y \in (I :_R x)$ and, consequently, $(I :_R x^2) \subseteq (I :_R x)$, as desired.

To establish the sufficiency part, assume that $x, y, z \in R$ such that $xyz \in I$ and $xy \notin I$. We show that either $xz \in I$ or $yz \in I$. From $xy \notin I$, it follows that $x \notin P$ or $y \notin P$, and so $(I :_R x) = (I :_R x^2)$ or $(I :_R y) = (I :_R y^2)$, respectively. Without loss of generality, suppose that $(I :_R x) = (I :_R x^2)$. To complete the proof as a contradiction, assume that $xz \notin I$ and $yz \notin I$. Consider $a \in \langle I + xz \rangle \cap \langle I + yz \rangle$, which follows that there are $a_1, a_2 \in I$ and $r_1, r_2 \in R$ such that $a \in (a_1 + r_1 xz) \cap (a_2 + r_2 yz)$. Consequently, $ax \in a_1 x + r_1 x^2 z$ and $ax \in a_2 x + r_2 yzx \subseteq I$, which lead to $r_1 x^2 z \in -a_1 x + ax \subseteq I$. Therefore, $r_1 z \in (I :_R x^2) = (I :_R x)$ by assumption. This implies that $a \in a_1 + r_1 xz \subseteq I$. Then, $< I + xz > \cap < I + yz > \subseteq I$, and so $< I + xz > \cap < I + yz > = I$, which contradicts the irreducibility of I.

(2) The proof is similar to that of Part (1). $\qquad\square$

In the process of proving the subsequent theorem, we need the following lemma.

Lemma 4.1. *Let $P_1, P_2, \ldots, P_n$, where $n \geq 2$ are hyperideals of R such that at most two of them are not prime. Furthermore, let S be an additive canonical subhypergroup of R which is closed under multiplication and $S \subseteq \bigcup_{i=1}^n P_i$. Then, there exists $1 \leq j \leq n$ such that $S \subseteq P_j$.*

Proof. We prove this statement by induction on n. First, consider for $n = 2$ that $S \subseteq P_1 \cup P_2$. As a contradiction, assume that $S \nsubseteq P_1$ and $S \nsubseteq P_2$. Then, there exists $a_j \in S \setminus P_j$, where $j = 1, 2$. Therefore, the hypothesis leads to $a_1 \in P_2$ and $a_2 \in P_1$. On the other hand, $a_1 + a_2 \in S \subseteq P_1 \cup P_2$, and so for all $t \in a_1 + a_2$, we have that t belongs to either P_1 or P_2. Since $a_1 \in \{a_1\} = a_1 + 0 \subseteq (a_1 + a_2) - a_2$, then there exists $t \in a_1 + a_2$ such that $a_1 \in t - a_2$. Based on the above results, if $t \in P_1$, then $a_1 \in P_1$. Also, if $t \in P_2$, then $a_2 \in -t - a_1 \subseteq P_2$, which is a contradiction in two cases. Thus, we must have $S \subseteq P_1$ or $S \subseteq P_2$. Now, suppose that $k \geq 2$, and our assertion is valid for $n = k$. To complete the proof, assume that $n = k + 1$, where $k \geq 2$. Thus, we have $S \subseteq \bigcup_{i=1}^{k+1} P_i$, and since at most two of P_i are not prime, we can assume that they have been indexed in such a way that P_{k+1} is prime. We claim that there is $1 \leq j \leq k$ such that $S \subseteq \bigcup_{\substack{i=1 \\ i \neq j}}^{k+1} P_i$. To prove this claim as a contradiction, suppose that $S \nsubseteq \bigcup_{\substack{i=1 \\ i \neq j}}^{k+1} P_i$, for all $1 \leq j \leq k$. This follows that for all $1 \leq j \leq k$, there exists $a_j \in S \setminus \bigcup_{\substack{i=1 \\ i \neq j}}^{k+1} P_i$, which implies that $a_j \in P_j$ by hypothesis. Moreover, since $P_{k+1} \in Spec(R)$, we conclude that $a_1 \cdots a_k \notin P_{k+1}$.

Consequently, $a_1 \cdots a_k \in \bigcap_{i=1}^{k} P_i \setminus P_{k+1}$ and $a_{k+1} \in P_{k+1} \setminus \bigcup_{i=1}^{k} P_i$. Now, consider the element $b \in a_1 \cdots a_k + a_{k+1}$. If $b \in P_{k+1}$, then $a_1 \cdots a_k \in b - a_{k+1} \subseteq P_{k+1}$, and this is a contradiction. Therefore, b does not belong to P_{k+1}. Also, we cannot have $b \in P_j$, where $1 \leq j \leq k$, as it would imply $a_{k+1} \in b - a_1 \cdots a_k \subseteq P_j$, which is again a contradiction. However, $b \in S$ since for all $1 \leq j \leq k$, we have $a_j \in S$, which leads to a contradiction to the hypothesis that $S \subseteq \bigcup_{i=1}^{k+1} P_i$. It follows that the statement is valid. In fact, there is $1 \leq j \leq k+1$ such that $S \subseteq \bigcup_{\substack{i=1 \\ i \neq j}}^{k+1} P_i$. By applying the inductive hypothesis, we deduce that $S \subseteq P_i$, where $1 \leq i \leq k+1$. $\qquad\square$

Theorem 4.10. *Let $I_1, I_2, \ldots, I_n$ be 2-absorbing hyperideals of R and I be a hyperideal of R such that $I \subseteq I_1 \cup I_2 \cup \cdots \cup I_n$. Then, there exists $1 \leq i \leq n$ such that $I^2 \subseteq I_i$.*

Proof. First, we show that there exists $1 \leq i \leq n$ such that $\sqrt{I} \subseteq \sqrt{I_i}$. By Theorem 4.1, we can assume that they have been indexed in such a way that $\sqrt{I_i} = p_i$ and $\sqrt{I_j} = p_{j,1} \cap p_{j,2}$, for all $1 \leq i \leq k$ and $k+1 \leq j \leq n$, where $p_i, p_{j,1}, p_{j,2}$ are prime hyperideals of R. Then, $\sqrt{I} \subseteq p_1 \cup p_2 \cup \cdots \cup p_k \cup (p_{k+1,1} \cap p_{k+1,2}) \cup \cdots \cup (p_{n,1} \cap p_{n,2})$, which follows that $\sqrt{I} \subseteq p_1 \cup p_2 \cup \cdots \cup p_k \cup p_{k+1,t_{k+1}} \cup \cdots \cup p_{n,t_n}$, where $t_{k+1}, \ldots, t_n \in \{1,2\}$. Therefore, by applying Lemma 4.1, we find that $\sqrt{I} \subseteq p_i$ or $\sqrt{I} \subseteq p_{j,t_s}$, for some $1 \leq i \leq k$, $k+1 \leq j \leq n$ and $t_s \in \{1,2\}$. If $\sqrt{I} \subseteq p_{j,t_s}$, where $k+1 \leq j \leq n$, $t_s \in \{1,2\}$, then $\sqrt{I} \subseteq p_{j,t_s} \subseteq \bigcup_{j=k+1}^{n} p_{j,t_s}$. We may assume that $\sqrt{I} \subseteq \bigcap_{j=k+1}^{s} p_{j,1}$ and $\sqrt{I} \not\subseteq \bigcup_{j=s+1}^{n} p_{j,1}$, where $k+1 \leq s \leq n$. On the other hand, $\sqrt{I} \subseteq p_{k+1,2} \cup \cdots \cup p_{s,2} \cup p_{s+1,1} \cup \cdots \cup p_{n,1}$. Therefore, $\sqrt{I} \subseteq p_{j,2}$, for some $k+1 \leq j \leq s$, by Lemma 4.1. Hence, $\sqrt{I} \subseteq p_{j,1} \cap p_{j,2} = \sqrt{I_j}$, where $k+1 \leq j \leq s$. Then, in general, there is $1 \leq i \leq n$ such that $\sqrt{I} \subseteq \sqrt{I_i}$, which leads to $I^2 \subseteq (\sqrt{I})^2 \subseteq (\sqrt{I_i})^2$. By applying Theorem 4.1, we get $I^2 \subseteq I_i$. $\qquad\square$

4.2 *r*-Hyperideals

In this section, some properties of r-hyperideals are examined in commutative Krasner hyperrings. r-hyperideals are compared with prime and maximal hyperideals. These results were obtained by Xu *et al.* [103].

Definition 4.4. ([73]) A proper hyperideal N of a commutative Krasner hyperring R is called an r-*hyperideal* (pr-*hyperideal*) if $a \cdot b \in N$ and $Ann(a) = 0$ implies that $b \in N$ ($b^n \in N$, for some $n \in \mathbb{N}$) and for all $a, b \in R$, where $Ann(a) = \{r \in R \mid ra = 0\}$.

$Min(N)$ is the set of all minimal prime hyperideals which contain N.

Let us mention now some important notations.

If $Ann(a) = 0$ (resp., $Ann(a) \neq 0$), a is said to be *regular* (resp., *zero divisor*), $r(R)$ denotes the set of all regular elements of R, while $zd(R)$ denotes the set of all zero divisors.

If N is a hyperideal of R and A is a subset of R, then

$$(N : A) = \{x \in R \mid x \cdot A \subseteq N\}.$$

Hence, $(0 : A) = Ann(A)$.

Let $f : R \to S$ be a homomorphism and M be a hyperideal of S. The hyperideal $f^{-1}(M)$ is called a contraction of M and denoted by M^c. The mapping $\varphi : R \to S^{-1}R$, given by $a = a/1$, is a homomorphism. If N is a hyperideal of R, then $\varphi(N) = S^{-1}N = \{i/s \in S^{-1}R \mid i \in N, s \in S\}$ is also a hyperideal of $S^{-1}R$.

If $S = r(R)$, then the hyperring $S^{-1}R$ is called a quotient hyperring, which is denoted by $Q(R)$.

Example 4.3. $\mathbb{Z}_6$ is a Krasner hyperring with the usual addition and multiplication, and $\{\hat{0}, \hat{2}, \hat{4}\} = N$ is a proper hyperideal:

$Ann(\hat{5}) = 0$ and $\hat{0} \cdot \hat{5} = \hat{0} \in N$ implies that $\hat{0} \in N$,
$Ann(\hat{5}) = 0$ and $\hat{2} \cdot \hat{5} = \hat{4} \in N$ implies that $\hat{2} \in N$,
$Ann(\hat{5}) = 0$ and $\hat{4} \cdot \hat{5} = \hat{2} \in N$ implies that $\hat{4} \in N$.

Then, N is an r-hyperideal of $\mathbb{Z}_6$.

Example 4.4. Let $R = \{0, a, b, c\}$ be a set with the hyperaddition $\oplus$ and the multiplication $\odot$ defined as follows:

$\oplus$	0	a	b	c
0	0	a	b	c
a	a	$\{0, b\}$	$\{a, c\}$	b
b	b	$\{a, c\}$	$\{0, b\}$	a
c	c	b	a	0

$\odot$	0	a	b	c
0	0	0	0	0
a	0	a	b	c
b	0	b	b	0
c	0	c	0	c

Then, $(R, \oplus, \odot)$ is a Krasner hyperring, and $I_1 = \{0, b\}, I_2 = \{0, c\}, I_3 = \{0, b, c\}$ are r-hyperideals.

Theorem 4.11. *Let R be a Krasner hyperring and N be a hyperideal of R. Then, the following statements are equivalent:*

(1) N *is an r-hyperideal;*
(2) $(a \cdot R) \cap N = a \cdot N$, *for all $a \in r(R)$;*
(3) $N = (N : a)$, *for all $a \in r(R)/N$;*
(4) $N = M^c$, *where M is a hyperideal of $Q(R)$.*

Proof. $(1 \Rightarrow 2)$ Let N be an r-hyperideal and a be a regular element. Suppose that $x \in (a \cdot R) \cap N$. Then, $x \in a \cdot R$ and $x \in N$. Thus, $x = a \cdot a'$, for $a' \in R$. Since $x = a \cdot a' \in N, Ann(a) = 0$ and N is an r-hyperideal, $a' \in N$. Hence, $a \cdot a' \in a \cdot N$ and $(a \cdot R) \cap N \subseteq a \cdot N$. Therefore, for every $a \in R, a \cdot N \subseteq (a \cdot R) \cap N, (a \cdot R) \cap N = a \cdot N$.

$(2 \Rightarrow 3)$ We have $N \subseteq (N : a)$, for every $a \in R$. Let a be regular, $a \notin N$, and $x \in (N : a)$. Hence, $Ann(a) = 0$ and $x \cdot a \in N$. From (b), $a \cdot x \in (a \cdot R) \cap N$ and $a \cdot x \in a \cdot N$. Hence, $a \cdot x = a \cdot y$, for $y \in N$. Since $Ann(a) = 0$, then $x = y \in N$. Hence, $x \in N$. Thus, $(N : a) \subseteq N$.

$(c \Rightarrow 4)$ Let S be a set of regular elements and $\varphi : R \to Q(R)$ be a natural homomorphism. We know that $N \subseteq \varphi^{-1}(M) = M^c$, for M a hyperideal of $Q(R)$. Suppose $x \in \varphi^{-1}(M)$. Since $\varphi(x) = x/s \in S^{-1}R$, $s \cdot x \in N$, for $s \in S$. From (c), $x \in (N : s) = N$.

$(4 \Rightarrow 1)$ Let $a \cdot x \in N$ and $Ann(a) = 0$. We have $(a \cdot x)/1 = a/1 \cdot x/1 \in M$, and since a is regular, there exists $1/a$, which is the inverse of $a/1$ in S. Thus, $(a/1) \cdot (x/1) \cdot (1/a) \in M$ and $(x/1) \in M$. Hence, $x \in N$, and so N is an r-hyperideal. $\qquad \square$

Corollary 4.2. *The following statements hold:*

(1) *The zero hyperideal is an r-hyperideal.*
(2) *The intersection of r-hyperideals is an r-hyperideal.*
(3) *When N is an r-hyperideal, $N \subseteq zd(R)$.*
(4) *Every r-hyperideal is a pr-hyperideal.*
(5) *A prime hyperideal is an r-hyperideal if and only if it consists all of zero divisors. As a result, every minimal prime hyperideal is an r-hyperideal.*

(6) *Let N be an r-hyperideal, $S \subseteq R$, and $S \not\subseteq N$. Then, $(N : S)$ is an r-hyperideal. Particularly, $Ann(x)$ and $Ann(S)$ are always r-hyperideals.*

Proof.   (2) Suppose that $I_1, I_2, \ldots, I_n$ are r-hyperideals of R. Let $r \cdot x \in \cap_{i=1}^{n} I_i$ and $Ann(r) = 0$. Then, $r \cdot x \in I_i$. Since every I_i is an r-hyperideal, it follows that $x \in I_i$. Hence, $x \in \cap_{i=1}^{n} I_i$, which means the intersection of r-hyperideals is also r-hyperideal.

(3) Let N be an r-hyperideal of R and $N \not\subseteq zd(R)$. Then, there exists a regular element r in N. Now, let e be identity element of R. In this way, $e \cdot r \in N$. Since N is an r-hyperideal, then $e \in N$, which is a contradiction.

(4) If a prime hyperideal N is an r-hyperideal, then it consists of all zero divisors from (c). For the converse, suppose that N is prime and $N \subseteq zd(R)$. Let $r \cdot x \in N$ and $Ann(r) = 0$. Since N is prime, then $r \in N$ or $x \in N$. Since we assume $N \subseteq zd(R)$, there is no regular element in N, and so $r \notin N$. Then, we get $x \in N$. We conclude that N is an r-hyperideal. $\square$

Remark 4.1. The sum of two r-hyperideals may not be an r-hyperideal.

In order to see that the sum of two r-hyperideals may not be an r-hyperideal, we consider the following example.

Example 4.5. Let $(G, \odot)$ be a group and $H = G \cup \{0, u, v\}$, where 0 is an absorbing element under multiplication and u, v are distinct orthogonal idempotents with

$u \odot v = v \odot u = 0; u \odot u = u$
$v \odot v = v; a \odot 0 = 0 \odot a = 0;$ for all $a \in H$
$u \odot g = g \odot u = u; v \odot g = g \odot v = v;$ for all $g \in G$.

Let us define the hyperaddition $\oplus$ on H as follows:

$a \oplus 0 = 0 \oplus a = \{a\}; a \oplus a = \{0, a\}$,   for all $a \in H$,
$a \oplus b = b \oplus a = H/\{0, a, b\}$,   for all $a, b \in H/\{0\}$ and $a \neq b$.

Then, $(H, \oplus, \odot)$ is a Krasner hyperring [41]. Consider $I = \{0, u\}$ and $J = \{0, v\}$. Hence, I, J are r-hyperideals, while $I \oplus J = H$ is not an r-hyperideal.

Lemma 4.2. *Let R be a Krasner hyperring and N be a hyperideal. Then, the following statements hold:*

(1) *N is an r-hyperideal if and only if for any J, K hyperideals of R with $J \cap r(R) \neq \emptyset$ and $J \cdot K \subseteq N$, then $K \subseteq N$.*

(2) *Assume that $N \subseteq zd(R)$ is not an r-hyperideal. There exist the hyperideals J and K such that $J \cap r(R) \neq \emptyset$, $N \subseteq J$, $N \neq J$, $N \subseteq K$, $N \neq K$ and $J \cdot K \subseteq N$.*

Proof. (1) Let N be an r-hyperideal, J and K be hyperideals of R such that $J \cap r(R) \neq \emptyset$, and $J \cdot K \subseteq N$. Let $a \in J \cap r(R)$ and $b \in K$. Since J and K are hyperideals, then we can take $a \cdot R \subseteq J$ and $b \cdot R \subseteq K$. By assumption, $a \cdot b \cdot R \subseteq N$. Since $J \cap r(R) \neq \emptyset$, $Ann(a) = 0$. Then, N is an r-hyperideal, and so $b \cdot R \subseteq N$. Thus, $K \subseteq N$. Conversely, assume that $J \cap r(R) \neq \emptyset$ and $J \cdot K \subseteq N$, $K \subseteq N$. Let $a \cdot R \subseteq J$ and $b \cdot R \subseteq K$. Then, $a \cdot b \cdot R \subseteq N$. Let $a \in J \cap r(R)$. So, $Ann(a) = 0$. At the same time, since $K \subseteq N$, $b \cdot R \subseteq N$.

(2) Assume that $N \subseteq zd(R)$ is not an r-hyperideal. Then, there exist $r \in r(R)$ and $x \notin N$ such that $r \cdot x \in N$. We have $Ann(r) = 0$. Let $J = (N : x)$ and $K = (N : J)$. It follows that $r \in J$. Since $Ann(r) = 0$, it follows that $J \cap r(R) \neq \emptyset$ and $r \notin N$. Since $N \subseteq zd(R)$, then $Ann(x) \neq 0$. Therefore, $x \in K$. Thus, there exist $b, c \in R$ such that $b \cdot x \in N$ and $c \cdot J \subseteq N$. Hence, $b \cdot x \cdot c \cdot J \subseteq N$. Then, $x \cdot J \cdot K \subseteq N$ and $J \cdot K \subseteq N$. □

Theorem 4.12. *Let R be a Krasner hyperring.*

(1) *Let N be a hyperideal of R with $N \cap r(R) \neq \emptyset$. If J and K are r-hyperideals of R such that $N \cdot J = N \cdot K$ or $N \cap J = N \cap K$, then $J = K$.*

(2) *Let N, M be hyperideals of R with $M \cap r(R) \neq \emptyset$. If $N \cdot M$ is an r-hyperideal of R, then $N = N \cdot M$. Moreover, N is an r-hyperideal.*

Proof. (1) Let J and K be r-hyperideals of R and $N \cdot J = N \cdot K$. Since $N \cdot J = N \cdot K \subseteq K$, $J \subseteq K$. Therefore, $K \subseteq J$. Then, $J = K$.

(2) Let $N \cdot M$ be an r-hyperideal of R and $M \cap r(R) \neq \emptyset$. Since $N \cdot M \subseteq M \cdot N$, $N \subseteq N \cdot M$. Therefore, $N \cdot M \subseteq N$. Then, $N = N \cdot M$. □

Theorem 4.13. *Let $I_1, I_2, \ldots, I_n$ be prime hyperideals of R, which are not comparable. If $\cap_{i=1}^{n} I_i$ is an r-hyperideal, then I_i is an r-hyperideal, for $i = 1, \ldots, n$.*

Proof. Let $r, x \in R$ such that $r \cdot x \in I_i$ and $Ann(r) = 0$. Let $y \in \bigcap_{j \neq i} I_j / I_i$. Then, $r \cdot x \cdot y \in \bigcap_{i=1}^{n} I_i$. We get $x \cdot y \in \cap_{i=1}^{n} I_i$ since $\bigcap_{i=1}^{n} I_i$ is an r-hyperideal and $Ann(r) = 0$. Thus, $x \cdot y \in I_i$. We conclude that $x \in I_i$ since $y \notin I_i$ and I_is are prime. Hence, I_i is an r-hyperideal. □

Let $\rho : R \to S$ be a good homomorphism. We investigate whether the image of an r-hyperideal and the inverse image of an r-hyperideal are r-hyperideals.

Theorem 4.14. *Let $\rho : R \to S$ be a good epimorphism such that $Ker\rho \subseteq N$. If N is an r-hyperideal of $(R, +, \cdot)$, then $\rho(N)$ is an r-hyperideal of the hyperring $(S, \oplus, \odot)$.*

Proof. Obviously, $\rho(N)$ is a hyperideal of S. Let $b_1 \odot b_2 \in \rho(N)$ and $Ann(b_1) = 0_S$, for $b_1, b_2 \in S$. Since ρ is onto, there exist $a_1, a_2 \in R$ such that $b_1 = \rho(a_1)$ and $b_2 = \rho(a_2)$. Then, $b_1 \odot b_2 = \rho(a_1) \odot \rho(a_2) = \rho(a_1 \cdot a_2) = \rho(x) \in f(N)$, for some $x \in N$. $0 \in \rho(a_1 \cdot a_2) \ominus \rho(x) = \rho(a_1 \cdot a_2 - x)$. Then, there exists $t \in a_1 \cdot a_2 - x$ such that $\rho(t) = 0$. We have $a_1 \cdot a_2 \in t + x \subseteq Ker\rho + N \subseteq N + N \subseteq N$. Thus, $a_1 \cdot a_2 \in N$. Let $Ann(a_1) \neq 0$. Then, there exists $0 \neq c \in R$ such that $a_1 \cdot c = 0_R$. So, $\rho(a_1 \cdot c) = \rho(a_1) \odot \rho(c) = \rho(0_R) = 0_S$. Since $\rho(c) \neq 0_S$, then $Ann(\rho(a_1)) = Ann(b_1) = 0_S$, which is a contradiction. This implies that $Ann(a_1) = 0$. Since N is an r-hyperideal, $a_2 \in N$. Therefore, $b_2 = \rho(a_2) \in \rho$. □

Theorem 4.15. *Let $\rho : R \to S$ be a good monomorphism. If M is an r-hyperideal of $(S, \oplus, \odot)$, then $\rho^{-1}(M) = M^c$ is an r-hyperideal of $(R, +, \cdot)$.*

Proof. $\rho^{-1}(M)$ is a hyperideal. Let $r \cdot x \in \rho^{-1}(M)$ and $Ann(r) = 0$. Then, $\rho(r) \odot \rho(x) = \rho(r \cdot x) \in M$. If $Ann(\rho(r)) \neq 0$, then there exists $0 \neq \rho(s) \in M$ such that $\rho(r) \odot \rho(s) = \rho(r \cdot s) = 0_S$. This means that there exists a $0 \neq s \in \rho^{-1}(M)$ such that $r \cdot s = 0_R$. Then, $Ann(r) \neq 0$. This is a contradiction, so $Ann(\rho(r)) = 0$. Since M is an r-hyperideal, $\rho(x) \in M$; therefore, $x \in \rho^{-1}(M)$. □

Theorem 4.16. *Let R be a Krasner hyperring. The following state-ments are equivalent:*

(1) *R is a hyperdomain.*
(2) *The only r-hyperideal of R is the zero hyperideal.*
(3) *$Ann(x \cdot y) = Ann(x) \cup Ann(y)$, for all $x, y \in R$.*

Proof. $(1 \Rightarrow 2)$ Let us suppose that R is a hyperdomain and $(0) \neq N$ is a proper hyperideal of R. Then, there exists $0 \neq r \in N$. Since R is a hyperdomain, we have $Ann(r) = 0$, which is a contradiction. Then, the zero hyperideal is the only r-hyperideal.

$(2 \Rightarrow 3)$ $Ann(x)$ is an r-hyperideal. Assume that the zero hyperideal is the only r-hyperideal of R. Hence, $Ann(x) = 0$. So, $Ann(x \cdot y) = Ann(x) \cup Ann(y) = 0$, for every $x, y \in R$.

$(3 \Rightarrow 1)$ Let $x \cdot y = 0$, $x, y \in R$. Thus, $Ann(x \cdot y) = Ann(0) = Ann(x) \cup Ann(y) = R$. This means $1_R \in Ann(x) \cup Ann$. Then, $x = \{0\}$ or $y = \{0\}$. Hence, R is a hyperdomain. $\square$

A Krasner hyperring R is called a *reduced hyperring* if there are no nilpotent elements in R. If $x^n = 0$, for $x \in R, n \in \mathbb{N}$, then $x = 0$.

Theorem 4.17. *Let R be a Krasner hyperring.*

(1) *Let $x, y \in R$ with $1_R \in x + y$. Then, $N = Ann(x) + Ann(y)$ is an r-hyperideal.*
(2) *Let R be a reduced hyperring, $I \in Min(R)$, and $e \in R$ be an idem-potent element. Moreover $N = I + Ann(e)$ is an r-hyperideal.*

Proof. (1) Let $a \cdot b \in N$ and $Ann(a) = 0$. Then, there exist $r \in Ann(x)$ and $s \in Ann(y)$ such that $a \cdot b \in r + s$. Obviously, $r \cdot x = 0$ and $s \cdot y = 0$. We get that $a \cdot b \cdot x \cdot y \in (r + s) \cdot (x \cdot y) = r \cdot (x \cdot y) + s \cdot (x \cdot y) = y \cdot (x \cdot r) + x \cdot (y \cdot s) = b \cdot 0 + x \cdot 0 = 0$, and by our assumption, $b \cdot x \cdot y = 0$. Thus, $b \cdot x \in Ann(y)$ or $b \cdot y \in Ann(x)$. Hence, $b = b \cdot 1_R \in b \cdot (x + y) = (b \cdot x)(b \cdot y) \subseteq Ann(x) + Ann(y) = N$ and therefore $b \in N$.

(2) Let $r \cdot x \in N$ and $Ann(r) = 0$. We have $r \cdot x \in a + b$ such that $a \in I$ and $b \in Ann(e)$. Since $I \in Min(R)$, I is an r-hyperideal. There is $y \notin I$ such that $a \cdot y = 0$. $e \cdot y \cdot r \cdot x \in e \cdot y \cdot (a + b) = (e \cdot y \cdot a) + (e \cdot y \cdot b) = 0$. Since R is a reduced hyperring, every idempotent element is 0. We have $e \cdot y \cdot x \in I$. Since $y \notin I$, it follows that $e \cdot x \in I$. Hence, $x \in x + (e \cdot x) - (e \cdot x)$, $x \in (e \cdot x) + x - (e \cdot x) = e \cdot x + x \cdot (1_R - e) \subseteq I + Ann(e) = N$. Then, N is an r-hyperideal. $\square$

Theorem 4.18. *Let N be a hyperideal of R. Then, N is a pr-hyperideal if and only if $\sqrt{N}$ is an r-hyperideal.*

Proof. Let N be a pr-hyperideal, $x \cdot y \in \sqrt{N}$ and $Ann(x) = 0$. There is $n \in \mathbb{N}$ such that $(x \cdot y)^n \in N$. Hence, $x^n \cdot y^n \in N$ and $Ann(x^n) = 0$. Since N is a pr-hyperideal, $\{y^{nm} \in N \mid m \in \mathbb{N}\} = \sqrt{0}$. Obviously, $\sqrt{0}$ is an r-hyperideal. Then, $nil(R)$ is also an r-hyperideal. $\qquad\square$

Theorem 4.19. *Every hyperideal which contains zero divisors is contained by a prime r-hyperideal.*

Proof. It is straightforward. $\qquad\square$

Theorem 4.20. *If N is a hyperideal and $I \in Min(N)$ in hyperring R, then I is an r-hyperideal.*

Proof. Let $x \cdot y \in I$ and $Ann(x) = 0$. There exist $a \notin I$ and $n \in N$ such that $a \cdot (x \cdot y)^n = a \cdot x^n \cdot y^n \in N$. Since N is an r-hyperideal and $Ann(x^n) = 0$, $a \cdot yn \in N \subseteq I$. Since $a \notin I$, $y^n \in I$. Thus, $y \in I$. $\qquad\square$

Theorem 4.21. *Let N be an r-hyperideal and $N \subseteq M$. If M/N is an r-hyperideal of R/N, then M is an r-hyperideal of R.*

Proof. Let M/N be an r-hyperideal of R/N, $(a + N) \cdot (b + N) \subseteq M/N$ and $Ann(a + N) = 0$ for $a + N, b + N \in R/N$. We have $(a+N) \cdot (b+N) \subseteq a \cdot (b+N) + N \cdot (b+N) \subseteq (a \cdot b) + (a \cdot N) + (N \cdot b) + N \subseteq (a \cdot b) + N \subseteq M + N$. Hence, $a \cdot b \in M$. Since $Ann(a) = 0$ and $b \in M$, M is an r-hyperideal. $\qquad\square$

Theorem 4.22. *Every maximal r-hyperideal is a prime hyperideal.*

Proof. It is straightforward. $\qquad\square$

Theorem 4.23. *If every prime hyperideal of R is an r-hyperideal, every maximal hyperideal of R is an r-hyperideal.*

Proof. Suppose that M is a maximal hyperideal of R. Then, it is a prime hyperideal. Let $a \cdot b \in M$ with $Ann(a) = 0$. Then, $b \in (M : a)$. Since M is maximal, it follows that $M = (M : a)$ and $b \in M$. $\qquad\square$

4.3   Generalizations of Prime and Primary Hyperideals

In this section, ϕ-prime, ϕ-primary, and ϕ–δ-primary hyperideals are presented, together with several of their properties. These results are obtained by Guan *et al.* [44]. Let $(R, \oplus, \circ)$ be a commutative Krasner hyperring with nonzero identity. We denote the set of all hyperideals of R by $L(R)$.

Definition 4.5. Let ϕ be a function such that $\phi : L(R) \to L(R) \cup \{\emptyset\}$ and N be a hyperideal of R. N is said to be a ϕ-*prime hyperideal* of R if $a \circ b \in N - \phi(N)$. Then, $a \in N$ or $b \in N$, for $a, b \in R$.

Example 4.6. Let R be a commutative Krasner hyperring. Consider the following functions $\phi : L(R) \to L(R) \cup \emptyset$, defined as follows. For all $N \in L(R)$:

(1) $\phi_{\emptyset}(N) = \emptyset$.
(2) $\phi_0(N) = 0$.
(3) $\phi_2(N) = N^2$.
(4) $\phi_n(N) = N^n$, (for all $n \geq 2$).
(5) $\phi_w(N) = \cap_{n=1}^{\infty} N^n$
(6) $\phi_1(N) = N$.

We have $\phi_{\emptyset} \leq \phi_0 \leq \phi_w \leq \cdots \leq \phi_{n+1} \leq \phi_n \leq \phi_{n-1} \leq \cdots \leq \phi_2 \leq \phi_1$.

Theorem 4.24. *Let* $\phi : L(R) \to L(R) \cup \emptyset$ *be a function and* T *be a proper hyperideal of* R *such that* T *is a* ϕ-*prime hyperideal of* R. *If* T *is not prime, then* $T^2 \subseteq \phi(T)$. *Hence, if* $T^2 \not\subseteq \phi(T)$, *then* T *is prime.*

Proof.   Assume that $T^2 \not\subseteq \phi(T)$. We prove that T is prime. Take $x \circ y \in T$, for $x, y \in R$. If $x \circ y \notin \phi(T)$, then $x \circ y \in T - \phi(T)$. Since T is ϕ-prime, then $x \in T$ or $y \in T$. Let us suppose $x \circ y \in \phi(T)$. We can suppose $x \circ T \not\subseteq \phi(K)$, so $x \circ m \notin \phi(T)$, for some $m \in T$. Then, $x \circ (y \oplus m) \subseteq T - \phi(T)$. So, $x \in T$ or $y \oplus m \subseteq T$ since T is a ϕ-prime hyperideal and hence $x \in T$ or $y \in T$. Now, we suppose that $x \circ T \subseteq \phi(T)$. (Similarly, $y \circ T \subseteq \phi(T)$.) We can find some $n, k \in T$ with $n \circ k \notin \phi(T)$ because $T^2 \not\subseteq \phi(T)$. Then, $(x \oplus n) \circ (y \oplus k) \subseteq T - \phi(T)$. Since T is a ϕ-prime hyperideal, it follows that $(x \oplus n) \subseteq T$ or $(y \oplus k) \subseteq T$. So, $x \in T$ or $y \in T$. Hence, T is prime hyperideal of R. □

Theorem 4.25. *Let N be a proper hyperideal of the commutative Krasner hyperring R, and let $\phi : L(R) \to L(R) \cup \emptyset$ be a function. Then, the following statements hold:*

(1) *N is a ϕ-prime hyperideal of R.*
(2) *For $a \in R - N$, $(N : a) = N \cup (\phi(N) : a)$.*
(3) *For $a \in R - N$, $(N : a) = N$ or $(N : a) = (\phi(N) : a)$.*
(4) *For each hyperideals K and L of R such that $K \circ L \subseteq N$ and $K \circ L \nsubseteq \phi(N)$, we have $K \subseteq N$ or $L \subseteq N$.*

Proof. $(1 \Rightarrow 2)$: Take $a \in R - N$. Suppose that $b \in (N : a)$. Then, $a \circ b \in N$. If $a \circ b \nsubseteq \phi(N)$, then $b \in N$ since N is a ϕ-prime hyperideal of R. If $a \circ b \in \phi(N)$, then $b \in (\phi(N) : a)$. Hence, $(N : a) \subseteq N \cup (\phi(N) : a)$. The other side holds since $\phi(N) \subseteq N$.

$(2 \Rightarrow 3)$: It is obvious because $(N : a)$ is a hyperideal of R.

$(3 \Rightarrow 4)$: Let K and L be hyperideals of R such that $K \circ L \subseteq N - \phi(N)$. Assume that $K \nsubseteq N$. Then, there exists an element $a \in K - N$. By (3), we have $(N : a) = N$ or $(N : a) = (\phi(N) : a)$. If $a \circ L \nsubseteq \phi(N)$, then $L \nsubseteq (\phi(N) : a)$. Since (3) holds, $L \subseteq (N : a) = N$. Now, assume that $a \circ L \subseteq \phi(N)$. Since $K \circ L \nsubseteq \phi(N)$, we can choose an element $x \in K$ such that $x \circ L \subseteq N - \phi(N)$. If $x \notin N$, then by (3), $(N : x) = N$ or $(N : x) = (\phi(N) : x)$. Since $L \subseteq (N : x)$, but $L \nsubseteq (\phi(N) : x)$, we conclude that $L \subseteq (N : x) = N$. Assume that $x \in N$. We have $a \oplus x \subseteq K - N$, and also note that $(a \oplus x) \circ L \subseteq N - \phi(N)$ since $a \circ L \subseteq \phi(N)$ and $x \circ L \nsubseteq \phi(N)$. Then, $L \subseteq (N : a \oplus x) = N$, which completes the proof.

$(4 \Rightarrow 1)$: Let $a \circ b \in N - \phi(N)$. Then, $(a) \circ (b) \subseteq N$, but $(a) \circ (b) \nsubseteq \phi(N)$. So, $(a) \subseteq N$ or $(b) \subseteq N$. Then, $a \in N$ or $b \in N$. $\qquad \square$

Let S be a multiplicatively closed subset of a Krasner hyperring R. Define $\phi : L(R) \to L(R) \cup \emptyset$ and $S : L(R_S) \to L(R_S) \cup \emptyset$, with $\phi_S(M) = (\phi(M \cap R))_S$. Also, $\phi_S(M) = (\phi(M \cap R)_S$, where $\phi(M \cap R) = \emptyset$.

Let N, M be hyperideals of R and $N \subseteq M$. We define $M : L(R/M) \to L(R/M) \cup \{\emptyset\}$, with $\phi_M(N/M) = (\phi(N) \oplus M)/M$. Also, $\phi_M(N/M) = \emptyset$, where $\phi(N) = \emptyset$.

Theorem 4.26. *Let $\phi : L(R) \to L(R) \cup \{\emptyset\}$ be a function and T be a ϕ-prime hyperideal of R. We have:*

(1) *If M is a hyperideal of R with $M \subseteq T$, then T/M is a ϕ_M-prime hyperideal of R/M.*

(2) *Let S be a multiplicatively closed subset of R, with $T \cap S = \emptyset$ and $\phi(T)_S \subseteq \phi_S(T_S)$. Then, T_S is a ϕ_S-prime hyperideal of R_S.*

Proof. (1) Let $x, y \in R$. Suppose that $(x \oplus M) \circ (y \oplus M) \subseteq T/M - \phi_M(T/M)$. So, $x \circ y \oplus M \subseteq T/M - (\phi(T \oplus M))/M$. Then, $x \circ y \in T - \phi(T \oplus M)$. We find that $x \circ y \in T - \phi(T)$, and $x \in T$ or $y \in T$. Therefore, $x \oplus M \subseteq T/M$ or $y \oplus M \subseteq T/M$. So, T/M is a ϕ_M-prime hyperideal of R/M.

(2) Let $a/s \circ b/t \in T_S - \phi_S(T_S)$, for some $a, b \in R$, $s, t \in S$. Then, we have $p \circ a \circ b \in T$, for some $p \in S$, but $q \circ a \circ b \notin \phi(T) \cap R$, for all $q \in S$. Now, if $q \circ a \circ b \in \phi(T)$, then $a/s \circ b/t \in \phi(T)_S \subseteq \phi_S(T_S)$, which is a contradiction. So, $p \circ a \circ b \in T - \phi(T)$, and since T is a ϕ-prime hyperideal of R, we obtain either $p \circ a \in T$ or $b \in T$. Hence, $a/s \in T_S$ or $b/t \in T_S$ since $T \cap S = \emptyset$. Thus, T_S is a ϕ_S-prime hyperideal of R_S. $\square$

Definition 4.6. Let R be a commutative hyperring and N be a proper hyperideal of R. Let $\phi : L(R) \to L(R) \cup \{\emptyset\}$. N is called a ϕ-*primary hyperideal* of R if $a \circ b \in N - \phi(N)$. Then, either $a \in N$ or $b^k \in N$, for some $a, b \in R, k \in \mathbb{N}$.

Example 4.7. Let R be a commutative Krasner hyperring. Then, we define the functions $\phi : L(R) \to L(R) \cup \{\emptyset\}$ as in the previous example. Also, we have the same order: $\phi_\emptyset \leq \phi_0 \leq \phi_w \leq \cdots \leq \phi_{n+1} \leq \phi_n \leq \phi_{n-1} \leq \cdots \leq \phi_2 \leq \phi_1$.

Theorem 4.27. *Let $\phi : L(R) \to L(R) \cup \{\emptyset\}$ be a function and T be a proper hyperideal of R such that T is a ϕ-primary hyperideal of R. If T is not primary, then $T^2 \subseteq \phi(T)$. Hence, if $T^2 \not\subseteq \phi(T)$, then T is a primary hyperideal of R.*

Proof. Assume that $T^2 \not\subseteq \phi(T)$. We check that T is primary. Take some $x \circ y \in T$ for $x, y \in R$. If $x \circ y \notin \phi(T)$, then since T is ϕ-primary, $x \in T$ or $y^k \in T$, for some $k \in \mathbb{N}$. If $x \circ y \in \phi(T)$, assuming that $x \circ T \not\subseteq \phi(T)$, we obtain $x \circ p_0 \notin \phi(T)$, where $p_0 \in T$. We have $x \circ (y \oplus p_0) \subseteq T - \phi(T)$. Thus, $x \in T$ or $(y \oplus p_0)^k \subseteq T$, for some $k \in \mathbb{N}$. It follows that $x \in T$ or $y^k \in T$. Now, we assume that

$x \circ T \subseteq \phi(T)$. (In the same way, we can assume that $y \circ T \subseteq \phi(T)$). Because $T^2 \not\subseteq \phi(T)$, there exist $p_1, q_1 \in T$, with $p_1 \circ q_1 \notin \phi(T)$. Then, $(x \oplus p_1) \circ (y \oplus q_1) \subseteq T - \phi(T)$. As T is ϕ-primary, $(x \oplus p_1) \subseteq T$ or $(y \oplus q_1)^m \subseteq T$, for some $m \in \mathbb{N}$. Consequently, T is a primary hyperideal of R. $\square$

Theorem 4.28. *Let N be a proper hyperideal of the commutative Krasner hyperring R, and let $\phi : L(R) \to L(R) \cup \{\emptyset\}$ be a function. The following statements hold:*

(1) *N is a ϕ-primary hyperideal of R.*
(2) *For $a \in R - \sqrt{N}$, $(N : a) = N \cup (\phi(N) : a)$.*
(3) *For $a \in R - \sqrt{N}$, $(N : a) = N$ or $(N : a) = (\phi(N) : a)$.*
(4) *For all hyperideals K and L of R, if $K \circ L \subseteq N$ and $K \circ L \not\subseteq \phi(N)$, then $K \subseteq N$ or $L \subseteq \sqrt{N}$.*

Proof. $(1 \Rightarrow 2)$: Let N be a ϕ-primary hyperideal of R. It is obvious that $N \cup (\phi(N) : a) \subseteq (N : a)$. To prove the other side, for all $b \in (N : a)$, we have $a \circ b \in N$. If $a \circ b \in N - \phi(N)$, then $b \in N$ since N is ϕ-primary and $a \in R - \sqrt{N}$. If $a \circ b \in \phi(N)$, then $b \in (\phi(N) : a)$. So, $(N : a) \subseteq N \cup (\phi(N) : a)$. Hence, $(N : a) = N \cup (\phi(N) : a)$.

$(2 \Rightarrow 3)$: It follows from the fact that $(N : a)$ is a hyperideal of R.

$(3 \Rightarrow 4)$: Let K and L be hyperideals of R, with $K \circ L \subseteq N - \phi(N)$. Assume that $L \not\subseteq \sqrt{N}$. Then, there exists an element $a \in L - \sqrt{N}$. By (3), we have $(N : a) = N$ or $(N : a) = (\phi(N) : a)$. If $K \circ a \not\subseteq \phi(N)$, then $K \not\subseteq (\phi(N) : a)$. Since (iii) holds, then $K \subseteq (N : a) = N$. Assume that $K \circ a \subseteq \phi(N)$. Since $K \circ L \not\subseteq \phi(N)$, then we can choose an element $x \in L$ such that $K \circ x \subseteq N - \phi(N)$. If $x \notin \sqrt{N}$, then by (3), $(N : x) = N$ or $(N : x) = (\phi(N) : x)$. Since $K \subseteq (N : x)$, but $K \not\subseteq (\phi(N) : x)$, then we conclude that $K \subseteq (N : x) = N$. So, assume that $x \in \sqrt{N}$. Then, we have $a \oplus x \subseteq L - \sqrt{N}$, and also note that $K \circ (a \oplus x) \subseteq N - \phi(N)$ since $K \circ a \subseteq \phi(N)$ and $K \circ x \not\subseteq \phi(N)$. Then, $K \subseteq (N : a \oplus x) = N$, which completes the proof.

$(4 \Rightarrow 1)$: Let $a \circ b \in N - \phi(N)$. Then, $(a) \circ (b) \subseteq N$, but $(a) \circ (b) \not\subseteq \phi(N)$. So, $(a) \subseteq N$ or $(b) \subseteq \sqrt{N}$. Therefore, $a \in N$ or $b^m \in N$, for some $m \in \mathbb{N}$. $\square$

Let N be a proper hyperideal of hyperring R. Denote the set of all hyperideals of R by $L(R)$, and denote the set of all proper hyperideals of R by $L^*(R)$.

Definition 4.7. The function $\phi : L(R) \to L(R) \cup \{\emptyset\}$ is said to be a *reduction function* if $\phi(N) \subseteq N$ and $N \subseteq M$ implies $\phi(N) \subseteq \phi(M)$, for all $N, M \in L(R)$, and similarly, δ is an *expansion function* if $\delta : L(R) \to L(R)$.

Example 4.8. Let R be a commutative Krasner hyperring with a non-zero identity. Let us consider the following functions δ on $L(R)$, for all $N \in L(R)$:

(1) $\delta_0(N) = N$, i.e., δ is the identity function.
(2) $\delta_1(N) = \sqrt{N}$, i.e., δ is the radical operation.
(3) $\delta_{res}(N) = (N : M)$, for a fixed $M \in L(R)$.
(4) $\delta_{Ann}(N) = Ann(Ann(N))$.
(5) $\delta_M(N) = N \oplus M$, for a fixed $M \in L(R)$.

All the above functions are examples of expansion on $L(R)$.

Example 4.9. Let R be a commutative Krasner hyperring with a nonzero identity. Consider the functions $\phi : L(R) \to L(R) \cup \{\emptyset\}$, defined as follows. For all $N \in L(R)$:

(1) $\phi_\emptyset(N) = \emptyset$.
(2) $\phi_0(N) = 0$.
(3) $\phi_1(N) = N$.
(4) $\phi_2(N) = N^2$.
(5) $\phi_k(N) = N^k$.
(6) $\phi_w(N) = \cap_{i=1}^{\infty} N^i$.

All the above functions are reductions on $L(R)$. Remember that $\phi_\emptyset \leq \phi_0 \leq \phi_w \leq \cdots \leq \phi_{n+1} \leq \phi_n \leq \phi_{n-1} \leq \cdots \leq \phi_2 \leq \phi_1$.

Definition 4.8. Let δ be a hyperideal expansion, ϕ be a hyperideal reduction, and N be a proper hyperideal of R. N is said to be a ϕ-δ-*primary hyperideal* if $a \circ b \in N - \phi(N)$. Then, either $a \in N$ or $b \in \delta(N)$, for each $a, b \in R$.

Theorem 4.29. *Let R be a Krasner hyperring and N be a proper hyperideal of R. Then, the following statements hold:*

(1) *N is a $\phi - \delta$-primary hyperideal.*
(2) *For each $a \in R - \delta(N), (N : a) = N \cup (\phi(N) : a)$.*
(3) *For each $a \in R - \delta(N), (N : a) = N$ or $(N : a) = (\phi(N) : a)$.*

(4) *For each hyperideal K, L of R, $K \circ L \subseteq N$ and $K \circ L \nsubseteq \phi(N)$ imply that $K \subseteq N$ or $L \subseteq \delta(N)$.*

(5) *For each hyperideal M of R such that $M \nsubseteq \delta(N)$, then $(N : M) = N$ or $(N : M) = (\phi(N) : M)$.*

Proof. (1⇒2): Suppose that N is a ϕ–δ-primary hyperideal and $a \in R - \phi(N)$. It is clear that $N \cup (\phi(N) : a)) \subseteq (N : a)$. Let $m \in (N : a)$. Then, we have $a \circ m \in N$. If $a \circ m \in \phi(N)$, then we obtain $m \in (\phi(N) : a) \subseteq N \cup (\phi(N) : a))$. Assume that $a \circ m \notin \phi(N)$. Since $a \circ m \in N - \phi(N)$ and $a \notin \delta(N)$, we get $m \in N \subseteq N \cup (\phi(N) : a))$. Hence, $(N : a) = N \cup (\phi(N) : a))$.

(2⇒3): It follows from the fact that a hyperideal is a union of two hyperideals. Then, it must be equal to one of them.

(3⇒4): Let $K \circ L \subseteq N$, for some hyperideals K and L of R. Assume that $L \nsubseteq \delta(N)$ and $K \nsubseteq N$. Then, there exists $m \in L - \delta(N)$. We have to check that $K \circ L \subseteq \phi(N)$. By (iii), we have either $(N : m) = N$ or $(N : m) = (\phi(N) : m)$. If $K \circ m \subseteq N$, then by (iii), we obtain $K \subseteq (N : m) = N$. For $a \in K - N$, it means that $a \in (N : m) - N$. Hence, by (iii), $(N : m) = (\phi(N) : m)$. Thus, $K \subseteq (N : m) = (\phi(N) : m)$, which implies that $K \circ m \subseteq \phi(N)$. On the other side, suppose that $m \in \delta(N)$. Then, $m \in L \cup \delta(N)$. Choose an element $m' \in L - \delta(N)$, so $m \oplus m' \subseteq L - \delta(N)$. Hence, $K \circ m' \subseteq \phi(N)$ and $K \circ (m \oplus m') \subseteq \phi(N)$. Let $a \in K$. Then, $a \circ (m \oplus m') \ominus a \circ m' = a \circ m \subseteq \phi(N)$. Thus, $K \circ m \subseteq \phi(N)$. Therefore, $K \circ L \subseteq \phi(N)$.

(4⇒5): Suppose that M is a hyperideal of R such that $M \nsubseteq \delta(N)$. Also, note that $M \circ (N : M) \subseteq N$. If $M \circ (N : M) \subseteq \phi(N)$, then $(N : M) \subseteq (\phi(N) : M) \subseteq (N : M)$. Assume that $M \circ (N : M) \nsubseteq \phi(N)$. Then, by (iv), $(N : M) \subseteq N \subseteq (N : M)$.

(5⇒1): Suppose that $a \circ m \in N - \phi(N)$, with $a \notin \delta(N)$, for $a, m \in R$. Set $R \circ a = M$, and note that $m \in (N : M)$. Then, by (v), we obtain either $m \in (N : M) = N$ or $m \in (N : M) = (\phi(N) : M)$. The latter case is impossible since $a \circ m \notin \phi(N)$. Therefore, N is a ϕ–δ-primary hyperideal. □

Theorem 4.30.

(1) *Let T be a ϕ–δ-primary hyperideal of R such that $(\phi(T):a) = (T : a)$, for all $a \in R$. Then, $(T : a)$ is a ϕ–δ-primary hyperideal of R.*

(2) *Let $\delta \leq \delta_1$ be hyperideal expansions and T be a ϕ–δ-primary hyperideal of R such that $\delta_1(\phi(T)) \subseteq \delta(T)$. Then, $\delta(T) = \delta_1(T)$.*

Proof. (1) Let T be a ϕ–δ-primary hyperideal of R such that $(\phi(T) : a) = \phi(T : a)$, for all $a \in R$. We check that $(T : a)$ is a ϕ–δ-primary hyperideal of R. For this, let us take $x, m \in R$ such that $x \circ m \in (T : a) - \phi(T : a)$. Then, we have $x \circ a \circ m \in T$. Since $(\phi(T) : a) = (T : a)$, we also have $x \circ a \circ m \notin \phi(T)$. Since T is a ϕ–δ-primary hyperideal, we get either $x \in \delta(T)$ or $a \circ m \in T$, whence either $x \in \delta(T : a)$ or $m \in (T : a)$. Therefore, $(T : a)$ is a ϕ–δ-primary hyperideal of R.

(2) Since $\delta \leq \delta_1$, we have $\delta(T) \subseteq \sqrt{T} = \delta_1(T)$.

For the converse, take $x \in \sqrt{T}$. Then, there exists a minimal $k \in \mathbb{N}$ such that $x^k \in T$, i.e., $x^{k-1} \notin T$. If $k = 1$, then we have $x \in T \subseteq \delta(T)$. Now, assume that $k > 1$. We have two cases.

Case 1: Let $x \in \delta_1(\phi(T)) = \sqrt{\phi(T)}$. Then, we have $x \in \delta(T)$.

Case 2: Let $x \notin \sqrt{\phi(T)}$. Then, we have $x^k \notin \phi(T)$, i.e., $x^{k-1} \circ (x \circ R) \subseteq T$ and $x \circ (x^{k-1} \circ R) \not\subseteq \phi(T)$. Since T is a ϕ–δ-primary hyperideal of R, $x \in \delta(T)$ or $x^{k-1} \circ R \subseteq T$. The latter case is impossible since $x^{k-1} \notin T$. Thus, in both cases, we have $x \in \delta(T)$. Therefore, $\delta(T) = \delta_1(T)$. □

Theorem 4.31. *Let T be a ϕ–δ-primary hyperideal of R such that $\delta(T) \circ T \not\subseteq \phi(T)$. Then, T is a δ-primary hyperideal of R.*

Proof. Let $a \circ m \in T$, for some $a, m \in R$. If $a \circ m \notin \phi(T)$, then we conclude that either $a \in \delta(T)$ or $m \in T$, as T is a ϕ–δ-primary hyperideal of R. Assume that $a \circ m \in \phi(T)$. If $a \circ T \not\subseteq \phi(T)$, then there exists $n \in T$ such that $a \circ n \notin \phi(T)$. Thus, we have $a \circ (m \oplus n) \subseteq T - \phi(T)$, which implies either $a \in \delta(T)$ or $m \oplus n \subseteq T$. Then, we get $a \in \delta(T)$ or $m \in T$, which completes the proof. Assume that $a \circ T \subseteq \phi(T)$. Similarly, we may assume that $\delta(T) \circ m \subseteq \phi(T)$. As $\delta(T) \circ T \not\subseteq \phi(T)$, we can find $b \in \delta(T)$ and $m' \in T$ such that $b \circ m' \notin \phi(T)$. Then, we conclude that $(a \oplus b) \circ (m \oplus m') \subseteq T - \phi(T)$. Since T is a ϕ–δ-primary hyperideal of R, we have either $a \oplus b \subseteq \delta(T)$ or $m \oplus m' \subseteq T$, which implies that $a \in \delta(T)$ or $m \in T$. Therefore, T is a δ-primary hyperideal of R. □

Chapter 5

Lower and Upper Approximations in Krasner Hyperrings

Rough set theory was proposed by Pawlak for knowledge discovery in databases and experimental data sets. The purpose of this chapter is to introduce and discuss the concept of lower and upper approximations in a Krasner hyperring. We consider a Krasner hyperring R as a universal set, and we introduce the notion of lower and upper approximations with respect to a hyperideal of R. The results of this chapter are contained in Refs. [25, 26, 55].

5.1 Introduction to Pawlak Approximations

Suppose that U is a non-empty set. A partition or classification of U is a family $\mathcal{P}$ of non-empty subsets of U such that each element of U is contained in exactly one element of $\mathcal{P}$. Recall that an equivalence relation ρ on a set U is a reflexive, symmetric, and transitive binary relation on U. Each partition $\mathcal{P}$ induces an equivalence relation ρ on U by setting

$$x\rho y \quad \Leftrightarrow \quad x \text{ and } y \text{ are in the same class of } \mathcal{P}.$$

Conversely, each equivalence relation ρ on U induces a partition $\mathcal{P}$ of U whose classes have the form $\rho(x) = \{y \in U \mid x\rho y\}$.

The following notation is used. Given a non-empty universe U, by $\mathcal{P}(U)$ we denote a power set on U. If ρ is an equivalence relation on U, then for every $x \in U$, $\rho(x)$ stands for the equivalence class of ρ with the representative x. For any $X \subseteq U$, we write X^c to denote the complement of X in U, i.e., the set $U \backslash X$.

Definition 5.1. A pair (U, ρ), where $U \neq \emptyset$ and ρ is an equivalence relation on U, is called an *approximation space*.

Definition 5.2. For an approximation space (U, ρ), by a rough approximation of (U, ρ), we mean a mapping $app : \mathcal{P}(U) \to \mathcal{P}(U) \times \mathcal{P}(U)$ defined as follows. For every $X \in \mathcal{P}(U)$,

$$app(X) = \big(\underline{app}(X), \overline{app}(X)\big),$$

where

$$\underline{app}(X) = \{x \in X \mid \rho(x) \subseteq X\}, \quad \overline{app}(X) = \{x \in X \mid \rho(x) \cap X \neq \emptyset\}.$$

$\underline{app}(X)$ is called the *lower approximation* of X in (U, ρ), whereas $\overline{app}(X)$ is called the *upper approximation* of X in (U, ρ).

Definition 5.3. Given an approximation space (U, ρ), a pair $(A, B) \in \mathcal{P}(U) \times \mathcal{P}(U)$ is called a *rough set* in (U, ρ) if and only if $(A, B) = app(X)$, for some $X \in \mathcal{P}(U)$.

The reader can find in Refs. [80,81] a deep discussion of the rough set theory.

Definition 5.4. A subset X of U is called a *definable* if $\underline{app}(X) = \overline{app}(X)$.

The properties of rough sets can be examined via either partition or equivalence classes. The objects of the given universe U can be divided into three classes with respect to any subset X of U, see Figure 5.1. Indeed, if $X \subseteq U$ is given by a predicate P and $x \in U$, then:

(1) $x \in \underline{app}(X)$ means that x certainly has property P;
(2) $x \in \overline{app}(X)$ means that x possibly has property P;
(3) $x \in U \backslash \overline{app}(X)$ means that x definitely does not have property P.

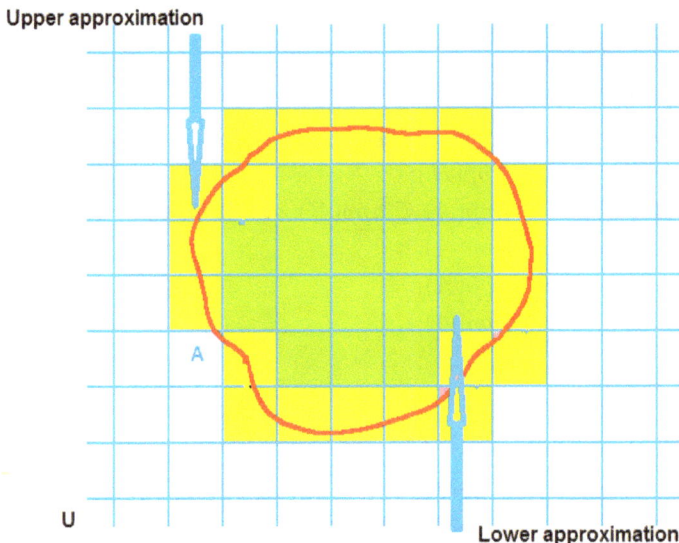

Fig. 5.1.   Lower and upper approximations.

Definition 5.5. Let $app(A) = \left(\underline{app}(A),\ \overline{app}(A)\right)$ and $app(B) = \left(\underline{app}(B),\ \overline{app}(B)\right)$ be any two rough sets in the approximation space $(\overline{U}, \rho)$. Then:

(1) $app(A) \sqcup app(B) = \left(\underline{app}(A) \cup \underline{app}(B),\ \overline{app}(A) \cup \overline{app}(B)\right)$;
(2) $app(A) \sqcap app(B) = \left(\underline{app}(A) \cap \underline{app}(B),\ \overline{app}(A) \cap \overline{app}(B)\right)$;
(3) $app(A) \sqsubseteq app(B) \Leftrightarrow app(A) \cap app(B) = app(A)$.

When $app(A) \sqsubseteq app(B)$, we say that $app(A)$ is a rough subset of $app(B)$. Thus, in the case of rough sets $app(A)$ and $app(B)$,

$$app(A) \sqsubseteq app(B) \quad \text{if and only if } \underline{app}(A) \subseteq \underline{app}(B)$$

$$\text{and} \quad \overline{app}(A) \subseteq \overline{app}(B).$$

This property of rough inclusion has all the properties of set inclusion. The rough complement of $app(A)$, denoted by $app^C(A)$, is defined by

$$app^C(A) = \left(U\backslash\overline{app}(A),\ U\backslash\underline{app}(A)\right).$$

Also, we can define $app(A)\backslash app(B)$ as follows:

$$app(A)\backslash app(B) = app(A) \cap app^C(B)$$

$$= \left(\underline{app}(A)\backslash\overline{app}(B),\ \overline{app}(A)\backslash\underline{app}(B)\right).$$

Theorem 5.1. *We have:*

(1) $\underline{app}(A) \subseteq A \subseteq \overline{app}(A)$;
(2) $\underline{app}(\emptyset) = \emptyset = \overline{app}(\emptyset)$;
(3) $\underline{app}(H) = H = \overline{app}(H)$;
(4) *if* $A \subseteq B$, *then* $\underline{app}(A) \subseteq \underline{app}(B)$ *and* $\overline{app}(A) \subseteq \overline{app}(B)$;
(5) $\underline{app}(\underline{app}(A)) = \underline{app}(A)$;
(6) $\overline{app}(\overline{app}(A)) = \overline{app}(A)$;
(7) $\overline{app}(\underline{app}(A)) = \underline{app}(A)$;
(8) $\underline{app}(\overline{app}(A)) = \overline{app}(A)$;
(9) $\underline{app}(A) = (\overline{app}(A^C))^C$;
(10) $\overline{app}(A) = (\underline{app}(A^C))^C$;
(11) $\underline{app}(A \cap B) = \underline{app}(A) \cap \underline{app}(B)$;
(12) $\overline{app}(A \cap B) \subseteq \overline{app}(A) \cap \overline{app}(B)$;
(13) $\underline{app}(A \cup B) \supseteq \underline{app}(A) \cup \underline{app}(B)$;
(14) $\overline{app}(A \cup B) = \overline{app}(A) \cup \overline{app}(B)$.

5.2   Rough Hyperideals

Biswas and Nanda [11] introduced the notion of rough subgroups. Kuroki [51], introduced the notion of a rough ideal in a semigroup. Davvaz [24] introduced the notion of rough subrings (resp., ideal) with respect to an ideal of a ring. In this section, we consider a Krasner hyperring as a universal set, and we introduce the notion of rough hyperideals with respect to a hyperideal of a hyperring R.

Throughout this chapter, R is a commutative Krasner hyperring. The relational notation $A \approx B$ is used to assert that the sets A and B have an element in common, i.e., $A \cap B \neq \emptyset$. Let I be a hyperideal of R and X be a non-empty subset of R. Then, the sets $\underline{Apr}_I(A) = \{x \in R \mid x + I \subseteq A\}$ and $\overline{Apr}_I(A) = \{x \in R \mid (x + I) \approx A\}$ are called, respectively, the *lower and upper approximations of the set A with respect to the hyperideal I.* In this case, we use the pair (R, I) instead of the approximation space (U, θ). Clearly, we have $\underline{Apr}_I(A) \subseteq A \subseteq \overline{Apr}_I(A)$. For every $A \subseteq R$ and $x \in R$, $\underline{Apr}_I(A)$, $\overline{Apr}_I(A)$, and $x + I$ are definable sets.

Theorem 5.2. *Let I be a hyperideal of R and A, B be non-empty subsets of R. Then:*

(1) $\overline{Apr}_I(A) + \overline{Apr}_I(B) = \overline{Apr}_I(A + B)$,

(2) $\underline{Apr}_I(A) + \underline{Apr}_I(B) \subseteq \underline{Apr}_I(A + B)$.

Proof. It is straightforward. $\qquad\square$

Lemma 5.1. *Let I be a hyperideal of R and A any non-empty subset in R. Then, $\overline{Apr}_I(A) = I + A$.*

Proof. For any $x \in \overline{Apr}_I(A)$, we have $x + I \cap A \neq \emptyset$, and so there exists $y \in R$ such that $y \in x + I \cap A$, i.e., $x \in y + I$ and $y \in A$, and so $x \in I + A$.

Conversely, if $x \in I + A$, then there exist $y \in I$ and $a \in A$ such that $x \in y + a$, i.e., $a \in -y + x \subseteq x + I$. Thus, $x \in \overline{Apr}_I(A)$. $\qquad\square$

Theorem 5.3. *Let I be a hyperideal of R and A, B be non-empty subsets of R. Then,*

$$\overline{Apr}_I(A) \cdot \overline{Apr}_I(B) \subseteq \overline{Apr}_I(A \cdot B).$$

Proof. By Lemma 5.1, we have

$$\overline{Apr}_I(A) \cdot \overline{Apr}_I(B) = (I + A) \cdot (I + B) = I^2 + IB + AI + AB$$
$$\subseteq I + AB = \overline{Apr}_I(A \cdot B).$$

This completes the proof. $\qquad\square$

Theorem 5.4. *Let I be a hyperideal of R and A, B be non-empty subsets of R. If $I^2 = I$, then:*

(1) $\overline{Apr}_I(A) \cdot \overline{Apr}_I(B) = \overline{Apr}_I(A \cdot B)$,

(2) $\underline{Apr}_I(A) \cdot \underline{Apr}_I(B) \subseteq \underline{Apr}_I(A \cdot B)$.

Proof. (1) We have

$$\overline{Apr}_I(A) \cdot \overline{Apr}_I(B) = (I + A) \cdot (I + B) = I^2 + IB + AI + AB$$
$$= I + AB = \overline{Apr}_I(A \cdot B).$$

(2) For any $x \in \underline{Apr}_I(A) \cdot \underline{Apr}_I(B)$, then $x = \sum_{i=1}^n a_i b_i$, for some $a_i \in \underline{Apr}_I(A)$ and $b_i \in \underline{Apr}_I(B)$. Hence, $a_i + I \subseteq A$ and $b_i + I \subseteq B$. Since $\sum_{i=1}^n (a_i + I)(b_i + I) \subseteq AB$, we have $\sum_{i=1}^n (a_i b_i + a_i I + I b_i + I^2) = \sum_{i=1}^n a_i b_i + I = x + I \subseteq AB$. This proves that $x \in \underline{Apr}_I(A \cdot B)$. $\qquad\square$

Example 5.1. Let $R = \{0, a, b, c\}$ be a set with the hyperoperation $+$ and the multiplication $\cdot$ defined as follows:

+	0	a	b	c
0	0	a	b	c
a	a	$\{0,b\}$	$\{a,c\}$	b
b	b	$\{a,c\}$	$\{0,b\}$	a
c	c	b	a	0

$\cdot$	0	a	b	c
0	0	0	0	0
a	0	c	b	c
b	0	b	0	b
c	0	c	b	c

Then, R is a commutative Krasner hyperring. Let $I = \{0, b\}, A = \{0, a, b\}$, and $B = \{0, b, c\}$. Then, $\overline{Apr}_I(A) = \{0, a, b, c\}$ and $\overline{Apr}_I(B) = \{0, a, b, c\}$. Thus,

$$\overline{Apr}_I(A) \cdot \overline{Apr}_I(B) = \{0, b, c\} \quad \text{and} \quad \overline{Apr}_I(A \cdot B) = \{0, a, b, c\}.$$

Example 5.2. Let $R = \{0, a, b, c\}$ be a set with the hyperoperation $(+)$ and multiplication $(\cdot)$ as follows:

+	0	a	b	c
0	0	a	b	c
a	a	$\{0,c\}$	c	$\{a,b\}$
b	b	c	$\{0,c\}$	$\{a,b\}$
c	c	c	$\{a,b\}$	$\{0,c\}$

$\cdot$	0	a	b	c
0	0	0	0	0
a	0	a	a	0
b	0	a	a	0
c	0	0	0	0

Then, R is a commutative Krasner hyperring. Let $I = \{0, c\}$, $A = \{0, a, c\}$, and $B = \{0, b, c\}$. Then, $\underline{Apr}_I(A) = \{0, c\}$ and $\underline{Apr}_I(B) = \{0, c\}$. Thus,

$$\underline{Apr}_I(A) \cdot \underline{Apr}_I(B) = \{0\} \quad \text{and} \quad \underline{Apr}_I(A \cdot B) = \emptyset.$$

Remark 5.1. Examples 5.1 and 5.2 show that Theorem 5.4 is not true if we remove the condition $I^2 = I$.

Lemma 5.2. *Let I, J be two hyperideals of R such that $I \subseteq J$, and let A be a non-empty subset of R. Then:*

(1) $\underline{Apr}_J(A) \subseteq \underline{Apr}_I(A)$,
(2) $\overline{Apr}_I(A) \subseteq \overline{Apr}_J(A)$.

Proof. It is straightforward. $\square$

Note that for the hyperideals I and J, set $I \cap J$ is also a hyperideal.

Corollary 5.1. *Let I, J be two hyperideals of R and A be a non-empty subset of R. Then,*

$$\underline{Apr}_I(A) \cap \underline{Apr}_J(A) \subseteq \underline{Apr}_{(I \cap J)}(A).$$

Proof. The result follows from Lemma 5.2. □

Under some conditions, equality will hold in Corollary 5.1, as shown in the following theorem.

Theorem 5.5. *Let I, J be two hyperideals of R and A be a non-empty subset of R.*

(1) *If A is a subhyperring of R and $I, J \subseteq A$, then $\underline{Apr}_I(A) \cap \underline{Apr}_J(A) = \underline{Apr}_{(I \cap J)}(A)$.*
(2) *If $\overline{Apr}_I(A) = A$ or $\overline{Apr}_J(A) + A$, then $\overline{Apr}_{(I \cap J)}(A) = \overline{Apr}_I(A) \cap \overline{Apr}_J(A)$.*

Proof. (1) Assume that x is an arbitrary element of $\underline{Apr}_{(I \cap J)}(A)$. Then, $x \in A$. Since $x \in A$, $I \subseteq A$ and A is a subhyperring of R, it follows that $x + I \subseteq A$, which implies that $x \in \underline{Apr}_I(A)$. Similarly, we obtain $x \in \underline{Apr}_J(A)$, and so $x \in \underline{Apr}_I(A) \cap \underline{Apr}_J(A)$.
(2) Assume that $\overline{Apr}_I(A) = A$, then

$$\overline{Apr}_I(A) \cap \overline{Apr}_J(A) = A \cap \overline{Apr}_J(A) = A.$$

On the other hand, we have $A \subseteq \overline{Apr}_{(I \cap J)}(A)$. Therefore, $\overline{Apr}_I(A) \cap \overline{Apr}_J(A) \subseteq \overline{Apr}_{(I \cap J)}(A)$. □

Theorem 5.6. *Let I be a hyperideal of R and A be a non-empty subset of R. Then:*

(1) $\underline{Apr}_I(A^c) = \left(\overline{Apr}_I(A)\right)^c$,
(2) $\overline{Apr}_I(A^c) = \left(\underline{Apr}_I(A)\right)^c$.

Proof. (1) We have

$$x \in \underline{Apr}_I(A^c) \Leftrightarrow x + I \subseteq A^c \Leftrightarrow (x + I) \cap A = \emptyset$$
$$\Leftrightarrow x \notin \overline{Apr}_I(A) \Leftrightarrow x \in (\overline{Apr}_I(A))^c.$$

(2) We have

$$x \in \underline{Apr}_I(A) \Leftrightarrow (x + I) \subseteq A \Leftrightarrow (x + I) \cap A^c = \emptyset$$
$$\Leftrightarrow x \notin \overline{Apr}_I(A^c) \Leftrightarrow x \in (\overline{Apr}_I(A^c))^c. \quad □$$

Theorem 5.7. *If I is a hyperideal of R and A a non-empty subset of R, then:*

(1) $\underline{Apr}_I(-A) = -\underline{Apr}_I(A)$,
(2) $\overline{Apr}_I(-A) = \overline{Apr}_I(A)$.

Proof.   (1) We have

$$
\begin{aligned}
x \in \underline{Apr}_I(-A) \quad &\Leftrightarrow \quad x + I \subseteq -A \\
&\Leftrightarrow \quad -x - I \subseteq A \\
&\Leftrightarrow \quad -x + I \subseteq A \\
&\Leftrightarrow \quad -x \in \underline{Apr}_I(A) \\
&\Leftrightarrow \quad x \in -\underline{Apr}_I(A).
\end{aligned}
$$

Thus, $\underline{Apr}_I(-A) = -\underline{Apr}_I(A)$.
  (2) We have

$$
\begin{aligned}
x \in \overline{Apr}_I(-A) \quad &\Leftrightarrow \quad (x + I) \approx (-A) \\
&\Leftrightarrow \quad (-x - I) \approx A \\
&\Leftrightarrow \quad (-x + I) \approx A \\
&\Leftrightarrow \quad -x \in \overline{Apr}_I(A) \\
&\Leftrightarrow \quad x \in -\overline{Apr}_I(A).
\end{aligned}
$$

Thus, $\overline{Apr}_I(-A) = -\overline{Apr}_I(A)$. □

Theorem 5.8. *If I, J are hyperideals of R, then $\overline{Apr}_I(J) = \overline{Apr}_J(I)$.*

Proof.   Suppose that x is an arbitrary element of $\overline{Apr}_I(J)$, then $(x + I) \approx J$. Hence, there exists $y \in (x + I) \cap J$. So, $y \in J$ and $y \in x + I$. From $y \in x + I$, we get $-y \in -x + I$, and so $x - y \subseteq I$. Since $y \in J$, then $-y \in J$. Hence, $x - y \subseteq x + J$. Therefore, we have $(x + J) \approx A$, which implies that $x \in \overline{Apr}_J(I)$. Thus, we conclude that $\overline{Apr}_I(J) \subseteq \overline{Apr}_J(I)$. Similarly, we get $\overline{Apr}_J(I) \subseteq \overline{Apr}_I(J)$. This completes the proof. □

Theorem 5.9. *Let I be a hyperideal and J a hyperideal of R. Then, $\overline{Apr}_I(J)$ is a hyperideal of R.*

Proof. Suppose that $a, b \in \overline{Apr}_I(J)$ and $r \in R$, then $(a + I) \approx J$ and $(b + I) \approx J$. So, there exist $x \in (a + I) \cap J$ and $y \in (b + I) \cap J$. Since J is a hyperideal of R, we have $x - y \subseteq J$ and $x - y \subseteq (a + I) - (b + I) = a - b + I$. Now, for every $c \in a - b$, we have $x - y \subseteq c + I$. Therefore, $(c + I) \approx J$, and so $c \in \overline{Apr}_I(J)$. Also, we have $rx \in J$ and $rx \in r(a + I) = ra + I$. So, $(ra + I) \approx J$, which implies that $ra \in \overline{Apr}_I(J)$. Therefore, $\overline{Apr}_I(J)$ is a hyperideal of R. $\square$

Similarly, if I is a hyperideal and J a subhyperring of R, then $\overline{Apr}_I(J)$ is a subhyperring of R.

Theorem 5.10. *Let I be a hyperideal and J a subhypergroup of R. If $\underline{Apr}_I(J)$ is a non-empty set, then I is a subset of J.*

Proof. Suppose that $\underline{Apr}_I(J) \neq \emptyset$, then there exists $x \in \underline{Apr}_I(J)$, which implies that $x + I \subseteq J$. Since $0 \in I$, we have $x + 0 = x \in J$, then $-x \in J$. Therefore, $I = -x + x + I \subseteq -x + J \subseteq J$. $\square$

Theorem 5.11. *Let I be a hyperideal and J a subhyperring of R. If $\underline{Apr}_I(J)$ is a non-empty set, then it is equal to J.*

Proof. We have $\underline{Apr}_I(J) \subseteq J$. Now, we show that $J \subseteq \underline{Apr}_I(J)$. Suppose that b is an arbitrary element of J. We have $I \subseteq J$. Since J is a subhyperring, we obtain $b + I \subseteq J$, which implies that $b \in \underline{Apr}_I(J)$. Therefore, $\underline{Apr}_I(J) = J$. $\square$

Let I be a hyperideal of R and $Apr_I(A) = (\underline{Apr}_I(A), \overline{Apr}_I(A))$ a rough set in the approximation space (R, I). If $\underline{Apr}_I(A)$ and $\overline{Apr}_I(A)$ are (prime) hyperideals (resp., subhyperrings) of R, then we call $Apr_I(A)$ a *rough (prime) hyperideal* (resp., *subhyperring*). Note that a rough subhyperring is also called a *rough hyperring*.

Theorem 5.12. *If I is a hyperideal and J a hyperideal (subhyperring) of R such that $I \subseteq J$, then $Apr_I(J)$ is a rough hyperideal (rough hyperring).*

Theorem 5.13. *If I is a hyperideal and P is a prime hyperideal of R such that $I \subseteq P$, then $Apr_I(P)$ is a rough prime hyperideal of R.*

Proof. By Theorems 5.11 and 5.12, it is enough to show that $\overline{Apr}_I(P)$ is prime. Suppose that $xy \in \overline{Apr}_I(P)$, for some $x, y \in R$, then $(xy + I) \approx P$. So, there exists $a \in I$ such that $(xy + a) \approx P$, which yields $a \in I$, $b \in P$ and $b \in xy + a$. Hence, $xy \in b - a \subseteq P + I \subseteq P$. Since P is prime, so $x \in P$ or $y \in P$. Therefore, $(x + I) \approx P$ or $(y + I) \approx P$, and so $x \in \overline{Apr}_I(P)$ or $y \in \overline{Apr}_I(P)$. $\square$

Theorem 5.14. *Let I, J be hyperideals of R and A a subhyperring of R. Then:*

(1) $\overline{Apr}_I(A) + \overline{Apr}_J(A) = \overline{Apr}_{(I+J)}(A)$,
(2) $\underline{Apr}_I(A) + \underline{Apr}_J(A) = \underline{Apr}_{(I+J)}(A)$.

Proof.   (1) Since $I \subseteq I + J$ and $J \subseteq I + J$, it follows that $\overline{Apr}_I(A) \subseteq \overline{Apr}_{(I+J)}(A)$ and $\overline{Apr}_J(A) \subseteq \overline{Apr}_{(I+J)}(A)$. Therefore, $\overline{Apr}_I(A) + \overline{Apr}_J(A) \subseteq \overline{Apr}_{(I+J)}(A)$. Conversely, let x be an arbitrary element of $\overline{Apr}_{(I+J)}(A)$. Then, $(x + I + J) \approx A$. So, there exists $a \in I$ such that $(x + a + J) \approx A$, which yields $b \in x + a$ such that $b \in \overline{Apr}_J(A)$. Now, we have $x \in b - a = -a + b \in \overline{Apr}_I(A) + \overline{Apr}_J(A)$ because A is a subhyperring and $0 \in (-a + I) \cap A$.

(2) Since $I \subseteq I + J$ and $J \subseteq I + J$, it follows that $\underline{Apr}_{(I+J)}(A) \subseteq \underline{Apr}_I(A)$ and $\underline{Apr}_{(I+J)}(A) \subseteq \underline{Apr}_J(A)$, and so $\underline{Apr}_{(I+J)}(A) \subseteq \underline{Apr}_I(A) + \underline{Apr}_J(A)$.

Now, let $x \in \underline{Apr}_I(A) + \underline{Apr}_J(A)$. Then, $x \in a + b$, for some $a \in \underline{Apr}_I(A)$ and $b \in \underline{Apr}_J(A)$. Hence, $a + I \subseteq A$ and $b + J \subseteq A$. So,

$$x + I + J \subseteq a + b + I + J = a + I + b + J \subseteq A + A = A$$

which yields $x \in \underline{Apr}_{(I+J)}(A)$. $\square$

Theorem 5.15. *Let R be an idempotent hyperring, I, J be two hyperideals of R, and K be a subhyperring of R. Then,*

$$\underline{Apr}_I(K) \cdot \underline{Apr}_J(K) = \underline{Apr}_{(I+J)}(K).$$

Proof.   Let x be any element of $\underline{Apr}_I(A) \cdot \underline{Apr}_J(A)$. Then, $x \in \sum_{i=1}^{n} a_i b_i$ for some $a_i \in \underline{Apr}_I(A)$ and $b_i \in \underline{Apr}_J(A)$. Therefore, $a_i + I \subseteq A$ and $b_i + J \subseteq A$. Now, we have

$$\sum_{i=1}^{n}(a_i + I)(b_i + J) \subseteq A \quad \text{or} \quad \sum_{i=1}^{n} a_i b_i + I + J \subseteq A,$$

which yields $\sum_{i=1}^{n} a_i b_i \subseteq \underline{Apr}_{(I+J)}(A)$, and so $x \in \underline{Apr}_{(I+J)}(A)$.

Conversely, since $I \subseteq I+J$ and $J \subseteq I+J$, we have $\underline{Apr}_{(I+J)}(K) \subseteq \underline{Apr}_I(K)$ and $\underline{Apr}_{(I+J)}(K) \subseteq \underline{Apr}_J(K)$. Thus, $\underline{Apr}_{(I+J)}(K) = \underline{Apr}_{(I+J)}(K) \cdot \underline{Apr}_{(I+J)}(K) \subseteq \underline{Apr}_I(K) \cdot \underline{Apr}_J(K)$. This completes the proof. $\qquad\square$

Theorem 5.16. *If I, J are hyperideals of R and A is a subhyperring of R, then*

$$\overline{Apr}_I(A) \cdot \overline{Apr}_J(A) \subseteq \overline{Apr}_{(I+J)}(A).$$

Proof. Let x be an element of $\overline{Apr}_I(A) \cdot \overline{Apr}_J(A)$. Then, $x \in \sum_{i=1}^n a_i b_i$, for some $a_i \in \overline{Apr}_I(A)$ and $b_i \in \overline{Apr}_J(A)$. Hence, $(a_i+I) \sim A$ and $(b_i + J) \sim A$, and so there exist elements $x_i, y_i \in R$ such that $x_i \in (a_i + I) \cap A$ and $y_i \in (b_i + J) \cap A$. Since A is a subhyperring of R, it follows that $\sum_{i=1}^n x_i y_i \subseteq A$. On the other hand, we have

$$x \in \sum_{i=1}^n a_i b_i \subseteq \sum_{i=1}^n x_i y_i + I + J \subseteq A + I + J.$$

So there exists $a \in A$ such that $x \in a + I + J$ or $a \in x + I + J$. Therefore, $(x + I + J) \sim A$, which implies $x \in \overline{Apr}_{(I+J)}(A)$. $\qquad\square$

Theorem 5.17. *Let R be a hyperring with an identity 1, I, J be two hyperideals of R, and K be a subhyperring of R such that $1 \in K$. Then,*

$$\overline{Apr}_I(K) \cdot \overline{Apr}_J(K) = \overline{Apr}_{(I+J)}(K).$$

Proof. Since $1 \in K$, it follows that $IK = I, KJ = J$ and $K^2 = K$. Thus, by Lemma 5.1, we have

$$\overline{Apr}_I(K) \cdot \overline{Apr}_J(K) = (I + K) \cdot (J + K) = IJ + IK + KJ + K^2$$
$$= I + J + K = \overline{Apr}_{(I+J)}(K),$$

as desired. $\qquad\square$

Example 5.3. Let $R = \{0, a, b, c\}$ be a set with the operation $+$ and multiplication $\cdot$ as follows:

+	0	a	b	c
0	0	a	b	c
a	a	0	c	b
b	b	c	0	a
c	c	b	a	0

$\cdot$	0	a	b	c
0	0	0	0	0
a	0	0	0	0
b	0	0	c	c
c	0	0	c	c

Then, R is a commutative Krasner hyperring (indeed, it is a ring).

Let $I = \{0, a\}$, $J = \{0, c\}$ and $A = \{0, a, b, c\}$. Then,

$$\underline{Apr}_I(A) = \{0, a, b, c\}, \quad \underline{Apr}_J(A) = \{0, a, b, c\},$$

$$\underline{Apr}_{(I+J)}(A) = \{0, a, b, c\} \quad \text{and} \quad \underline{Apr}_I(A) \cdot \underline{Apr}_J(A) = \{0, c\}.$$

5.3   Approximations and Homomorphisms

Now, let R and R' be commutative Krasner hyperrings and $f : R \longrightarrow R'$ a good homomorphism from R onto R'. It is well known that $Ker f$ is a hyperideal of R. If A is a subset of R and $Ker f$ is normal, clearly we have $f(\underline{Apr}_{Ker f}(A)) \subseteq f(A)$. However, in general, $f(\underline{Apr}_{Ker f}(A)) \neq f(A)$.

Theorem 5.18. *Let R and R' be two hyperrings and f a good homomorphism from R onto R' such that $Ker f$ is normal. If A is a nonempty subset of R, then*

$$f(\overline{Apr}_{Ker f}(A)) = f(A).$$

Proof.   Since $A \subseteq \overline{Apr}_{Ker f}(A)$, it follows that $f(A) \subseteq f(\overline{Apr}_{Ker f}(A))$. Conversely, let $y \in f(\overline{Apr}_{Ker f}(A))$. Then, there exists an element $x \in \overline{Apr}_{Ker f}(A)$ such that $f(x) = y$, so we have $(x + Ker f) \approx A$. Thus, there exists an element $a \in (x + Ker f) \cap A$. Then, $a \in x + b$, for some $b \in Ker f$, i.e., $x \in a - b$. Then, we have $y = f(x) \in f(a - b) = f(a) - f(b) = f(a)$, and so $y \in f(A)$. Therefore, $f(\overline{Apr}_{Ker f}(A)) \subseteq f(A)$.   □

Let R and R' be two hyperrings and f a good homomorphism from R onto R'. Let I be a normal hyperideal of R and J a normal hyperideal of R'. Then:

(1) If f is an epimorphism, then $f(I)$ is a hyperideal of R'. If f is not surjective, $f(I)$ is not necessarily a hyperideal of R'.

(2) $f^{-1}(J)$ is a hyperideal of R.

Theorem 5.19. *Let R and R' be two hyperrings and f be an isomorphism from R onto R'. If I is a hyperideal of R and A a non-empty subset of R, then:*

(1) $\overline{Apr}_{f(I)}(f(A)) = f\left(\overline{Apr}_I(A)\right),$

(2) $\underline{Apr}_{f(I)}(f(A)) = f\left(\underline{Apr}_I(A)\right).$

Proof. (1) Suppose that y is an arbitrary element of $f(\overline{Apr}_I(A))$, then there exists $x \in \overline{Apr}_I(A)$ such that $f(x) = y$. Hence, $(x + I) \approx A$, and so there exists $a \in (x + I) \cap A$. Then, $f(a) \in f(A)$ and $f(a) \in f(x) + f(I)$. So, $(f(x) + f(I)) \approx f(A)$, which yields $y = f(x) \in \underline{Apr}_{f(I)}(f(A))$. So, we get $f(\overline{Apr}_I(A)) \subseteq \overline{Apr}_{f(I)}(f(A))$.

Conversely, let $y \in \overline{Apr}_{f(I)}(f(A))$. Then, there exists $x \in R$ such that $f(x) = y$ and $(f(x) + f(I)) \approx f(A)$. So, there exists $a \in A$ such that $f(a) \in f(A)$ and $f(a) \in f(x) + f(I)$. Then, $a \in x + I$. Hence, $(x + I) \approx A$, which implies that $x \in \overline{Apr}_I(A)$. So, $y = f(x) \in f(\overline{Apr}_I(A))$. It means that $\overline{Apr}_{f(I)}(f(A)) \subseteq f(\overline{Apr}_I(A))$.

(2) Suppose that y is an arbitrary element of $f(\underline{Apr}_I(A))$. Then, there exists $x \in \underline{Apr}_I(A)$ such that $f(x) = y$. Then, we have $(x + I) \subseteq A$. Let $z \in (y + f(I))$; then, there exists $c \in R$ such that $f(c) = z$ and $f(c) \in f(x) + f(I)$. So, $c \in x + I \subseteq A$, which implies that $z = f(c) \in f(A)$. Thus, $y + f(I) \subseteq f(A)$, which yields $y \in \underline{Apr}_{f(I)}(f(A))$, so we have $f(\underline{Apr}_I(A)) \subseteq \underline{Apr}_{f(I)}(f(A))$.

Conversely, assume that $y \in \underline{Apr}_{f(I)}(f(A))$, then there exists $x \in R$ such that $f(x) = y$ and $(f(x) + f(I)) \subseteq f(A)$. Let $a \in x + I$; then, $f(a) \in (f(x) + f(I)) \subseteq f(A)$. Since f is one to one, $a \in A$. Thus, $x + I \subseteq A$, which yields $x \in \underline{Apr}_I(A)$. Then, $y = f(x) \in f(\underline{Apr}_I(A))$, and so $\underline{Apr}_{f(I)}(f(A)) \subseteq f(\underline{Apr}_I(A))$. $\square$

Corollary 5.2. *Let R and R' be two hyperrings and f an isomorphism from R to R'. If J is a normal hyperideal of R and B a nonempty subset of R', then:*

(1) $\overline{Apr}_{f^{-1}(J)}(f^{-1}(B)) = f^{-1}\left(\overline{Apr}_J(B)\right),$

(2) $\underline{Apr}_{f^{-1}(J)}(f^{-1}(B)) = f^{-1}\left(\underline{Apr}_J(B)\right).$

Proof. The result follows from Theorem 5.19. ☐

Corollary 5.3. *Let R and R' be two hyperrings and f an isomorphism from R onto R'. If I, J are normal hyperideals of R and A a nonempty subset of R, then:*

(1) $f\left(\overline{Apr}_{(I+J)}(A)\right) = \overline{Apr}_{f(I)}(f(A)) + \overline{Apr}_{f(J)}(f(A))$,

(2) $f\left(\underline{Apr}_{(I+J)}(A)\right) = \underline{Apr}_{f(I)}(f(A)) + \underline{Apr}_{f(J)}(f(A))$.

Proof. The result follows from Theorem 5.19. ☐

Theorem 5.20. *Let R and R' be two hyperrings and f an isomorphism from R to R'. If J is a hyperideal of R' and A a non-empty subset of R, then:*

(1) $\overline{Apr}_{f^{-1}(J)}(A)$ *is a hyperideal of R $\Leftrightarrow$ $\overline{Apr}_J(f(A))$ is a hyperideal of R',*

(2) $\overline{Apr}_{f^{-1}(J)}(A)$ *is prime $\Leftrightarrow$ $\overline{Apr}_J(f(A))$ is prime.*

Proof. (1) Suppose that $\overline{Apr}_{f^{-1}(J)}(A)$ is a hyperideal of R. We show that $\overline{Apr}_J(f(A))$ is a hyperideal of R'. Let $x, y \in \overline{Apr}_J(f(A))$. Then, $f^{-1}(x)$, $f^{-1}(y) \in f^{-1}(\overline{Apr}_J(f(A)))$. By Theorem 5.19, we have $f^{-1}(x)$, $f^{-1}(y) \in \overline{Apr}_{f^{-1}(J)}(A)$, and so $f^{-1}(x) - f^{-1}(y) = f^{-1}(x-y) \subseteq \overline{Apr}_{f^{-1}(J)}(A)$, which implies that $x - y \subseteq \overline{Apr}_J(f(A))$.

Also, for every $r \in R$, we have $r \cdot \overline{Apr}_{f^{-1}(J)}(A) \subseteq \overline{Apr}_{f^{-1}(J)}(A)$, and so $f(r \cdot \overline{Apr}_{f^{-1}(J)}(A)) \subseteq f(\overline{Apr}_{f^{-1}(J)}(A))$. Hence, $f(r) \cdot f(\overline{Apr}_{f^{-1}(J)}(A)) \subseteq f(\overline{Apr}_{f^{-1}(J)}(A))$. By Theorem 5.19, we get $r' \cdot \overline{Apr}_J(f(A)) \subseteq \overline{Apr}_J(f(A))$. Therefore, $\overline{Apr}_J(f(A))$ is a hyperideal of R'.

Similarly, we can prove the converse.

(2) Suppose that $\overline{Apr}_{f^{-1}(J)}(A)$ is prime. We show that $\overline{Apr}_J(f(A))$ is prime. Let $x \cdot y \in \overline{Apr}_J(f(A))$, for some $x, y \in R'$. Then, there exist $a, b \in R$ such that

$$f(a) = x, \quad f(b) = y \quad \text{and} \quad (f(ab) + J) \cap f(A) \neq \emptyset.$$

Hence, $(f(a) + J)(f(b) + J) \cap f(A) \neq \emptyset$, and so there exist elements

$$f(r) \in f(a) + J \quad \text{and} \quad f(s) \in f(b) + J$$

such that $f(r)f(s) = f(rs) \in f(A)$, which implies that $rs \in A$. Also, we get

$$r \in a + f^{-1}(J) \quad \text{and} \quad s \in b + f^{-1}(J),$$

and so $rs \in ab + f^{-1}(J)$. Therefore, $(ab + f^{-1}(J)) \cap f(A) \neq \emptyset$, which implies $ab \in \overline{Apr}_{f^{-1}(J)}(A)$. Since $\overline{Apr}_{f^{-1}(J)}(A)$ is prime, we conclude that

$$a \in \overline{Apr}_{f^{-1}(J)}(A) \quad \text{or} \quad b \in \overline{Apr}_{f^{-1}(J)}(A).$$

Now, by Theorem 5.19, we get

$$x \in \overline{Apr}_J(f(A)) \quad \text{or} \quad y \in \overline{Apr}_J(f(A)).$$

Therefore, $x \in \overline{Apr}_J(f(A))$ is prime.

Similarly, we can prove the converse. $\qquad \square$

Corollary 5.4. *Let R and R' be two hyperrings and f be an isomorphism from R to R'. If J is a hyperideal of R' and A a non-empty subset of R, then:*

(1) $\underline{Apr}_{f^{-1}(J)}(A)$ *is a hyperideal of R* $\Leftrightarrow$ $\underline{Apr}_J(f(A))$ *is a hyperideal of R',*

(2) $\underline{Apr}_{f^{-1}(J)}(A)$ *is prime* $\Leftrightarrow$ $\underline{Apr}_J(f(A))$ *is prime.*

Proof. It is straightforward. $\qquad \square$

5.4 Rough Sets in a Quotient Hyperring

Let I be a hyperideal of R. It is important to note that the equivalence class $x + I$ containing x plays dual roles. It is a subset of R if it is considered in relation to the hyperring R and is an element of R/I if it is considered in relation to the quotient hyperring. Therefore, the lower and upper approximations can be presented in an equivalent form, as shown in the following.

Let I be a hyperideal of R and A a non-empty subset of R. Then, we define $\underline{Apr}_I(A) = \{x + I \in R/I \mid x + I \subseteq A\}$ and $\overline{Apr}_I(A) = \{x + I \in R/I \mid (x + I) \approx A\}$. Now, we discuss these sets as subsets of the quotient hyperring R/I.

Theorem 5.21. *Let I be a hyperideal and J a hyperideal of R. Then, $\overline{Apr}_I(J)$ is a hyperideal of R/I.*

Proof.   Assume that $a+I, b+I \in \overline{Apr}_I(J)$ and $r+I \in R/I$. Then, $(a+I) \approx J$ and $(b+I) \approx J$, so there exist $x \in (a+I) \cap J$ and $y \in (b+I) \cap J$. Since J is a hyperideal of R, we have $x - y \subseteq J$ and $rx \in J$. Also, we have

$$x - y \subseteq (a+I) \oplus (-b+I),$$
$$rx \in (r+I) \odot (a+I) = ra+I.$$

Therefore, $(a+I) \oplus (-b+I) \approx J$ and $(ra+I) \approx J$, which imply that $(a+I) \oplus (-b+I) \in \overline{Apr}_I(J)$ and $(r+I) \odot (a+I) \in \overline{Apr}_I(J)$. Therefore, $\overline{Apr}_I(J)$ is a hyperideal of R/I.   □

Theorem 5.22. *Let I be a hyperideal and J a hyperideal of R such that $I \subseteq J$. Then, $\underline{Apr}_I(J)$ is a hyperideal of R/I.*

Proof.   It is straightforward.   □

Similarly, if I is a normal hyperideal and J is a subhyperring of R, then $\underline{Apr}_I(J)$ and $\overline{Apr}_I(J)$ are subrings of R/I.

Corollary 5.5. *Let R and R' be two hyperrings and f an isomorphism from R onto R'. If J is a hyperideal of R' and A a non-empty subset of R, then:*

(1) $\overline{Apr}_{f^{-1}(J)}(A)$ *is a hyperideal of* $R/f^{-1}(J)$ $\Leftrightarrow$ $\overline{Apr}_J(f(A))$ *is a hyperideal of* R'/J,

(2) $\overline{Apr}_{f^{-1}(J)}(A)$ *is prime* $\Leftrightarrow$ $\overline{Apr}_J(f(A))$ *is prime.*

Proof.   It is straightforward.   □

Corollary 5.6. *Let R and R' be two hyperrings and f an isomorphism from R onto R'. If J is a normal hyperideal of R' and A a nonempty subset of R, then:*

(1) $\underline{Apr}_{f^{-1}(J)}(A)$ *is a hyperideal of* $R/f^{-1}(J)$ $\Leftrightarrow$ $\underline{Apr}_J(f(A))$ *is a hyperideal of* R'/J,

(2) $\underline{Apr}_{f^{-1}(J)}(A)$ *is prime* $\Leftrightarrow$ $\underline{Apr}_J(f(A))$ *is prime.*

Proof.   It is straightforward.   □

Chapter 6

Derived Hyperstructures from Hyperconics

6.1 Conics

An elliptical curve is a curve of the form $y^2 = p(x)$, where $p(x)$ is a cubic polynomial with no repeat roots over the field F. This kind of curve is considered and extended over Krasner's hyperfields in Ref. [97]. Now, let $g(x, y) = ax^2 + bxy + cy^2 + dx + ey + f \in F[x, y]$ and $g(x, y) = 0$ be the quadratic equation of two variables in the field F. If $a = c = 0$ and $b \neq 0$, then the equation $g(x, y) = 0$ is called a homographic transformation. Vahedi *et al.* [98] extended this particular quadratic equation on Krasner's quotient hyperfield $\frac{F}{G}$. The motivation of this chapter aligns with that of Ref. [98]. If, in the general form of the quadratic equation, we suppose that $ae \neq 0$ and $b = 0$, then we obtain an important quadratic equation known as a conic. Note that the conditions which are considered for the coefficients of the equations of a conic curve and a homographic curve are completely different. Until now, the study of conic curves has been focused on fields. In recent works, authors have investigated some main classes of curves, including elliptic curves and homographics over Krasner's hyperfields Refs. [97,98]. Vahedi *et al.* [99] investigated the conic curves over some quotients of Krasner's hyperfields. They introduced generalized quadratic forms and hyperconics over quotient hyperfields as a generalization of the notion of conics on fields.

In the following, we recall some basic notions, which can be found in Refs. [37, 48].

Definition 6.1. A *conic* is an affine plane curve of degree 2. Irreducible conics C come in three types: we say that C is a *hyperbola*, a *parabola*, or an *ellipse* if the number of points at infinity on (the projective closure of) C equals $2, 1$, or 0, respectively. Over an algebraically closed field, every irreducible conic is a hyperbola.

Let d be a square-free integer that is not equal to 1, and set

$$\Delta = \begin{array}{ll} d & \text{if } d \equiv 1 \pmod 4, \\ 4d & \text{if } d \equiv 2, 3 \pmod 4. \end{array}$$

The conic $C : Q_0(x, y) = 1$ associated with the principal quadratic form of discriminant Δ,

$$Q_0(x, y) = \begin{cases} x^2 + xy + \frac{1-d}{4}y^2 & \text{if } d \equiv 1 \pmod 4, \\ x^2 - dy^2 & \text{if } d \equiv 2, 3 \pmod 4, \end{cases}$$

is called the *Pell conic* of discriminant. Pell conics are irreducible nonsingular affine curves with a distinguished integral point $N = (1, 0)$. The problem corresponding to the determination of $E(Q)$ pertains to finding the integral points on a Pell conic. The idea that certain sets of points on curves can be given a group structure is relatively modern. For elliptic curves, the group structure became well known only in the 1920s; implicitly, it can be found in the work of Clebsch. In a rarely cited article, Juel wrote down the group law for elliptic curves defined over $\mathbb{R}$ and $\mathbb{C}$ at the end of the 19th century. The group law on Pell conics is defined over a field F. For two rational points $p, q \in Q(F)$, draw a line through $\mathcal{O}$ parallel to the line joining p and q. Denote its second point of intersection as $p * q$, which is the sum of p and q, where $\mathcal{O}$ is some arbitrary point in the Pell conic, perhaps at infinity, and is the identity element of the group.

Example 6.1. Consider $f(x) = x^{-1} + x$ over a finite field $F = \mathbb{Z}_7$. Then, we have a Cayley table of points as follows:

$\bullet_{11}$	$(0,\infty)$	$(1,2)$	$(2,-1)$	$(3,1)$	$(-3,-1)$	$(-2,1)$	$(-1,-2)$	(∞,∞)
$(0,\infty)$	$(0,\infty)$	$(1,2)$	$(2,-1)$	$(3,1)$	$(-3,-1)$	$(-2,1)$	$(-1,-2)$	(∞,∞)
$(1,2)$	$(1,2)$	$(-1,-2)$	$(-3,-1)$	$(-2,1)$	$(3,1)$	$(2,-1)$	$(0,\infty)$	(∞,∞)
$(2,-1)$	$(2,-1)$	$(-3,-1)$	$(1,2)$	$(-1,-2)$	(∞,∞)	$(0,\infty)$	$(-2,1)$	$(3,1)$
$(3,1)$	$(3,1)$	$(-2,1)$	$(-1,-2)$	$(1,2)$	$(0,\infty)$	(∞,∞)	$(-3,-1)$	$(2,-1)$
$(-3,-1)$	$(-3,-1)$	$(3,1)$	(∞,∞)	$(0,\infty)$	$(-1,-2)$	$(1,2)$	$(2,-1)$	$(-2,1)$
$(-2,1)$	$(-2,1)$	$(2,-1)$	$(0,\infty)$	(∞,∞)	$(1,2)$	$(-1,-2)$	$(3,1)$	$(-3,-1)$
$(-1,-2)$	$(-1,-2)$	$(0,\infty)$	$(-2,1)$	$(-3,-1)$	$(2,-1)$	$(3,+1)$	(∞,∞)	$(1,2)$
(∞,∞)	(∞,∞)	$(-1,-2)$	$(3,1)$	$(2,-1)$	$(-2,+1)$	$(-3,-1)$	$(1,2)$	$(0,\infty)$

Also, see Fig. 6.1. The associativity of the group law is induced from a special case of Pascal's theorem. In the following, we recall this theorem, which is in turn a special case of Bezout's theorem.

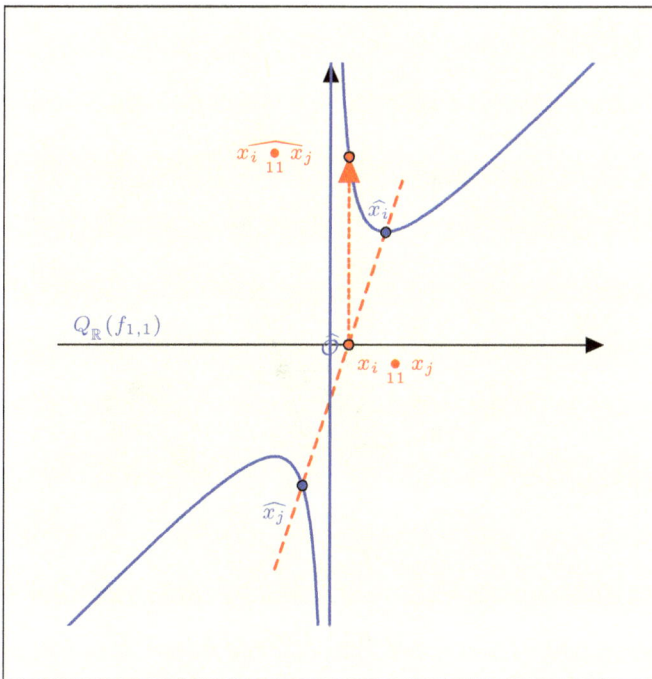

Fig. 6.1.   The curve for $f(x) = x^{-1} + x$.

Theorem 6.1 ([94]). *For any conic and any six points $p_1, p_2, \ldots,$ p_6 on it, the opposite sides of the resulting hexagram, extended if necessary, intersect at points lying on some straight line. More specifically, let $L(p, q)$ denote the line through the points p and q. Then, the points $L(p_1, p_2) \cap L(p_4, p_5)$, $L(p_2, p_3) \cap L(p_5, p_6)$, and $L(p_3, p_4) \cap L(p_6, p_1)$ lie on a straight line, called the Pascal line of the hexagon, see Fig. 6.2.*

6.2 Hyperconic

Since this chapter deals only with Krasner's hyperfields, we simply write "quotient hyperfields" instead of "Krasner's quotient hyperfields."

The notion of hyperconics in a quotient hyperfield is studied by Vahedi *et al.* [99]. Using the hyperconic $Q_{\bar{F}}(f_{\bar{A}, \bar{B}})$, they introduced

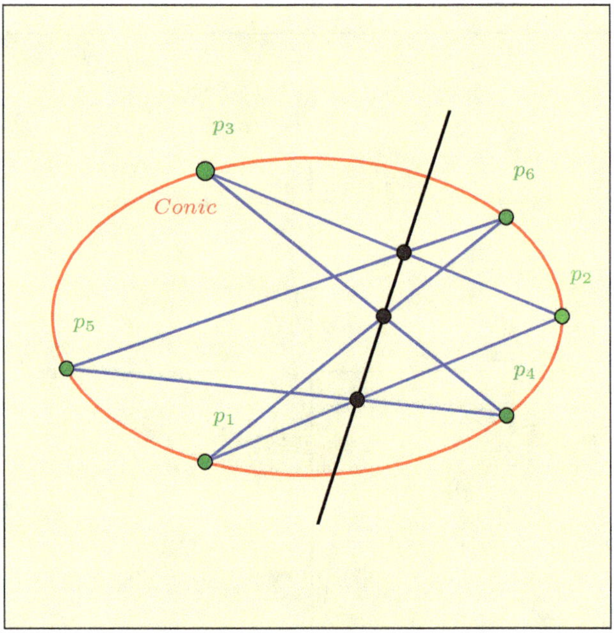

Fig. 6.2.   Conic.

certain hyperoperations as a generalization of group operations on fields. The main reference for this section is Ref. [99].

Let $g(x, y) = ax^2 + bxy + cy^2 + dx + ey + f \in F[x, y]$ and $g(x, y) = 0$ be the quadratic equation of two variables in the field F. If $c = 0$ and the equation $g(x, y) = 0$ remains quadratic in two variables, or, specifically, $c = 0$ and $(a, b) \neq 0 \neq (e, b)$, then it can be calculated as an explicit function y in terms of x and, through a change of variables, can also be expressed in the form of $Y = AX^2 + BX$ or $AX^{-1} + BX$, where $A, B \in F$.

For this purpose, if $a, e \neq 0 = b$ and setting $x = X$ and $y = Y - fe^{-1}$, then $Y = AX^2 + BX$, where $A = -ae^{-1}$ and $B = -de^{-1}$. If $b \neq 0 = a$ and setting $x = X - eb^{-1}$ and $y = Y - db^{-1}$, then $Y = AX^{-1}$, where $A = edb^{-2} - fb^{-1}$. If $b \neq 0 \neq a$ and setting $x = X - eb^{-1}$ and $y = Y + 2aeb^{-2}\alpha_F - db^{-1}$ so that

$$\alpha_F = \begin{cases} 0 & \text{if } char(F) = 2, \\ 1 & \text{if } char(F) \neq 2, \end{cases}$$

then $Y = AX^{-1} + BX$, where $A = -ae^2b^{-3} + edb^{-2} - fb^{-1}$ and $B = -ab^{-1}$. The reduced quadratic equation in two variables, $ax^2 + bxy + dx + ey + f = 0$, in the field F can be generalized to the quotient hyperfield $\bar{F}$.

Definition 6.2. Let $\bar{F}$ be a quotient hyperfield, and let $(\bar{A}, \bar{B}) \in \bar{F}^2$ and $f_{\bar{A}, \bar{B}}(\bar{x})$ be equal to $\bar{A}\bar{x}^{-1} \oplus \bar{B}\bar{x}$ or $\bar{A}\bar{x}^2 \oplus \bar{B}\bar{x}$. Then, the relation $\bar{y} \in f_{\bar{A}, \bar{B}}(\bar{x})$ is called the generalized reduced two-variable quadratic equation in $\bar{F}^2$. Moreover, the set $Q(f_{\bar{A}, \bar{B}}, \bar{F}) = \{(\bar{x}, \bar{y}) \in \bar{F}^2 | \bar{y} \in f_{\bar{A}, \bar{B}}(\bar{x})\}$ is called the conic hypersection, and if $\bar{A} \neq 0$, $Q(f_{\bar{A}, \bar{B}}, \bar{F})$ is called the nondegenerate conic hypersection. Also, for all $a \in \bar{A}$ and $b \in \bar{B}$, $Q(f_{a,b}, F) = \{(x, y) \in F^2 | y = f_{a,b}(x)\}$ is a conic section, and for $a \neq 0$, it is a nondegenerate conic section, in which $f_{a,b}(z) = az^2 + bz$ or $az^{-1} + bz$ corresponding to $f_{\bar{A}, \bar{B}}$. Additionally, $Q(f_{\bar{A}, \bar{B}}, F) = \bigcup_{(\bar{x}, \bar{y}) \in Q(f_{\bar{A}, \bar{B}}, \bar{F})} \bar{x} \times \bar{y}$ is said to be a conic hypersection, as a subset of F^2, and $Q(f_{a,b}, \bar{F}) = \overline{Q(f_{a,b}, F)} = \left\{ \overline{(x, y)} | (x, y) \in Q(f_{a,b}, F) \right\}$, where $\overline{(x, y)} = (\bar{x}, \bar{y})$ for all $(x, y) \in Q(f_{\bar{A}, \bar{B}}, F)$.

Theorem 6.2. *Using the above notions, we have* $Q(f_{\bar{A},\bar{B}}, F) = \bigcup_{a\in\bar{A}, b\in\bar{B}} Q(f_{a,b}, F)$.

Proof. Let $(x,y) \in Q(f_{\bar{A},\bar{B}}, F)$ and without loss of generality $f(x) = Ax^2 + Bx$. Then,

$$
\begin{aligned}
(\bar{x}, \bar{y}) \in Q(f_{\bar{A},\bar{B}}, \bar{F}) &\Leftrightarrow \bar{y} \in \bar{A}\bar{x}^2 \oplus \bar{B}\bar{x} \\
&\Leftrightarrow \bar{y} \in \overline{\bar{A}x^2 + \bar{B}x} \\
&\Leftrightarrow \bar{y} = \overline{ax^2 + bx}, \quad \text{for some } (a,b) \in \bar{A} \times \bar{B} \\
&\Leftrightarrow y = agx^2 + bgx \quad \text{for some } g \in G \\
&\Leftrightarrow y = a'x^2 + b'x, \quad \text{where } a' = ag, \ b' = bg \\
&\Leftrightarrow (x,y) \in Q(f_{a',b'}, F), \text{for some} \\
&\qquad (a',b') \in \bar{A} \times \bar{B} \\
&\Leftrightarrow (x,y) \in \bigcup_{a\in\bar{A}, b\in\bar{B}} Q(f_{a,b}, F).
\end{aligned}
$$

Consequently, $Q(f_{\bar{A},\bar{B}}, F) = \bigcup_{a\in\bar{A}, b\in\bar{B}} Q(f_{a,b}, F)$. $\qquad\square$

Example 6.2. Let $F = \mathbb{Z}_5$ be a field of order 5, $G = \{\pm 1\} \leqslant F^*$, and $f_{\bar{1},\bar{0}}(\bar{x}) = \bar{x}^2$. Then, we have $\bar{F} = \{\bar{0}, \bar{1}, \bar{2}\}$, $Q(f_{\bar{1},\bar{0}}, \bar{F}) = \overline{Q(f_{1,0}, F)} \cup \overline{Q(f_{(-1),0}, F)}$, where

$$Q(f_{1,0}, F) = \{(0,0), (1,1), (-1,1), (2,-1), (-2,-1)\},$$
$$Q(f_{(-1),0}, F) = \{(0,0), (1,-1), (-1,-1), (2,1), (-2,1)\},$$

and $Q(f_{1,0}, \bar{F}) = \overline{Q(f_{1,0}, F)} = \{(\bar{0},\bar{0}), (\bar{1},\bar{1}), (\bar{2},\bar{1})\} = \overline{Q(f_{(-1),0}, F)} = Q(f_{(-1),0}, \bar{F})$. In this case, $Q(f_{\bar{1},\bar{0}}, \bar{F})$ is a nondegenerate conic hypersection because $\bar{A} = \bar{1} \neq \bar{0}$.

Definition 6.3. Let F be a field, $x \in F$ and G be a subgroup in F^*. We consider

$$
\mathcal{O} = \begin{cases} 0^{-1} & \text{if } f_{a,b}(z) = az^2 + bz, \\ 0 & \text{if } f_{a,b}(z) = az^{-1} + bz; \end{cases}
$$

$$
\mathcal{G}_x(f_{a,b}) = \begin{cases} \{x\} & \text{if } G = \{1\}, \\ \{z \in F | f_{a,b}(z) & \text{if } G \neq \{1\}, f_{a,b}(z) = az^2 + bz, \\ \quad = f_{a,b}(x)\} & \\ \{-x, x\} & \text{if } G \neq \{1\}, f_{a,b}(z) = az^{-1} + bz. \end{cases}
$$

Obviously, 0^{-1} is an element outside of F. We denote $0^{-1} = \frac{1}{0}$ by ∞, where $\infty \notin F$, and $\overline{\infty} = \infty$. Suppose that $\mathcal{G}_\infty(f_{a,b}) = \{\infty\}$, $f_{a,b}(\infty) = \infty$, $f_{a,b}(\mathcal{O}) = \infty$, for all $a \in \bar{A}, b \in \bar{B}$, and also $\hat{X} = \{\hat{x} | x \in X\}$, where $\hat{x} = (x, f_{a,b}(x))$ and $X \subseteq F \cup \{\mathcal{O}\}$. Moreover, $\mathcal{O} \cdot \mathcal{O} =$

$$\mathcal{O} = \mathcal{O} + \mathcal{O}, \ x \cdot \mathcal{O} = \begin{cases} \mathcal{O} & \text{if } x \neq 0, \\ 0 & \text{if } x = 0, \end{cases} \text{ and } x + \mathcal{O} = \begin{cases} \mathcal{O} & \text{if } \mathcal{O} = 0^{-1}, \\ x & \text{if } \mathcal{O} = 0, \end{cases}$$

for all x in the field $(F, +, \cdot)$.

Remark 6.1. It should be noted that associativity, as defined by adding $\mathcal{O}$ to the field $(F, +, \cdot)$ for the two operations "+" and "$\cdot$", remains preserved.

Definition 6.4. Let $Q(f_{\bar{A},\bar{B}}, \bar{F})$ be a nondegenerate conic hypersection, $F_\infty = F \cup \{\infty\}$, and

$$Q_F(f_{a,b}) = \{\hat{x} : x \in F_\infty, \hat{x} \notin L_0\},$$

$$Q_F(f_{\bar{A},\bar{B}}) = \bigcup_{a \in \bar{A}, b \in \bar{B}} Q_F(f_{a,b}),$$

where $L_0 = \{(x, 0) | x \in F_o\}$. Also, for all $\hat{x}_i, \hat{x}_i \in Q_F(f_{a,b})$,

$$\widehat{x_i \bullet_{ab} x_j} = (x_i \bullet_{ab} x_j, f_{a,b}(x_i \bullet_{ab} x_j)), \text{ in which } \{(x_i \bullet_{ab} x_j, 0)\}$$
$$= L_0 \cap L_{a,b}(\hat{x}_i, \hat{x}_j),$$

and

$$L_{a,b}(\hat{x}_i, \hat{x}_j)$$
$$= \begin{cases} \{(x, y) \in F^2 | y - f_{a,b}(x_i) & x_i \neq x_j, \mathcal{O} \notin \{x_i, x_j\}, \\ \quad = \frac{f_{a,b}(x_j) - f_{a,b}(x_i)}{(x_j - x_i)}(x - x_i)\}, \\ \{(x, y) \in F^2 | y - f_{a,b}(x_i) & x_i = x_j \notin \{\mathcal{O}\}, \\ \quad = f'_{a,b}(x_i)(x - x_i)\}, \\ \{(x, y) \in F^2 | \mathcal{O} \neq x \in \{x_i, x_j\}\} & x_i \neq x_j, \mathcal{O} \in \{x_i, x_j\}, \\ \quad \cup \{\hat{\mathcal{O}}\}, \\ \{(\mathcal{O}, y) | y \in F_\infty = F \cup \{\infty\}\}, & (x_i, x_j) = (\mathcal{O}, \mathcal{O}), \end{cases}$$

and $f'_{a,b}$ represent the formal derivative of $f_{a,b}$.

We denote $\overline{Q_F(f_{a,b})}$ by $Q_{\bar{F}}(f_{a,b})$ and $\overline{Q_F(f_{\bar{A},\bar{B}})}$ by $Q_{\bar{F}}(f_{\bar{A},\bar{B}})$. Further, we consider $\bar{\mathcal{O}} = \{\mathcal{O}\} = \mathcal{O}, \overline{f(\mathcal{O})} = \{f(\mathcal{O})\} = f(\mathcal{O})$, and

$\overline{(\mathcal{O}, f(\mathcal{O}))} = (\bar{\mathcal{O}}, \overline{f(\mathcal{O})}) = (\mathcal{O}, f(\mathcal{O}))$. Moreover, $\mathcal{O} \odot \mathcal{O} = \mathcal{O}$ and $\mathcal{O} \oplus \mathcal{O} = \mathcal{O}$. Also, for all $\bar{x}$ in the hyperfield $(\bar{F}, \oplus, \odot)$, $\bar{x} \odot \mathcal{O} = \begin{cases} \mathcal{O} & \text{if } x \neq 0, \\ \bar{0} & \text{if } x = 0, \end{cases}$ and $\bar{x} \oplus \mathcal{O} = \begin{cases} \mathcal{O} & \text{if } \mathcal{O} = 0^{-1}, \\ \bar{x} & \text{if } \mathcal{O} = 0, \end{cases}$ and we agree that $L_o \cap L(\hat{x}_i, \hat{x}_j) = \{(\infty, 0)\}$ if $f_{a,b}(x_i) = f_{a,b}(x_j)$. In addition, let $L_{a,b}(\hat{x}_i, \hat{x}_j)$ denote the line passing through $\hat{x}_i, \hat{x}_j$. Intuitively, we can call any line passing through $(\mathcal{O}, \infty)$ a vertical line, and every vertical line passes through $(\mathcal{O}, \infty)$. $\hat{\mathcal{O}}$ plays the role of asymptotically extending the function $f_{a,b}$.

Remark 6.2. On adding $\mathcal{O}$ to the hyperfield $(F, \oplus, \odot)$, the associativity for the two hyperoperations "$\oplus$" and "$\odot$" remains preserved.

Suppose that $\hat{x} \in Q(f_{a,b}, F)$ and $\tilde{x} = \begin{cases} \{x\} & f_{a,b}(x) = ax^2 + bx, \\ \{x, -x\} & f_{a,b}(x) = ax^{-1} + bx. \end{cases}$ Hence, we have the following theorem.

Theorem 6.3. *If* $|Q_F(\tilde{f}_{a_1,b_1}) \cap Q_F(\tilde{f}_{a_2,b_2})| \geq 2$, *then* $Q(f_{a_1,b_1}, F) = Q(f_{a_2,b_2}, F)$.

Proof. Let $\{\tilde{x}_1, \tilde{x}_2\} \subseteq Q_F(\tilde{f}_{a_1,b_1}) \cap Q_F(\tilde{f}_{a_2,b_2})$, $\tilde{x}_1 \neq \tilde{x}_2$ and $i, j = 1, 2$. Then,

$$y_i = a_j x_i^2 + b_j x_i \Rightarrow x_1 \neq x_2 \Rightarrow \begin{cases} a_1 = a_2 = \dfrac{x_2 y_1 - x_1 y_2}{x_1^2 x_2 - x_2^2 x_1}, \\ b_1 = b_2 = \dfrac{-x_2^2 y_1 + x_1^2 y_2}{x_1^2 x_2 - x_2^2 x_1}, \end{cases}$$

$$y_i = a_j x_i^{-1} + b_j x_i \Rightarrow x_1 \neq \pm x_2 \Rightarrow \begin{cases} a_1 = a_2 = \dfrac{x_2 y_1 - x_1 y_2}{x_2 x_1^{-1} - x_1 x_2^{-1}}, \\ b_1 = b_2 = \dfrac{-x_2^{-1} y_1 + x_1^{-1} y_2}{x_2 x_1^{-1} - x_1 x_2^{-1}}. \end{cases}$$

Hence, $Q(f_{a_1,b_1}, F) = Q(f_{a_2,b_2}, F)$, as we expected. □

Definition 6.5. Let $Q(f_{\bar{A},\bar{B}}, \bar{F})$ be a nondegenerate conic hypersection. It is called a *hyperconic* and denoted $Q_{\bar{F}}(f_{\bar{A},\bar{B}})$ if the following implication for all $a, c \in \bar{A}$ and $b, d \in \bar{B}$ holds:

$$Q_F(f_{a,b}) \cap Q_F(f_{c,d})$$

$$\neq \begin{cases} \{\hat{O}\} & f_{a,b}(x) = ax^2 + bx, \\ \{\hat{O}\widehat{\infty}\} & f_{a,b}(z) = ax^{-1} + bx \end{cases} \Rightarrow Q_F(f_{a,b}) = Q_F(f_{c,d}).$$

Theorem 6.4. *Let* $\hat{x}_i = (x_i, f(x_i))$ *and* $\hat{x}_j = (x_j, f(x_j))$ *belong to* $Q_F(f_{a,b})$, *then*

$$x_i \bullet_{ab} x_j = \begin{cases} \frac{x_i f_{a,b}(x_j) - x_j f_{a,b}(x_i)}{f_{a,b}(x_j) - f_{a,b}(x_i)} & x_i \neq x_j, \mathcal{O} \notin \{x_i, x_j\}, \\ x_i - \frac{f_{a,b}(x_i)}{f'_{a,b}(x_i)} & x_i = x_j \notin \{\mathcal{O}\}, \\ x_i & x_i \neq \mathcal{O} = x_j, \\ x_j & x_j \neq \mathcal{O} = x_i, \\ \mathcal{O} & (x_i, x_j) = (\mathcal{O}, \mathcal{O}). \end{cases}$$

Proof. The proof is straightforward for the first two cases. If $f_{a,b}(x_i) = f_{a,b}(x_j)$, then

$$x_i \bullet_{ab} x_j = \frac{x_i f_{a,b}(x_j) - x_j f_{a,b}(x_i)}{f_{a,b}(x_j) - f_{a,b}(x_i)} = \frac{x_i f_{a,b}(x_j) - x_j f_{a,b}(x_i)}{0} = \infty,$$

$$\{(x_i \bullet_{ab} x_j, 0)\} = L_o \cap L(\hat{x}_i, \hat{x}_j) = \{(\infty, 0)\} \Rightarrow x_i \bullet_{ab} x_j = \infty.$$

Suppose that $(x_i, x_j) \in Q_F^2(f_{a,b})$ by regarding Definition 6.4 if $x_i \neq \mathcal{O} = x_j$, then

$$\{(x_i \bullet_{ab} \mathcal{O}, 0)\} = L_0 \cap L_{a,b}(\hat{x}_i, \hat{O}) = \{(x_i, 0)\} \Rightarrow x_i \bullet_{ab} \mathcal{O} = x_i.$$

If $x_j \neq \mathcal{O} = x_i$, then the proof is similar to the previous case. Ultimately, if $x_i = x_j = \mathcal{O}$, then

$$\{(\mathcal{O} \bullet_{a,b} \mathcal{O}, 0)\} = L_0 \cap L(\hat{O}, \hat{O}) = \{(\mathcal{O}, 0)\} \Rightarrow \mathcal{O} \bullet_{a,b} \mathcal{O} = \mathcal{O}. \qquad \square$$

Remark 6.3. $\left(Q_F(f_{a,b}), \bullet_{ab}\right)$ is a conic group, for all $(a, b) \in \bar{A} \times \bar{B}$. Note that $\bullet_{ab}$ is the group operation on the conic $Q_F(f_{a,b})$.

Example 6.3. Let $F = \mathbb{Z}_5$ be a field of order 5, $G = \{\pm 1\} \leqslant F^*$, and $f_{\bar{1},\bar{0}}(\bar{x}) = \bar{x}^2$. Then, we have $\bar{F} = \{\bar{0}, \bar{1}, \bar{2}\}$, $Q_{\bar{F}}(f_{\bar{1},\bar{0}}) = \overline{Q_F(f_{1,0})} \cup$ $\overline{Q_F(f_{(-1),0})}$, where $Q_F(f_{1,0}) = \{\hat{O}, (1,1), (-1,1), (2,-1), (-2,-1)\}$

and $\overline{Q_F(f_{1,0})} = \{\hat{O}, (\bar{1}, \bar{1}), (\bar{2}, \bar{1})\}, Q_F(f_{(-1),0}) = \{\hat{O}, (1, -1),$ $(-1, -1), (2, 1), (-2, 1)\}$, and $\overline{Q_F(f_{(-1),0})} = \{\hat{O}, (\bar{1}, \bar{1}), (\bar{2}, \bar{1})\}$. In this case $Q_{\bar{F}}(f_{\bar{1},\bar{0}})$ is a hyperconic because $Q_F(f_{1,0}) \cap Q_F(f_{(-1),0}) = \hat{O}$.

Definition 6.6. We introduce the hyperoperation "$\circ$" on $Q_F(f_{\bar{A},\bar{B}})$ as follows.

Let $(x, y), (x', y') \in Q_F(f_{\bar{A},\bar{B}})$. If $(x, y) \in Q_F(f_{a,b})$ and $(x', y') \in Q_F(f_{a',b'})$ for some $a, a' \in \bar{A}$ and $b, b' \in \bar{B}$:

$$
(x, y) \circ (x', y')
$$
$$
= \begin{cases} \{\widehat{x_i \bullet_{ab} x_j} | (x_i, x_j) \in \mathcal{G}_x(f_{a,b}) & \text{if } Q_F(f_{a,b}) = Q_F(f_{a',b'}), \\ \times \mathcal{G}_{x'}(f_{a',b'})\} & \\ Q_F(f_{a,b}) \cup Q_F(f_{a',b'}) & \text{otherwise.} \end{cases}
$$

Theorem 6.5. $(Q_F(f_{\bar{A},\bar{B}}), \circ)$ *is a hypergroup.*

Proof. Suppose that $\{X, Y, Z\} \subseteq Q_F(f_{\bar{A},\bar{B}})$, by Bezout's theorem $(x, y) \circ (x', y') \subseteq P^*(Q_F(f_{a,b}))$. Now, let $X = (x, y) \in Q_F(f_{a,b}), Y = (x', y') \in Q_F(f_{a',b'}), Z = (x'', y'') \in Q_F(f_{a'',b''})$, where $J = \{(a, b), (a', b'), (a'', b'')\} \subseteq \bar{A} \times \bar{B}$. If $(x, y) = (x_1, y_1)$ and $(x', y') = (x_1', y_1')$, then $x = x_1$ and $x' = x_1'$. Because $\mathcal{G}_x(f_{a,b}) = \mathcal{G}_{x_1}(f_{a,b})$ and $\mathcal{G}_{x'}(f_{a,b}) = \mathcal{G}_{x_1'}(f_{a,b})$, we have $\mathcal{G}_x(f_{a,b}) \times \mathcal{G}_{x'}(f_{a,b}) = \mathcal{G}_{x_1}(f_{a,b}) \times \mathcal{G}_{x_1'}(f_{a,b})$ and thus

$$
\{\widehat{z \bullet_{ab} w} | (z, w) \in \mathcal{G}_x(f_{a,b}) \times \mathcal{G}_{x'}(f_{a,b})\}
$$
$$
= \{\widehat{z \bullet_{ab} w} | (z, w) \in \mathcal{G}_{x_1}(f_{a,b}) \times \mathcal{G}_{x_1'}(f_{a,b})\}.
$$

That is, $(x, y) \circ (x', y') = (x_1, y_1) \circ (x_1', y_1')$; consequently, "$\circ$" is well defined. If $X = (\mathcal{O}, \infty)$, $Y = (\mathcal{O}, \infty)$, or $Z = (\mathcal{O}, \infty)$, associativity is evident. If this property is not satisfied, the following cases may occur:

Case 1. $|J| = 1$. In this case, we have $Q_F(f_{a,b}) = Q_F(f_{a',b'}) = Q_F(f_{a'',b''})$.

$$
[(x, y) \circ (x', y')] \circ (x'', y'')
$$
$$
= \left\{ (\widehat{x_i \bullet_{ab} x_j'}) | (x_i, x_j') \in \mathcal{G}_x(f_{a,b}) \times \mathcal{G}_{x'}(f_{a,b}) \right\} \circ (x'', y'')
$$
$$
= \left\{ (\widehat{x_i \bullet_{ab} x_j'}) \bullet_{ab} x_k'' | (x_i, x_j', x_k'') \in \mathcal{G}_x(f_{a,b}) \times \mathcal{G}_{x'}(f_{a,b}) \times \mathcal{G}_{x''}(f_{a,b}) \right\}.
$$

Similarly,

$$(x, y) \circ [(x', y') \circ (x'', y'')]$$
$$= \left\{ x_i \bullet_{ab} \widehat{(x'_j \bullet_{ab} x''_k)} \mid (x_i, x'_j, x''_k) \in \mathcal{G}_x(f_{a,b}) \times \mathcal{G}_{x'}(f_{a,b}) \times \mathcal{G}_{x''}(f_{a,b}) \right\}.$$

On the other hand, we have

$$L(\hat{x}_i, \hat{x'_j}) \cap L(\widehat{x_i \bullet_{ab} x'_j}, \hat{\mathcal{O}}) = \{(x_i \bullet_{ab} x_j, 0)\} \subseteq L_0,$$
$$L(\hat{x'_j}, \hat{x''_k}) \cap L(\hat{\mathcal{O}}, \widehat{x'_j \bullet_{ab} x''_k}) = \{(x_j \bullet_{ab} x_k, 0)\} \subseteq L_0.$$

Therefore, by Pascal's theorem, we have

$$L(\hat{x''_k}, \widehat{x_i \bullet_{ab} x'_j}) \cap L(\hat{x'_j} \bullet_{ab} \widehat{x''_k}, \hat{x}_i) \subseteq L_0,$$

and in addition,

$$\{((x_i \bullet_{ab} x'_j) \bullet_{ab} x''_k, 0)\} = L_0 \cap L(\widehat{x_i \bullet_{ab} x'_j}, \hat{x''_k}),$$
$$\{(x_i \bullet_{ab} (x'_j \bullet_{ab} x''_k), 0)\} = L_0 \cap L(\hat{x}_i, \widehat{x'_j \bullet_{ab} x''_k}),$$
$$L_0 \cap L(\widehat{x_i \bullet_{ab} x'_j}, \hat{x''_k}) = L(\hat{x''_k}, \widehat{x_i \bullet_{ab} x'_j}) \cap L(\widehat{x'_j \bullet_{ab} x''_k}, \hat{x}_i)$$
$$= L_0 \cap L(\hat{x}_i, \widehat{x'_j \bullet_{ab} x''_k}).$$

In other words,

$$(x_i \bullet_{ab} x'_j) \bullet_{ab} x''_k = x_i \bullet_{ab} (x'_j \bullet_{ab} x''_k).$$

So,

$$\left(\widehat{(x_i \bullet_{ab} x'_j) \bullet_{ab} x''_k} \right) = \left(x_i \bullet_{ab} \widehat{(x'_j \bullet_{ab} x''_k)} \right),$$

for all $(x_i, x'_j, x''_k) \in \mathcal{G}_x(f_{a,b}) \times \mathcal{G}_{x'}(f_{a,b}) \times \mathcal{G}_{x''}(f_{a,b})$.

Case 2. $|J| = 2$. (i) If $Q_F(f_{a,b}) = Q_F(f_{a',b'}) \neq Q_F(f_{a'',b''})$, we have

$$[(x, y) \circ (x', y')] \circ (x'', y'') = [\{\widehat{z \bullet_{ab} w} \mid (z, w) \in \mathcal{G}_x(f_{a,b}) \times \mathcal{G}_{x'}(f_{a,b})\}]$$
$$\circ (x'', y'')$$
$$= \bigcup_{(u,v) \in (x,y) \circ (x',y')} (u, v) \circ (x'', y'')$$
$$= Q_F(f_{a,b}) \cup Q_F(f_{a'',b''}).$$

Otherwise,

$$\begin{aligned}
(x,y) \circ [(x',y') \circ (x'',y'')] &= (x,y) \circ \big(Q_F(f_{a',b'}) \cup Q_F(f_{a'',b''})\big) \\
&= Q_F(f_{a',b'}) \cap Q_F(f_{a,b}) \cup Q_F(f_{a'',b''}) \\
&= Q_F(f_{a,b}) \cup Q_F(f_{a'',b''}).
\end{aligned}$$

(ii) If $Q_F(f_{a,b}) \neq Q_F(f_{a',b'}) = Q_F(f_{a'',b''})$, this case is similar to (i).
(iii) If $Q_F(f_{a,b}) = Q_F(f_{a'',b''}) \neq Q_F(f_{a',b'})$, we have

$$\begin{aligned}
[(x,y) \circ (x',y')] \circ (x'',y'') &= \big(Q_F(f_{a,b}) \cup Q_F(f_{a',b'})\big) \circ (x'',y'') \\
&= Q_F(f_{a,b}) \cup Q_F(f_{a',b'}) \cup Q_F(f_{a'',b''}) \\
&= Q_F(f_{a,b}) \cup Q_F(f_{a',b'}).
\end{aligned}$$

On the other hand,

$$\begin{aligned}
(x,y) \circ [(x',y') \circ (x'',y'')] &= (x,y) \circ \big(Q_F(f_{a',b'}) \cup Q_F(f_{a'',b''})\big) \\
&= Q_F(f_{a,b}) \cup Q_F(f_{a',b'}) \cup Q_F(f_{a'',b''}) \\
&= Q_F(f_{a,b}) \cup Q_F(f_{a',b'}).
\end{aligned}$$

Case 3. If $|J| = 3$, we have

$$\begin{aligned}
[(x,y) \circ (x',y')] \circ (x'',y'') &= \big(Q_F(f_{a,b}) \cup Q_F(f_{a',b'})\big) \circ (x'',y'') \\
&= Q_F(f_{a,b}) \cup Q_F(f_{a',b'}) \cup Q_F(f_{a'',b''}).
\end{aligned}$$

On the other hand,

$$\begin{aligned}
(x,y) \circ [(x',y') \circ (x'',y'')] &= (x,y) \circ \big(Q_F(f_{a',b'}) \cup Q_F(f_{a'',b''})\big) \\
&= Q_F(f_{a,b}) \cup Q_F(f_{a',b'}) \cup Q_F(f_{a'',b''}).
\end{aligned}$$

To prove the validity of the law of reproduction axiom for "$\circ$", let us consider two cases:

Case 1. If $|\bar{A} \times \bar{B}| = 1$, then $\bar{F} = F$ and $Q_F(f_{\bar{A},\bar{B}}) = Q_F(f_{a,b})$, where $a \in \bar{A}, b \in \bar{B}$, and also, $(Q_F(f_{a,b}), \circ)$ is a conic group. Hence, there is nothing to prove.

Case 2. If $|\bar{A} \times \bar{B}| > 1$, consider an arbitrary element $\hat{x} \in Q_F(f_{a,b}) \subseteq Q_F(f_{\bar{A},\bar{B}})$, then

$$\hat{x} \circ Q_F(f_{\bar{A},\bar{B}}) = \left(\hat{x} \circ \bigcup_{a \neq i \in \bar{A}, b \neq j \in \bar{B}} Q_F(f_{i,j}) \right) \cup (\hat{x} \circ Q_F(f_{a,b})),$$

$$= \left(\bigcup_{a \neq i \in \bar{A}, b \neq j \in \bar{B}} \hat{x} \circ Q_F(f_{i,j}) \right) \cup Q_F(f_{a,b}),$$

$$= \left(\bigcup_{i \in \bar{A}, j \in \bar{B}} Q_F(f_{i,j}) \right) \cup Q_F(f_{a,b}),$$

$$= Q_F(f_{\bar{A},\bar{B}}).$$

Similarly, $Q_F(f_{\bar{A},\bar{B}}) \circ \hat{x} = Q_F(f_{\bar{A},\bar{B}})$, and the reproduction axiom is established. Therefore, $\left(Q_F(f_{\bar{A},\bar{B}}), \circ \right)$ is a hypergroup. $\qquad \square$

Remark 6.4. The hyperconic and the associated hypergroup are a conic and a conic group, respectively, if $G = \{1\}$.

Example 6.4. Let $F = \mathbb{Z}_5$ be a field of order 5 and $G = \{\pm 1\} \leqslant F^*$. We have $\bar{F} = \{\bar{0}, \bar{1}, \bar{2}\}$. In addition, if we go back to Example 6.3, then $Q_{\bar{F}}(f_{\bar{1},\bar{0}}) = \overline{Q_F(f_{1,0})} \cup \overline{Q_F(f_{(-1),0})}$ is a hyperconic, where $Q_F(f_{1,0}) = \{\hat{O}, (1,1), (-1,1), (2,-1), (-2,-1)\}$ $Q_F(f_{(-1),0}) = \{\hat{O}, (1,-1), (-1,-1), (2,1), (-2,-1)\}$ $Q_{\bar{F}}(f_{\bar{1},\bar{0}}) = \{\hat{O}, (\bar{1},\bar{1}), (\bar{2},\bar{1})\}$. Now, let $H = Q_F(f_{1,0})$ and $K = Q_F(f_{(-1),0})$. Then, H and K are reversible subhypergroups of $Q_F(f_{\bar{1},\bar{0}})$, which are defined as follows, respectively:

$\circ$	$\hat{O}$	$(1,1)$	$(-1,1)$	$(2,-1)$	$(-2,-1)$
$\hat{O}$	$\hat{O}$	$(\pm 1, 1)$	$(\pm 1, 1)$	$(\pm 2, -1)$	$,(\pm 2, -1)$
$(1,1)$	$(\pm 1, 1)$	$\hat{O}, (\pm 2, -1)$	$\hat{O}, (\pm 2, -1)$	$(\pm 1, 1), (\pm 2, -1)$	$(\pm 1, 1), (\pm 2, -1)$
$(-1,1)$	$(\pm 1, 1)$	$\hat{O}, (\pm 2, -1)$	$\hat{O}, (\pm 2, -1)$	$(\pm 1, 1), (\pm 2, -1)$	$(\pm 1, 1), (\pm 2, -1)$
$(2,-1)$	$(\pm 2, -1)$	$(\pm 1, 1), (\pm 2, -1)$	$(\pm 1, 1), (\pm 2, -1)$	$(\pm 1, 1), \hat{O}$	$(\pm 1, 1), \hat{O}$
$(-2,-1)$	$(\pm 2, -1)$	$(\pm 1, 1), (\pm 2, -1)$	$(\pm 1, 1), (\pm 2, -1)$	$(\pm 1, 1), \hat{O}$	$(\pm 1, 1), \hat{O}$

∘	$\hat{O}$	$(1,-1)$	$(-1,-1)$	$(2,1)$	$(-2,1)$
$\hat{O}$	$\hat{O}$	$(\pm 1,-1)$	$(\pm 1,-1)$	$(\pm 2,1)$	$(\pm 2,1)$
$(1,-1)$	$(\pm 1,-1)$	$\hat{O},(\pm 2,1)$	$\hat{O},(\pm 2,1)$	$(\pm 1,-1),(\pm 2,1)$	$(\pm 1,-1),(\pm 2,1)$
$(-1,-1)$	$(\pm 1,-1)$	$\hat{O},(\pm 2,1)$	$\hat{O},(\pm 2,1)$	$(\pm 1,-1),(\pm 2,1)$	$(\pm 1,-1),(\pm 2,1)$
$(2,1)$	$(\pm 2,1)$	$(\pm 1,-1),(\pm 2,1)$	$(\pm 1,-1),(\pm 2,1)$	$(\pm 1,-1),\hat{O}$	$(\pm 1,-1),\hat{O}$
$(-2,-1)$	$(\pm 2,-1)$	$(\pm 1,1),(\pm 2,-1)$	$(\pm 1,1),(\pm 2,-1)$	$(\pm 1,1),\hat{O}$	$(\pm 1,-1),\hat{O}$

Theorem 6.6. H *is a subhypergroup of* $Q_F(f_{\bar{A},\bar{B}})$ *if and only if* $H = \bigcup_{(i,j)\in I\subseteq \bar{A}\times\bar{B}} Q_F(f_{i,j})$, *or* $H \leqslant Q_{i,j}(F)$, *for some* $(i,j) \in \bar{A}\times\bar{B}$.

Proof. ($\Rightarrow$) Let us assume that $H \not\leqslant Q_F(f_{i,j})$ for every (i,j) in $\bar{A}\times\bar{B}$. Then, in $\bar{A}\times\bar{B}$, we have $(i,j) \neq (s,t)$ such that $H\cap Q_F(f_{i,j}) \neq \emptyset \neq H \cap Q_F(f_{s,t})$. Now, let $I=\{(i,j) \in \bar{A} \times \bar{B}|H \cap Q_F(f_{i,j}) \neq \emptyset\}$. Thus, we have $H \subseteq \bigcup_{(i,j)\in I} Q_F(f_{i,j}) \subseteq \bigcup_{(s,t),(i,j)\in I}(Q_F(f_{i,j}) \cap H) \circ (Q_F(f_{s,t}) \cap H) \subseteq H$. Accordingly, $H = \bigcup_{(i,j)\in I} Q_F(f_{i,j})$. ($\Leftarrow$). It is obvious. $\square$

Theorem 6.7. *A subhypergroup* H *of* $Q_F(f_{\bar{A},\bar{B}})$ *is a reversible hypergroup if and only if* $H \leqslant Q_F(f_{i,j})$, *for some* $(i,j) \in \bar{A} \times \bar{B}$.

Proof. ($\Leftarrow$) First, we prove that $H \leqslant Q_F(f_{i,j})$ is a regular reversible hypergroup for all $(i,j) \in \bar{A} \times \bar{B}$. Let (x,y) and (x',y') be elements in $Q_F(f_{i,j})$.

Case 1. If $x' \notin \mathcal{G}_x(f_{a,b})$, then

$$(x'',y'') \in (x,y) \circ (x',y')$$
$$\Rightarrow (x'',y'') = \widehat{z \bullet_{ij} w}, \text{ for some } (z,w) \in \mathcal{G}_x(f_{a,b}) \times \mathcal{G}_{x'}(f_{a,b})$$
$$\Rightarrow x'' = z \bullet_{ij} w$$
$$\Rightarrow z = x'' \bullet_{ij} h, \text{ where } w \bullet_{ij} h = \mathcal{O}$$
$$\Rightarrow (z, f_{a,b}(z)) = \widehat{x'' \bullet_{ij} h} \text{ and } h \in \mathcal{G}_w(f_{a,b}) = \mathcal{G}_{x'}(f_{a,b})$$
$$\Rightarrow (z, f_{a,b}(z)) \in (x'', f_{a,b}(x'')) \circ (x', f_{a,b}(x')).$$

Case 2. If $x' \in \mathcal{G}_x(f_{a,b})$, then $z, w \in \mathcal{G}_x(f_{a,b})$ and $\hat{m} \in \hat{z} \circ \hat{w} = \{\widehat{x \bullet_{ab} x}, \widehat{n \bullet_{ab} n}, \hat{\mathcal{O}}\}$, where $x \bullet_{ab} n = \mathcal{O}$. Then, $\hat{z} \in \hat{m} \circ \hat{w}$ and $\hat{w} \in \hat{z} \circ \hat{m}$.

Case 3. If $Y \in \infty \circ X = X \circ \infty$, then $\infty \in Y \circ X$ and $X \in \infty \circ Y$. Note the fact that $\infty \in X \circ X$, for all $X \in Q_F(f_{i,j})$ (i.e., every element is one of its inverses).

($\Rightarrow$) Assume that $(x, y) \in H \cap Q_F(f_{i,j})$ and $(x', y') \in H \cap Q_F(f_{s,t})$, in which $(i, j) \neq (s, t)$, $(x, y) \neq \hat{\mathcal{O}} \neq (x', y')$, and $(x'', y'') \in (x, y) \circ (x', y') \cap Q_F(f_{i,j})$. Then, $(x', y') \in (z, w) \circ (x'', y'') \subseteq Q_F(f_{i,j})$, where $z \in \mathcal{G}_x(f_{a,b})$. Hence, $Q_F(f_{i,j}) = Q_F(f_{s,t})$, which means that the reversibility conditions do not hold. $\quad\square$

The class of H_v-*groups* is more general than the class of hypergroups, which was introduced by Vougiouklis [102]. The hyperstructure $(H, \circ)$ is called a H_v−group if $x \circ H = H = H \circ x$, and also the weak associativity condition holds, i.e., $x \circ (y \circ z) \cap (x \circ y) \circ z \neq \emptyset$, for all $x, y, z \in H$. In Refs. [97, 98], the authors have investigated certain hyperoperations denoted by $\bar{\circ}$ and $\diamond$ on some main classes of curves, elliptic curves, and homographics over Krasner's hyperfields. In the following, we study them on a hyperconic. Consider the following hyperoperation on the hyperconic $(Q_{\bar{F}}(f_{\bar{A},\bar{B}})$:

$$(\bar{x}, \bar{y}) \bar{\circ} (\bar{x}', \bar{y}') = \{(\bar{v}, \bar{w}) | (v, w) \in (\bar{x} \times \bar{y}) \circ (\bar{x}' \times \bar{y}')\},$$

for all $(\bar{x}, \bar{y})$, $(\bar{x}', \bar{y}')$ in $Q_{\bar{F}}(f_{\bar{A},\bar{B}})$, see Fig. 6.3.

Theorem 6.8. $(Q_{\bar{F}}(f_{\bar{A},\bar{B}}), \bar{\circ})$ *is an* H_v-*group.*

Proof. The proof is straightforward. $\quad\square$

Theorem 6.9. *If* $\psi_{\bar{A},\bar{B}} : Q_F(f_{\bar{A},\bar{B}}) \to Q_{\bar{F}}(f_{\bar{A},\bar{B}})$, $\psi_{\bar{A},\bar{B}}(x, y) = (\bar{x}, \bar{y})$, *then* $\psi_{\bar{A},\bar{B}}$ *is an epimorphism of* H_v-*groups.*

Proof. The basis of the proof is similar to that of the proof of Proposition 3 in Ref. [98]. $\quad\square$

Example 6.5. Let $G = \{\pm 1\}$ be a subgroup of F^*, where $F = \mathbb{Z}_5$. Consider $f_{\bar{1},\bar{1}}(\bar{x}) = \bar{x}^2 \oplus \bar{x}$ on $\bar{F} = \{\bar{0}, \bar{1}, \bar{2}\}$. Consequently, $Q_{\bar{F}}(f_{\bar{1},\bar{1}}) = \{\hat{\mathcal{O}}, (\bar{1}, \bar{2}), (\bar{2}, \bar{1}), (\bar{2}, \bar{2})\}$ is a hyperconic. A calculation gives us the following table of H_v-group.

$\bar{\circ}$	$\hat{O}$	$(\bar{1},\bar{2})$	$(\bar{2},\bar{1})$	$(\bar{2},\bar{2})$
$\hat{O}$	$\hat{O}$	$(\bar{1},\bar{2}),(\bar{2},\bar{2})$	$(\bar{2},\bar{1})$	$(\bar{2},\bar{2}),(\bar{1},\bar{2})$
$(\bar{1},\bar{2})$	$(\bar{1},\bar{2}),(\bar{2},\bar{2})$	$(\bar{1},\bar{2}),(\bar{2},\bar{1}),(\bar{2},\bar{2})$	$(\bar{1},\bar{2}),(\bar{2},\bar{1}),(\bar{2},\bar{2})$	$(\bar{1},\bar{2}),(\bar{2},\bar{1}),(\bar{2},\bar{2})$
$(\bar{2},\bar{1})$	$(\bar{2},\bar{1})$	$(\bar{1},\bar{2}),(\bar{2},\bar{1}),(\bar{2},\bar{2})$	$Q_{\bar{F}}(f_{\bar{1},\bar{1}})$	$(\bar{1},\bar{2}),(\bar{2},\bar{1}),(\bar{2},\bar{2})$
$(\bar{2},\bar{2})$	$(\bar{1},\bar{2}),(2,2)$	$(\bar{1},\bar{2}),(\bar{2},\bar{1}),(\bar{2},\bar{2})$	$(\bar{1},\bar{2}),(\bar{2},\bar{1}),(\bar{2},\bar{2})$	$(\bar{1},\bar{2}),(\bar{2},\bar{1}),(\bar{2},\bar{2})$

Let A be a finite set called *alphabet*, let K be a nonempty subset of A, called *key-set*, and let "$\cdot$" be a hyperoperation on A. Berardi *et al.* [10] utilized such hyperoperations with the following condition: $k \cdot x = k \cdot y \Rightarrow x = y$, for all $(x,y) \in A^2$ and $k \in K$. Let $\big(Q_F(f_{m,n}),\circ\big)$ be a subhypergroup of $\big(Q_F(f_{\bar{m},\bar{n}}),\circ\big)$ and $A = \{a_x | \hat{x} \in Q_F(f_{m,n})\}$, where $a_x = i(x)$. Note that $i(x) = \widehat{\mathcal{G}_x(f_{a,b})}$ is the set of all inverses

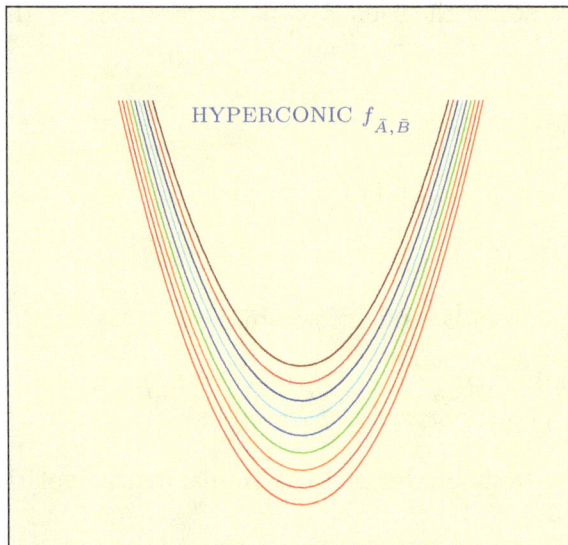

Fig. 6.3.   Hyperconic.

of $\hat{x}$, for all $\hat{x} \in Q_F(f_{m,n})$. We define the hyperoperation $\diamond$ on A as follows:

$$a_u \diamond a_v = \{a_w | w \in u \circ v\}.$$

Theorem 6.10. $(A, \diamond)$ *is a canonical hypergroup which satisfies Berardi's condition.*

Proof. The proof is similar to the one of Theorem 4.1 in Ref. [97]. $\qquad\qquad\qquad\qquad\qquad\qquad\qquad\qquad\qquad\qquad\qquad\square$

Conic curve cryptography (CCC) renders efficient digital signature schemes (CCDLP), which have a high level of security with small key sizes. Let $g(x, y) = ax^2 + bxy + cy^2 + dx + ey + f \in F[x, y]$ and $g(x, y) = 0$ be the quadratic equation in two variables in the field F. If $a = c = 0$ and $b \neq 0$, then the equation $g(x, y) = 0$ is called a homographic transformation. Vahedi *et al.* extended this particular quadratic equation to the quotient hyperfield $\frac{F}{G}$. Now, suppose that $ae \neq 0$ and $b = 0$ in $g(x, y)$. Then, the curve is called a conic. The motivation of this chapter aligns with that of Ref. [98]. In fact, in a similar way, the notion of a conic on a field is extended to a hyperconic over a quotient hyperfield. Note that, as one can see that the group structures of these two classes of curves have different applications, the associated hyperstructures can be different in different applications. To conclude this chapter, a canonical hypergroup, which is expressed as $\left(Q_F(f_{n,m}), \diamond\right)$, is investigated.

Chapter 7

Fundamental Relations on Krasner Hyperrings

7.1 γ^*-Relation on Krasner Hyperrings

Vougiouklis [101], introduced the *fundamental relation* γ^* on a general hyperring R as the smallest equivalence relation on R such that the quotient R/γ^* is a *fundamental ring*. We apply the relation γ^* on Krasner hyperrings. Let $(R, +, \cdot)$ be a Krasner hyperring; we can define the relation γ as follows:

$$a\gamma b \quad \Leftrightarrow \quad \{a, b\} \subseteq u,$$

where u is a finite sum of finite products of the elements of R. Suppose that γ^* is the transitive closure of γ. Then, both $\oplus$ and $\odot$ on R/γ^* are defined as follows:

$$\gamma^*(a) \oplus \gamma^*(b) = \gamma^*(c), \quad \text{for all } c \in \gamma^*(a) + \gamma^*(b),$$
$$\gamma^*(a) \odot \gamma^*(b) = \gamma^*(ab).$$

Let $\mathcal{U}$ be the set of all finite sums of products of the elements of R. We can rewrite the definition of γ^* on R as follows:

$a\gamma^* b$ if and only if $\exists z_1, \ldots, z_{n+1} \in R$, with $z_1 = a, z_{n+1} = b$, and $u_1, \ldots, u_n \in \mathcal{U}$ such that $\{z_i, z_{i+1}\} \subseteq u_i$, for $i \in \{1, \ldots, n\}$.

According to the distributive law, every set which is the value of a polynomial in the elements of R is a subset of a sum of products in R.

Also, we can define the relation β_+^* as the smallest equivalence relation such that the quotient R/β_+^* is a group. The equivalence relation β_+^* was introduced by Koskas [49] on hypergroups and was studied by many others. We denote by β_+ the relation defined in R as follows:

$$a\beta_+ b \Leftrightarrow \text{ there exists } (c_1, \ldots, c_n) \in R^n \text{ such that } \{a, b\} \subseteq c_1 + \cdots + c_n.$$

For hypergroups, we have $\beta_+^* = \beta_+$.

Theorem 7.1. *Let $(R, +, \cdot)$ be a Krasner hyperring. Then, the relation γ^* is the smallest equivalence relation in R such that the quotient R/γ^* is a ring.*

Proof. First, we prove that R/γ^* is a ring. The product $\odot$ and the sum $\oplus$ in R/γ^* are defined as follows:

$$\gamma^*(a) \oplus \gamma^*(b) = \{\gamma^*(c) \mid c \in \gamma^*(a) + \gamma^*(b)\},$$
$$\gamma^*(a) \odot \gamma^*(b) = \gamma^*(bc).$$

Let $a' \in \gamma^*(a), b' \in \gamma^*(b)$. Hence, we have

$a'\gamma^* a$ implies that $\exists x_1, \ldots, x_{m+1}$, with $x_1 = a'$, $x_{m+1} = a$, and $u_1, \ldots, u_m \in \mathcal{U}$ such that $\{x_i, x_{i+1}\} \subseteq u_i$, for $i \in \{1, \ldots, m\}$;

$b'\gamma^* b$ implies that $\exists y_1, \ldots, y_{n+1}$, with $y_1 = b', y_{n+1} = b$, and $v_1, \ldots, v_n \in \mathcal{U}$ such that $\{y_j, y_{j+1}\} \subseteq v_j$, for $j \in \{1, \ldots, n\}$.

Then, we obtain

$$\{x_i, x_{i+1}\} + y_1 \subseteq u_i + v_1, \quad i \in \{1, \ldots, m-1\},$$
$$x_{m+1} + \{y_j, y_{j+1}\} \subseteq u_m + v_j, \quad j \in \{1, \ldots, n\}.$$

The sets $u_i + v_1 = t_i$, $i \in \{1, \ldots, m-1\}$ and $u_m + v_j = t_{m+j-1}$, $j \in \{1, \ldots, n\}$ are the elements of $\mathcal{U}$. We choose the elements $z_1, \ldots, z_{m+n}$ such that $z_i \in x_i + y_1$, $i \in \{1, \ldots, m\}$ and $z_{m+j} \subseteq x_{m+1} + y_{j+1}$, $j \in \{1, \ldots, n\}$. We obtain $\{z_k, z_{k+1}\} \subseteq t_k$, $k \in \{1, \ldots, m+n-1\}$. Hence, any element $z_1 \in x_1 + y_1 = a' + b'$ is γ^* equivalent to any element $z_{m+n} \in x_{m+1} + y_{n+1} = a + b$. Thus, $\gamma^*(a) \oplus \gamma^*(b)$ is a singleton, and we have $\gamma^*(a) \oplus \gamma^*(b) = \gamma^*(c)$, for all $c \in \gamma^*(a) + \gamma^*(b)$. According to the distributive law, we have $u \cdot v \in \mathcal{U}$, for all $u, v \in \mathcal{U}$. Similarly, we obtain $\gamma^*(a) \odot \gamma^*(b) = \gamma^*(ab)$. Therefore, it follows immediately that R/γ^* is a ring.

Now, let ρ be an equivalence relation in R such that R/ρ is a ring, and let $\rho(a)$ be the equivalence class of the element a. Then, $\rho(a) \oplus \rho(b)$ and $\rho(a) \odot \rho(b)$ are singletons, for all $a, b \in R$, which means that for all $a, b \in R$ and for all $c \in \rho(a) + \rho(b)$, we have

$$\rho(a) \oplus \rho(b) = \rho(c), \quad \rho(a) \odot \rho(b) = \rho(ab).$$

The above equalities are called the *fundamental properties* in $(R/\rho, \oplus, \odot)$. Hence, for all $a, b \in R$ and $A \subseteq \rho(a)$, $B \subseteq \rho(b)$, we have

$$\rho(a) \oplus \rho(b) = \rho(a + b) = \rho(A + B) \quad \text{and}$$
$$\rho(a) \odot \rho(b) = \rho(a \cdot b) = \rho(A \cdot B).$$

By induction, we extend the above equalities to finite sums and products. Now, set $u \in \mathcal{U}$, which means that there exist the finite sets of indices J and I_j and the elements $x_i \in R$ such that

$$u = \sum_{j \in J} \left(\prod_{i \in I_j} x_i \right).$$

For all I_j, the set $\prod_{i \in I_j} x_i$ is a subset of one class, say $\rho(a_j)$. Thus, for all $a \in \sum_{j \in J} a_j$, we have

$$u \subseteq \sum_{j \in J} \rho(a_j) = \rho \left(\sum_{j \in J} a_j \right) = \rho(a).$$

Therefore, for all $x, y \in R$, $x\gamma y$ implies $x\rho y$, whence $x\gamma^* y$ implies $x\rho y$. Hence, for all $a \in R$, $\gamma^*(a) \subseteq \rho(a)$, which means that γ^* is the smallest equivalence relation in R such that the quotient R/γ^* is a ring. $\qquad\square$

We denote the transitive closures of the relations β_+ by β_+^*, and we call β_+^* the *fundamental relations* with respect to addition. Let us consider the following canonical maps:

$$\varphi_+ : R \to R/\beta_+^*, \ \varphi_+(x) = \beta_+^*(x),$$
$$\varphi^* : R \to R/\gamma^*, \ \varphi^*(x) = \gamma^*(x).$$

We note that the maps $\varphi_+ : (R, +) \to (R/\beta_+^*, \oplus)$ and $\varphi^* : (R, +, \cdot) \to (R/\gamma^*, \oplus, \odot)$ are good homomorphisms.

We denote the kernels of φ_+, φ^* by ω_+, ω^*, respectively. If $\bar{0}$ is the zero element of R/β_+^* or R/γ^*, then

$$\omega_+ = ker\varphi_+ = \{x \in R : \varphi_+(x) = \bar{0}\},$$
$$\omega^* = ker\varphi^* = \{x \in R : \varphi^*(x) = \bar{0}\}.$$

We have $\omega_+ \subseteq \omega^*$.

Theorem 7.2. *Let $(R, +, \cdot)$ be a Krasner hyperring. Then:*

(1) *$R\omega^* \subseteq \omega^*$, $\omega^* R \subseteq \omega^*$;*
(2) *if $(R, +)$ is a regular hypergroup, then ω^* is a hyperideal of R.*

Proof.   (1) For all $r \in R$, $x \in \omega^*$, $a \in rx$, we have

$$\varphi^*(a) = \varphi^*(rx) = \varphi^*(r) \odot \varphi^*(x) = \varphi^*(r) \odot \bar{0} = \bar{0}.$$

(2) If $a, b \in \omega^*$, i.e., $\varphi^*(a) = \varphi^*(b) = \bar{0}$, then

$$\varphi^*(a + b) = \varphi^*(a) \oplus \varphi^*(b) = \bar{0}^* \oplus \bar{0} = \bar{0},$$

so $a + b \in \omega^*$. Suppose that $(R, +)$ is regular, and let e be an identity of it. Then, $e \in \omega_+ \subseteq \omega^*$. Set $x \in \omega^*$. Then, for all $x' \in R$, such that $e \in x + x'$, we have

$$\bar{0} = \varphi^*(e) = \varphi^*(x + x') = \varphi^*(x) \oplus \varphi^*(x') = \bar{0} \oplus \varphi^*(x').$$

So, $\varphi^*(x') = \bar{0}$, whence $x' \in \omega^*$. Therefore, for all $y \in \omega^*$, we have

$$y \in e + y \subseteq (x + x') + y = x + (x' + y),$$

and from $x' + y \in \omega^*$, we obtain $y \in x + \omega^*$. Hence, $\omega^* \subseteq x + \omega^*$, whence ω^* is a subhypergroup of $(R, +)$. $\square$

Theorem 7.3. *For all Krasner hyperrings we have $\gamma^* = \beta_+^*$.*

Proof.   In a Krasner hyperring R, any product of the elements of R is a singleton. Hence, for all

$$u = \sum_{j \in J} \left(\prod_{i \in I_j} x_i \right) \in \mathcal{U},$$

we consider the elements $y_j = \prod_{i \in I_j} x_i$, for all $j \in J$, and we have $u = \sum_{j \in J} y_j$. This means that $a\gamma^* b$ if and only if $a\beta_+^* b$. $\square$

7.2 α^*-Relation on Krasner Hyperrings

In this section, we define the relation α^* as the smallest equivalence relation on R such that the quotient R/α^*, the set of all equivalence classes, is a commutative ring. In this case, R/α^* is a commutative fundamental ring. Suppose that $\alpha^*(a)$ is the equivalence class containing $a \in R$. Then, both the sum $\uplus$ and the product $\otimes$ in R/α^* are defined as follows: $\alpha^*(a) \uplus \alpha^*(b) = \alpha^*(c)$, for all $c \in \alpha^*(a) + \alpha^*(b)$, and $\alpha^*(a) \otimes \alpha^*(b) = \alpha^*(ab)$. This relation was introduced and studied by Davvaz and Vougiouklis on general hyperrings [29]. In Ref. [63], Mirvakili and Davvaz applied this relation to Krasner hyperrings. The relation α is defined as follows.

Definition 7.1. $x \, \alpha \, y \iff \exists n \in \mathbb{N}, \exists (k_1, \ldots, k_n) \in \mathbb{N}^n, \exists \sigma \in \mathbb{S}_n, \exists (x_{i1}, \ldots, x_{ik_i}) \in R^{k_i}$ and $\exists \sigma_i \in \mathbb{S}_{k_i}, (i = 1, \ldots, n)$ such that

$$x \in \sum_{i=1}^{n} \left(\prod_{j=1}^{k_i} x_{ij} \right) \quad \text{and} \quad y \in \sum_{i=1}^{n} A_{\sigma(i)},$$

where $A_i = \prod_{j=1}^{k_i} x_{i\sigma_i(j)}$.

Also, we can define the relation α_+^* as the smallest equivalence relation such that the quotient R/α_+^* is an abelian group. The equivalence relation α_+^* was introduced by Freni [35] on hypergroups with a different name. We denote by α_+ the relation defined in R as follows: $a\alpha_+b$ if and only if there exists $(x_1, \ldots, x_n) \in R^n$, $\sigma \in S_n$ such that $a \in \sum_{i=1}^{n} x_i$ and $b \in \sum_{i=1}^{n} x_{\sigma(i)}$. The relational notation $A \approx B$ is used to assert that the sets A and B have an element in common, i.e., $A \cap B \neq \emptyset$.

Now, we present some properties of α^*-relation on Krasner hyperrings. The main reference is Ref. [63].

Theorem 7.4. *If $(R, +, \cdot)$ is a commutative Krasner hyperring, then we have $\alpha^* = \alpha_+^*$.*

Proof. In a Krasner hyperring R, every product of the elements of R is a singleton. Thus, for every

$$A = \sum_{i=1}^{n} \left(\prod_{j=1}^{k_i} x_{ij} \right),$$

we can consider the elements $y_i = \prod_{j=1}^{k_i} x_{ij}$ $(i = 1, \ldots, n)$ of R, for which we have $A = \sum_{i=1}^{n} y_i$. Since the semigroup $(R, \cdot)$ is commutative, this means that $a \; \alpha^* \; b$ if and only if $a \; \alpha_+^* \; b$. $\qquad\square$

The kernel of the canonical map $\varphi : R \to R/\gamma^*$ is called the *core* of R and is denoted by ω_R. Here, we also denote by ω_R the zero element of R/γ^*. We have the following statements. We set $D(R)$ the kernel of the canonical map $\phi : R \to R/\alpha^*$ and $D(R)$ is the zero element of commutative ring R/α^*. Thus, we have the following.

Lemma 7.1.

(1) $\omega_R = \varphi^{-1}(0_{R/\gamma^*})$ and $D(R) = \phi^{-1}(0_{R/\alpha^*})$;
(2) $\gamma^*(-x) = -\gamma^*(x)$ and $\alpha^*(-x) = -\alpha^*(x)$, for all $x \in R$;
(3) $\gamma^*(0) = 0_{R/\gamma^*}$ and $\alpha^*(-x) = 0_{R/\alpha^*}$.

Proof. It is straightforward. $\qquad\square$

Lemma 7.2. *Let R be a Krasner hyperring. Then,*

$$R \cdot D(R) \subseteq D(R) \quad \text{and} \quad D(R) \cdot R \subseteq D(R).$$

Proof. For all $a \in R \cdot D(R)$, there exist $r \in R$ and $x \in D(R)$ such that $a \in rx$. So, $\alpha^*(a) = \alpha^*(rx) = \alpha^*(r) \otimes \alpha^*(x) = \alpha^*(r) \otimes \overline{0} = \overline{0}$. $\qquad\square$

Theorem 7.5. *If R is a Krasner hyperring, then $D(R)$ is a hyperideal of R.*

Proof. We have $0 \in D(R)$. Let $x, y \in D(R)$. Then, for every $z \in x + y$, we have $\alpha^*(z) = \alpha^*(x) \oplus \alpha^*(y) = D(R) \oplus D(R) = D(R)$, which yields $z \in D(R)$, and so $x + y \subseteq D(R)$. Since $x \in D(R)$, there exists $-x \in R$ such that $0 \in x - x$. So,

$$D(R) = \alpha^*(0) = \alpha^*(x - x) = \alpha^*(x) \oplus \alpha^*(-x)$$
$$= D(R) \oplus \alpha^*(-x) = \alpha^*(-x),$$

and hence $-x \in D(R)$. $\qquad\square$

Theorem 7.6. *If R is a Krasner hyperring, then $D(R)$ is a normal hyperideal of R.*

Proof. If $y \in x + D(R) - x$, then there exists $a \in D(R)$ such that $y \in x + a - x$. Thus, $\phi(y) = \phi(x + a - x) = \phi(x) + \phi(a) + \phi(-x) = \phi(x) + 0_{R/\alpha^*} - \phi(x) = 0_{R/\alpha^*}$. Therefore, we have $y \in \phi^{-1}(0_{R/\alpha^*}) = D(R)$. $\qquad\square$

Theorem 7.7. *If R is a Krasner hyperring and $A = D(R)$, then $A^* = \alpha^*$.*

Proof. $x \; A^* \; y$ if and only $x - y \cap A \neq \emptyset$. Thus, there exists $z \in A = D(R)$ such that $z \in x - y$ and so $\alpha^*(x - y) = \alpha^*(z) = 0_{R/\alpha^*}$. Hence, $\alpha^*(x) = \alpha^*(y)$ and $x \; \alpha^* \; y$. Thus, $A^* \subseteq \alpha^*$. For the converse, if $\alpha^*(x) = \alpha^*(y)$, then $\alpha^*(x - y) = 0_{R/\alpha^*}$. So, $x - y \subseteq D(R) = A$, and hence $x \; A^* \; y$. Thus, $A^* = \alpha^*$. $\qquad\square$

If R is a Krasner hyperring and $B = \omega_R$, then:

(1) B is a normal hyperideal of R,
(2) equivalence relation B^* is equal to fundamental relation Γ^*.

Remark 7.1. If $A = D(R)$, then $R/A = R/\alpha^*$, and so R/A is a commutative ring. If $B = \omega_R$, then $R/B = R/\gamma^*$ and R/B is a ring.

Theorem 7.8. *Let R be a Krasner hyperring. Then, $I = D(R)/\gamma^*$ is an ideal of the ring $S = R/\gamma^*$, and we have $S/I \cong R/\alpha^*$.*

Proof. Set $A = D(R)$ and $B = \omega_R$. Then, we have $R/B = R/\gamma^*$, $R/A = R/\alpha^*$, and $D(R)/B = D(R)/\gamma^*$. Now, by the third isomorphism theorem, A/B is a hyperideal of R/B and $(R/B)/(A/B) \cong R/A$. However, $R/B = R/\gamma^*$ is a ring, and so $A/B = D(R)/\gamma^*$ is an ideal of R/γ^*. Therefore, $\big((R/\Gamma^*)/D(R)\big)/\gamma^*] \cong R/\alpha^*$. $\qquad\square$

We say that $x \; \alpha_{n,k_1^n} \; y$ if and only if there exist $\sigma \in \mathbb{S}_n$, $(x_{i1}, \dots, x_{ik_i}) \in R^{k_i}$, and $\sigma_i \in \mathbb{S}_{k_i}$, where $i = 1, \dots, n$ such that

$$x \in \sum_{i=1}^{n} \left(\prod_{j=1}^{k_i} x_{ij} \right) \quad \text{and} \quad y \in \sum_{i=1}^{n} \left(\prod_{j=1}^{k_{\sigma(i)}} x_{\sigma(i)\sigma_{\sigma(i)}(j)} \right),$$

Lemma 7.3. *Let $(R, +, \cdot)$ be a Krasner hyperring. If $m \geq n$, then $x \; \alpha_{n,k_1^n} \; y$ implies that $x \; \alpha_{m,k_1^m} \; y$, where for every $i \geq n$, $k_i = k_n$. Moreover, if $(R, +, \cdot)$ is a Krasner hyperring with an identity element for the hyperoperation $\cdot$ or if $(R, +, \cdot)$ is a hyperfield (where $(R, \cdot)$ is a hypergroup), $m \geq n$ and $l_1^m \in \mathbb{N}$, then $x \; \alpha_{n,k_1^n} \; y$ implies that $x \; \alpha_{m,q_1^m} \; y$, where $q_i = \max\{k_i, l_i\}$, such that for every $i \geq n$, $k_i = k_n$.*

Proof. Let $x \; \alpha_{n,k_1^n} \; y$. Then, there exist $\sigma \in \mathbb{S}_n$, $(x_{i1}, \ldots, x_{ik_i}) \in R^{k_i}$, and $\sigma_i \in \mathbb{S}_{k_i}$, where $i = 1, \ldots, n$, such that

$$x \in \sum_{i=1}^{n} \left(\prod_{j=1}^{k_i} x_{ij} \right) \quad \text{and} \quad y \in \sum_{i=1}^{n} \left(\prod_{j=1}^{k_{\sigma(i)}} x_{\sigma(i)\sigma_{\sigma(i)}(j)} \right).$$

If $n < m$, then there exist $x'_{ik_i} \in R$, where $i = n, \ldots, m$ and $k_i = k_n$, such that $x_{nk_n} \in x'_{nk_n} + \cdots + x'_{mk_m}$, and so

$$x \in \sum_{i=1}^{n} \left(\prod_{j=1}^{k_i} x_{ij} \right)$$

$$\subseteq \sum_{i=1}^{n-1} \left(\prod_{j=1}^{k_i} x_{ij} \right) + \left(\prod_{j=1}^{k_n-1} x_{nj} \right) \cdot (x'_{nk_n} + \cdots + x'_{mk_m})$$

$$= \sum_{i=1}^{n-1} \left(\prod_{j=1}^{k_i} x_{ij} \right) + \left(\prod_{j=1}^{k_n-1} x_{nj} \right) \cdot x'_{nk_n} + \cdots + \left(\prod_{j=1}^{k_n-1} x_{nj} \right) \cdot x'_{mk_m}$$

$$= u.$$

Now, there exist $t \in \{1, \ldots, n\}$ and $s \in \{1, \ldots, k_t\}$ such that $x_{\sigma(t)\sigma_{\sigma(t)}(s)} = x_{nk_n}$, so $\sigma(t) = n$ and $\sigma_n(s) = k_n$, and we have

$$y \in \sum_{i=1}^{n} \left(\prod_{j=1}^{k_{\sigma(i)}} x_{\sigma(i)\sigma_{\sigma(i)}(j)} \right)$$

$$= \underbrace{\sum_{i=1}^{t-1} \left(\prod_{j=1}^{k_{\sigma(i)}} x_{\sigma(i)\sigma_{\sigma(i)}(j)} \right) + \left(\prod_{j=1}^{k_{\sigma(t)}} x_{\sigma(t)\sigma_{\sigma(t)}(j)} \right)}_{A}$$

$$+ \underbrace{\sum_{i=t+1}^{n} \left(\prod_{j=1}^{k_{\sigma(i)}} x_{\sigma(i)\sigma_{\sigma(i)}(j)} \right)}_{B}$$

$$= A + \left(\prod_{j=1}^{s-1} x_{n\sigma_n(j)} \right) \cdot x_{nk_n} \cdot \left(\prod_{j=s+1}^{k_n} x_{n\sigma_n(j)} \right) + B$$

$$\subseteq A + \prod_{j=1}^{s-1} x_{n\sigma_n(j)} \cdot (x'_{nk_n} + \cdots + x'_{mk_m}) \cdot \prod_{j=s+1}^{k_n} x_{n\sigma_n(j)} B$$

$$= A + \prod_{j=1}^{s-1} x_{n\sigma_n(j)} \cdot x'_{nk_n} \cdot \prod_{j=s+1}^{k_n} x_{n\sigma_n(j)}$$

$$+ \cdots + \prod_{j=1}^{s-1} x_{n\sigma_n(j)} \cdot x'_{mk_m} \cdot \prod_{j=s+1}^{k_n} x_{n\sigma_n(j)} + B$$

$$= u'.$$

Therefore, $x \alpha_{m,k_1^m} y$.

Let $(R, +, \cdot)$ be a Krasner hyperring with an identity element for the hyperoperation $\cdot$ or let $(R, +, \cdot)$ be a hyperfield ($(R, \cdot)$ is a hypergroup), $m \geq n$, $l_1^m \in \mathbb{N}$ and $x \, \alpha_{n,k_1^n} \, y$. Then, $x \, \alpha_{n,k_1^n} \, y$ where $k_i = k_n$ for every $i \geq n$. So, there exist $\sigma \in \mathbb{S}_m$, $(x_{i1}, \ldots, x_{ik_i}) \in R^{k_i}$ and $\sigma_i \in \mathbb{S}_{k_i}$, where $i = 1, \ldots, m$ such that

$$x \in \sum_{i=1}^{m} \left(\prod_{j=1}^{k_i} x_{ij} \right) \quad \text{and} \quad y \in \sum_{i=1}^{m} \left(\prod_{j=1}^{k_{\sigma(i)}} x_{\sigma(i)\sigma_{\sigma(i)}(j)} \right).$$

Now, suppose that $\sigma(t_i) = i$ and $\sigma_i(s_i) = k_i$ and $q_i = \max\{k_i, l_i\}$. Then, there exists $x'_{ik_i}, \ldots, x'_{iq_i}$ such that $x_{ik_i} \in x'_{ik_i} \cdots x'_{iq_i}$. Therefore,

$$x \in \sum_{i=1}^{m} \left(\prod_{j=1}^{k_i} x_{ij} \right) = \sum_{i=1}^{m} \left(\prod_{j=1}^{k_i-1} x_{ij} \cdot x_{ik_i} \right)$$

$$\subseteq \sum_{i=1}^{m} \left(\prod_{j=1}^{k_i-1} x_{ij} \cdot x'_{ik_i} \cdots x'_{iq_i} \right) = u$$

and

$$y \in \sum_{i=1}^{m} \left(\prod_{j=1}^{k_{\sigma(i)}} x_{\sigma(i)\sigma_{\sigma(i)}(j)} \right)$$

$$= \sum_{i=1}^{m} \left(\prod_{j=1}^{t_i-1} x_{\sigma(i)\sigma_{\sigma(i)}(j)} \cdot x_{\sigma(t_i)\sigma_{\sigma(t_i)}(s_i)} \cdot \prod_{j=t_i+1}^{k_{\sigma(i)}} x_{\sigma(i)\sigma_{\sigma(i)}(j)} \right)$$

$$\subseteq \sum_{i=1}^{m} \left(\prod_{j=1}^{t_i-1} x_{\sigma(i)\sigma_{\sigma(i)}(j)} \cdot x'_{ik_i} \cdots \cdots x'_{iq_i} \cdot \prod_{j=t_i+1}^{k_{\sigma(i)}} x_{\sigma(i)\sigma_{\sigma(i)}(j)} \right) = u'.$$

Therefore, $x \alpha_{m,q_1^m} y$. □

Theorem 7.9. *Let $(R,+,\cdot)$ be a Krasner hyperring with an identity element for the hyperoperation $\cdot$ or let $(R,+,\cdot)$ be a hyperfield $((R,\cdot)$ is a hypergroup). If $a \; \alpha_{n,k_1^n} \; b$ and $c \; \alpha_{m,l_1^m} \; d$, then $a \; \alpha_{p,q_1^p} \; b$ and $c \; \alpha_{p,q_1^p} \; d$, where $p = \max\{m,n\}$, and for every $i = 1,\ldots,p$, set*

$$q_i = \begin{cases} \max\{l_i,k_i\} & \text{if } i \le \min\{k_i,l_i\} \\ \max\{l_i,k_n\} & \text{if } i > \min\{k_i,l_i\} \quad \text{and} \quad n \le m \\ \max\{l_m,k_i\} & \text{if } i > \min\{k_i,l_i\} \quad \text{and} \quad m \le n. \end{cases}$$

Theorem 7.10. *Let R_1 and R_2 be two Krasner hyperrings with an identity element for the hyperoperation $\cdot$ or let them be two hyperfields, and let $\alpha^*_{R_1}$, $\alpha^*_{R_2}$, and $\alpha^*_{R_1 \times R_2}$ be the α^*-relations on R_1, R_2, and $R_1 \times R_2$, respectively. Then,*

$$(a,b) \; \alpha^*_{R_1 \times R_2} \; (c,d) \quad \text{if and only if } a \; \alpha^*_{R_1} \; c \quad \text{and} \quad b \; \alpha^*_{R_2} \; d.$$

Proof. It is clear that $(a,b) \; \alpha^*_{R_1 \times R_2} \; (c,d)$ imply $a \; \alpha^*_{R_1} \; c$ and $b \; \alpha^*_{R_2} \; d$.

Conversely, let $a \; \alpha^*_{R_1} \; c$ and $b \; \alpha^*_{R_2} \; d$. Then, there exists $h,g \in \mathbb{N}$, $a = u_1,\ldots,u_g = b \in R$ and $c = v_1,\ldots,v_h = d \in R$ such that

$$u_1 \; \alpha_{n_1,k_{11}^{1n_1}} \; u_2 \; \alpha_{n_2,k_{21}^{2n_2}} \cdots \alpha_{n_{(g-1)},k_{(g-1)1}^{(g-1)n_{(g-1)}}} \; u_g$$

and

$$v_1 \; \alpha_{m_1,l_{11}^{1m_1}} \; u_2 \; \alpha_{m_2,l_{21}^{2m_2}} \cdots \alpha_{m_{(h-1)},l_{(h-1)1}^{(h-1)m_{(h-1)}}} \; v_h.$$

Suppose that $h \geq g$. Then, set $u_{g+1} = \cdots = u_h := u_g$, and we obtain

$$u_1 \, \alpha_{n_1, k_{11}^{1n_1}} \, u_2 \, \alpha_{n_2, k_{21}^{2n_2}} \cdots \alpha_{n_{(g-1)}, k_{(g-1)1}^{(g-1)n_{(g-1)}}} \, u_g \, \alpha_{1,1} \, u_{g+1} \cdots \alpha_{1,1} \, u_h$$

and

$$v_1 \, \alpha_{m_1, l_{11}^{1m_1}} \, u_2 \, \alpha_{m_2, l_{21}^{2m_2}} \cdots \alpha_{m_{(h-1)}, l_{(h-1)1}^{(h-1)m_{(h-1)}}} \, v_h.$$

Now, by Theorem 7.9, there exist $p, q_1, \ldots, q_p \in \mathbb{N}$ such that

$$u_1 \, \alpha_{p,q_1^p} \, u_2 \, \alpha_{p,q_1^p} \cdots \alpha_{p,q_1^p} \, u_h \quad \text{and} \quad v_1 \, \alpha_{p,q_1^p} \, u_2 \, \alpha_{p,q_1^p} \cdots \alpha_{p,q_1^p} \, v_h.$$

So,

$$(a, c) = (u_1, v_1) \, \alpha_{p,q_1^p} \, (u_2, v_2) \, \alpha_{p,q_1^p} \cdots \alpha_{p,q_1^p} \, (u_h, v_h) = (c, d).$$

Therefore, $(a, b) \, \alpha_{R_1 \times R_2}^* \, (c, d)$. $\qquad\qquad\qquad\qquad\qquad \square$

Theorem 7.11. *Let R_1 and R_2 be two Krasner hyperrings with an identity element for the hyperoperation $\cdot$ or let them be two hyper-fields, and let $\alpha_{R_1}^*$, $\alpha_{R_2}^*$, and $\alpha_{R_1 \times R_2}^*$ be the commutative fundamental equivalence relations on R_1, R_2, and $R_1 \times R_2$, respectively. Then,*

$$R_1 \times R_2 / \alpha_{R_1 \times R_2}^* \cong R_1 / \alpha_{R_1}^* \times R_2 / \alpha_{R_2}^*.$$

Proof. We define the relation $\tilde{\alpha}$ on $R_1 \times R_2$ as follows:

$$(a_1, b_1) \, \tilde{\alpha} \, (a_2, b_2) \quad \Longleftrightarrow \quad a_1 \, \alpha_{R_1}^* \, a_2 \quad \text{and} \quad b_1 \, \alpha_{R_2}^* \, b_2.$$

Then, $\tilde{\alpha} = \alpha_{R_1 \times R_2}^*$. Now, we consider the map $\varphi : R_1 / \alpha_{R_1}^* \times R_2 / \alpha_{R_2}^* \to R_1 \times R_2 / \alpha_{R_1 \times R_2}^*$ by

$$\varphi(\alpha_{R_1}^*(a), \alpha_{R_2}^*(b)) = \alpha_{R_1 \times R_2}^*(a, b).$$

It is easy to see that φ is an isomorphism. $\qquad\qquad\qquad\qquad \square$

Lemma 7.4. *If A, B are hyperideals of R_1, R_2, respectively, then*

$$(R_1 \times R_2) / (A \times B) \cong R_1 / A \times R_2 / B.$$

Proof. The proof is straightforward, and we omit it. $\qquad\qquad \square$

Corollary 7.1. *If A, B are hyperideals of R_1, R_2, respectively, and if α_1^*, α_2^*, and α^* are fundamental equivalence relations on R_1/A, R_2/B, and $(R_1 \times R_2)/(A \times B)$, respectively, then,*

$$((R_1 \times R_2)/(A \times B))/\alpha^* \cong (R_1/A)/\alpha_1^* \times (R_2/B)/\alpha_2^*.$$

Proof.   The proof is obtained exactly from Theorem 7.11 and Lemma 7.4.   □

Definition 7.2. Let f be a good homomorphism from R_1 onto R_2, and let α_1^*, α_2^* be the fundamental relations on R_1, R_2, respectively, then we define

$$\overline{ker f} = \{\alpha_1^*(x) \mid x \in R_1, \ \alpha_2^*(f(x)) = D(R_2)\}.$$

Lemma 7.5. *$\overline{ker f}$ is an ideal of the commutative fundamental ring R_1/α_1^*.*

Proof.   Assume that $\alpha_1^*(x), \alpha_1^*(y) \in \overline{ker f}$. Then, for every $z \in x - y$, we have $\alpha_1^*(z) = \alpha_1^*(x) \uplus \alpha_1^*(-y)$. On the other hand, we have

$$\alpha_2^*(f(z)) = \alpha_2^*(f(x) + f(-y)) = \alpha_2^*(f(x)) \uplus \alpha_2^*(f(-y))$$
$$= \alpha_2^*(f(x)) \uplus (-\alpha_2^*(f(y)))$$
$$= D(R_2) \uplus D(R_2) = D(R_2).$$

Therefore, $\alpha_1^*(z) \in \overline{ker f}$. For $\alpha_1^*(r) \in R_1/\alpha_1^*$ and $\alpha_1^*(x) \in \overline{ker f}$, we have

$$\alpha_2^*(f(r \cdot x)) = \alpha_2^*(f(r) \cdot f(x)) = \alpha_2^*(f(r)) \otimes \alpha_2^*(f(x))$$
$$= \alpha_2^*(f(r)) \otimes D(R_2) = D(R_2),$$

and so $\alpha_1^*(r) \otimes \alpha_1^*(x) \in \overline{ker f}$. Therefore, $\overline{ker f}$ is an ideal of R_1/α^*.
                                                                       □

Theorem 7.12. *Let $(R+, \cdot)$ be a Krasner hyperring such that $(R, \cdot)$ be commutative. Let A, B be two hyperideals of R, with $A \subseteq B$, and $\phi : R/A \to R/B$ be the canonical map. Suppose that α_A^* and α_B^* are the fundamental equivalence relations on R/A and R/B, respectively. Then,*

$$((R/A)/\alpha_A^*))/\overline{ker \phi} \cong (R/B)/\alpha_B^*.$$

Proof.   We define the map

$$\rho : (R/A)/\alpha_A^* \to (R/B)/\alpha_B^*$$

by

$$\rho : \alpha_A^*(A + x) \mapsto \alpha_B^*(B + x) \quad \text{(for all } x \in R\text{)}.$$

We must check that ρ is well defined, i.e., if $x, y \in R$ and $\alpha_A^*(A+x) = \alpha_A^*(A+y)$, then $\alpha_B^*(B+x) = \alpha_B^*(B+y)$. Using Theorem 7.4, we have $\alpha_A^* = (\alpha_+^*)_A$ and $\alpha_B^* = (\alpha_+^*)_B$. Now, $(\alpha_+^*)_A(A + x) = (\alpha_+^*)_A(A + y)$ if and only if there exist $(A + x_1, A + x_2, \ldots, A + x_n) \in (R/A)^n$ and $\sigma \in S_n$ such that $A+x \subseteq \oplus \sum_{i=1}^n A+x_i$ and $A+y \subseteq \oplus \sum_{i=1}^n A+x_{\sigma(i)}$. We have

$$\oplus \sum_{i=1}^n A + x_i = \left\{ A + z \mid z \in \sum_{i=1}^n x_i \right\}.$$

Therefore, for some $z_1 \in \sum_{i=1}^n x_i$, $z_2 \in \sum_{i=1}^n x_{\sigma(i)}$, we have $A + x = A + z_1$ and $A + y = A + z_2$. So, there exist $a \in x - z_1 \cap A$ and $b \in y - z_2 \cap A$. Then, $x \in a + z_1$ and $y \in b + z_2$. Hence, $B + x \in (B + a) \oplus (B + z_1)$ and $B + y \in (B + b) \oplus (B + z_2)$. Since $a, b \in A \subseteq B$, $B+a = B$, $B+b = B$. Since $B \oplus (B+z_1) = B+z_1$ and $B \oplus (B + z_2) = B + z_2$, we have $B + x = B + z_1$ and $B + y = B + z_2$. Since

$$B + z_1 \subseteq \left\{ B + z \mid z \in \sum_{i=1}^n x_i \right\}$$

and

$$B + z_2 \subseteq \left\{ B + z \mid z \in \sum_{i=1}^n x_{\sigma(i)} \right\},$$

we get

$$B + x \subseteq \left\{ B + z \mid z \in \sum_{i=1}^n x_i \right\} = \oplus \sum_{i=1}^n (B + x_i),$$

$$B + y \subseteq \left\{ B + z \mid z \in \sum_{i=1}^n x_{\sigma(i)} \right\} = \oplus \sum_{i=1}^n (B + x_{\sigma(i)}).$$

Therefore, $(\alpha_+^*)_B(B + x) = (\alpha_+^*)(B + y)$. This follows that ρ is well defined. Moreover, ρ is a good homomorphism if $x, y \in R_1$. Hence,

$$\rho(\alpha_A^*(A + x) \uplus \alpha_A^*(A + y)) = \rho(\alpha_A^*(A + x + y)) = \alpha_B^*(B + x + y)$$

$$= \alpha_B^*(B + x) \uplus \alpha_B^*(B + y)$$

$$= \rho(\alpha_A^*(A + x)) \uplus \rho(\alpha_A^*(A + y)),$$

$$\rho(\alpha_A^*(A + x) \otimes \alpha_A^*(A + y)) = \rho(\alpha_A^*(A + xy)) = \alpha_B^*(B + xy)$$

$$= \alpha_B^*(B + x) \otimes \alpha_B^*(B + y)$$

$$= \rho(\alpha_A^*(A + x)) \otimes \rho(\alpha_A^*(A + y)),$$

and $\rho(D([R : A^*])) = \rho(\alpha_A^*(A)) = \alpha_B^*(B) = D([R : B^*])$. Clearly, ρ is onto. Now, we show that $ker\rho = \overline{ker\phi}$. We have

$$ker\rho = \{\alpha_A^*(A + x) \mid \rho(\alpha_A^*(A + x)) = D(R/B)\}$$

$$= \{\alpha_A^*(A + x) \mid \alpha_B^*(B + x) = D(R/B)\}$$

$$= \{\alpha_A^*(A + x) \mid \gamma_B^*(\phi(A + x)) = D(R/B)\}$$

$$= \overline{ker\phi}. \qquad \square$$

Theorem 7.13. *Let R be a Krasner hyperring and $a_1, \ldots,$ $a_m, b_1, \ldots, b_m \in R$ such that $a_j \ \alpha \ b_j$, for all $j = 1, \ldots m$. Then, for all $x \in \sum_{i=1}^m \delta_i a_i$ and for all $y \in \sum_{i=1}^m \delta_i b_i$, where $\delta_i \in \{1, -1\}$ $(i = \{1, \ldots, m\})$, we have $x \ \alpha \ y$.*

Proof. Suppose that $a_j \ \alpha \ b_j$, for all $j = 1, \ldots, m$. Then, there exist $n_j \in \mathbb{N}$, $(k_{j1}, \ldots, k_{jn_j}) \in \mathbb{N}^{n_j}$ and $(z_{ji1}, \ldots, z_{jik_{ji}}) \in R^{k_{ji}}$ and there exist $\sigma_j \in \mathbb{S}_{n_j}$ and $\sigma_{ji} \in \mathbb{S}_{k_{ji}}$ when $(i = 1, \ldots, n_j)$ such that

$$a_j \in \sum_{i=1}^{n_j} \left(\prod_{l=1}^{k_{ji}} x_{jil} \right) \quad \text{and} \quad b_j \in \sum_{i=1}^{n_j} \left(\prod_{l=1}^{k_{j\sigma(i)}} x_{j\sigma(i)\sigma_{j\sigma(i)}(l)} \right).$$

Therefore,

$$\sum_{j=1}^{m} a_j \subseteq \sum_{j=1}^{m} \left(\sum_{i=1}^{n_j} \left(\prod_{l=1}^{k_{ji}} x_{jil} \right) \right),$$

$$\sum_{i=1}^{n} \left(\prod_{j=1}^{k_i} x_{ij} \right) \subseteq \sum_{j=1}^{m} \left(\sum_{i=1}^{n_j} \left(\prod_{l=1}^{k_{j\sigma(i)}} x_{j\sigma(i)\sigma_{j\sigma(i)}(l)} \right) \right).$$

If we rename x_{ijl}s, then we conclude that there exist $n, k_i \in \mathbb{N}$, $\tau \in \mathbb{S}_n$, and $\tau_i \in \mathbb{S}_{k_i}$ such that

$$\sum_{j=1}^{m} a_j \subseteq \sum_{i=1}^{n} \left(\prod_{j=1}^{k_i} x_{ij} \right), \quad \sum_{j=1}^{m} b_j \subseteq \sum_{i=1}^{n} A_{\sigma(i)},$$

where $A_i = \prod_{j=1}^{k_i} x_{i\sigma_i(j)}$, and so for all $x \in \sum_{j=1}^{m} a_j$ and for all $y \in \sum_{j=1}^{m} b_j$, we get $x \, \alpha \, y$. Now, note that if $a_j \, \alpha \, b_j$, then $(-a_j) \, \alpha \, (-b_j)$. $\qquad \square$

Theorem 7.14. *Let R be a Krasner hyperring. Then, $x, y \in \alpha^*(0)$ if and only if there exist $A, A' \subseteq \alpha^*(z)$ and $B, B' \subseteq \alpha^*(-z)$, for some $z \in R$, such that $(x + A) \approx B$ and $(y + A') \approx B'$.*

Proof. Suppose that there exist $A, A' \subseteq \alpha^*(z)$ and $B, B' \subseteq \alpha^*(-z)$ for some $z \in R$ such that $(x + A) \approx B$ and $(y + A') \approx B'$. Then, we have

$$(\alpha^*(x) \oplus \{\alpha^*(a) \mid a \in A\}) \approx \{\alpha^*(b) \mid b \in B\},$$

$$(\alpha^*(y) \oplus \{\alpha^*(a') \mid a' \in A'\}) \approx \{\alpha^*(b') \mid b' \in B'\}.$$

Therefore, we obtain $\alpha^*(x) \oplus \alpha^*(z) = \alpha^*(-z)$ and $\alpha^*(y) \oplus \alpha^*(z) = \alpha^*(-z)$, which imply $\alpha^*(x) = \alpha^*(y) = \alpha^*(z) \oplus \alpha^*(-z) = \alpha^*(0)$.

For the converse, take $A = A' = B = B' = \alpha^*(0)$. Then, $\alpha^*(0) \subseteq \alpha^*(0)$ and $\{x\} \subseteq \alpha^*(0)$, which imply that $\alpha^*(x + \alpha^*(0)) = \alpha^*(0)$. Hence, $(x + \alpha^*(0)) \subseteq \alpha^*(0)$, or $(x + A) \approx A$. Similarly, we get $(y + A) \approx A$. This completes the proof. $\qquad \square$

If $r \in \mathbb{N}$ and $a \in R$, then we set $ra = \sum_{i=1}^{n} a$.

Theorem 7.15. *Let R be a finite Krasner hyperring. For every $a \in R$, there exist $r, s \in \mathbb{N}$ such that $0 < s < r$, $ra \approx sa$ and $(r - s)a \subseteq \alpha^*(0)$.*

Proof. Since R is finite, it follows that for every $r \in \mathbb{N}$, $ra \subseteq R$, and there exist $r, s \in \mathbb{N}$ such that $0 < s < r$, and $ra \approx sa$. From $ra \approx sa$, we have $\phi(ra) = \phi(sa)$, and so $r\alpha^*(a) = s\alpha^*(a)$. Since $r\alpha^*(a)$ and $s\alpha^*(a)$ are the elements of the commutative ring R/α^*,

$$(r - s)\alpha^*(a)) = 0_{R/\alpha^*} = \alpha^*(0),$$

which implies that $\phi((r - s)a) = \alpha^*(e)$, and so $(r - s)a \subseteq \alpha^*(0)$. $\square$

Let M be a non-empty subset of a Krasner hyperring R. We say that M is an α-*part* of R (see [62]) if for every $n \in \mathbb{N}$, $i = 1, 2, \ldots, n$, $\forall k_i \in \mathbb{N}$, $\forall (z_{i1}, z_{i2}, \ldots, z_{ik_i}) \in R^{n_i}, \forall \sigma \in \mathbb{S}_n, \forall \sigma_i \in \mathbb{S}_{k_i}$, we have

$$\sum_{i=1}^{n} \left(\prod_{j=1}^{k_i} z_{ij} \right) \approx M \quad \Rightarrow \quad \sum_{i=1}^{n} A_{\sigma(i)} \subseteq M,$$

where $A_i = \prod_{j=1}^{k_i} z_{i\sigma_i(j)}$.

Let A be a non-empty subset of R. The intersection of α-parts of R which contains A is called the α-*closure* of A in R. It is denoted $C_\alpha(A)$. We have the following:

(1) If A is an α-part of R, then $A + B$, $B + A$, AB, and BA are α-parts of R, for every $B \in \wp^*(R)$.
(2) $A \in \wp^*(R)$, then A is a α-part of R if and only if $A + D(R) = A$.
(3) $A \in \wp^*(R)$, one has $D(R) + A = A + D(R) = C_\alpha(A)$.
(4) $D(R)$ is an α-part of R.

Let R be a Krasner hyperring and $X = \langle \wp^*(P), \uplus \rangle$ be the set of non-empty subsets of R endowed with the hyperoperation $\uplus$, defined as follows:

$$A \uplus B = \{C \in \wp^*(R) \mid C \subseteq A + B\} \quad \text{for all } (A, B) \in \wp^*(R)^2.$$

Let $\sum(R)$ be the set of hyper-sums of the elements of R.

Theorem 7.16. *If $(R, +, \cdot)$ is a Krasner hyperring, then $(\sum(R), \uplus, \cdot)$ is a hyperring.*

Proof. It is clear that $\uplus$ is associative. Let $E = \{0\}$. Then, for all $A \in \sum(R)$, we have $A \uplus E = E \uplus A = A$. We define the function $-I$ as follows:

$$-I : \sum(R) \to \sum(R)$$

$$-I\left(\sum_{i=1}^{n} x_i\right) = \sum_{i=1}^{n}(-x_i).$$

Now, let $X = \sum_{i=1}^{m} x_i$, $Y = \sum_{i=1}^{n} y_i$, and $Z = \sum_{i=1}^{p} z_i$ be elements of $\sum(R)$ such that $X \in Y \uplus Z$. Let $x \in X$ and $y \in Y$ be arbitrary. Then, there exists $z \in Z$ such that $x \in y + z$. In other words, for every $z \in Z$, we have $x \notin y + z$, and so $x \notin Y + Z$, or $\{x\} \notin Y \uplus Z$, which is a contradiction. Since $x \in y + z$, $y \in x - z$, and so $y \in \sum_{i=1}^{n} x_i + \sum_{i=1}^{p}(-z_i)$, for every $y \in Y$. Therefore, $Y \subseteq \sum_{i=1}^{n} x_i + \sum_{i=1}^{p}(-z_i)$, or $Y \in X \uplus -I(Z)$. Similarly, we get $Z \in -I(Y) \uplus X$. Now, we prove that $(\sum(R), \cdot)$ is a semigroup. Suppose that $X = \sum_{i=1}^{m} x_i$, $Y = \sum_{i=1}^{n} y_i$. Then,

$$X \cdot Y = \left(\sum_{i=1}^{m} x_i\right) \cdot \left(\sum_{i=1}^{n} y_i\right) = \sum_{j=1}^{m}\sum_{i=1}^{n} x_i y_j \in \sum(R).$$

Also, $0 \cdot X = X \cdot 0 = 0$. The operation $\cdot$ is distributive on the hyperoperation $\uplus$ because if $X = \sum_{i=1}^{m} x_i$, $Y = \sum_{i=1}^{n} y_i$ and $Z = \sum_{i=1}^{p} z_i$ are elements of $\sum(R)$, then $a \in X \cdot (Y + Z)$ if there exist $x \in X, y \in Y$ and $z \in Z$ such that $a \in x(y+z) = xy+xz$. Therefore, $a \in X \cdot Y + X \cdot Z$. Thus, $X \cdot (Y+Z) \subseteq X \cdot Y + X \cdot Z$. For the converse, let $a \in X \cdot Y + X \cdot Z$. Then, there exist $x, x' \in X$, $y \in Y$, and $z \in Z$ such that $a \in xy + x'z$. However,

$$X \cdot Y + X \cdot Z = \sum_{i=1}^{m} x_i \sum_{i=1}^{n} y_i + \sum_{i=1}^{m} x_i \sum_{i=1}^{p} z_i$$

$$= \sum_{i=1}^{m} x_i \left(\sum_{i=1}^{n} y_i + \sum_{i=1}^{p} z_i\right) = X \cdot (Y + Z).$$

Hence, $a \in xy + x'z \subseteq X \cdot Y + X \cdot Z = X \cdot (Y+Z)$, and so $a \in X \cdot (Y \uplus Z)$. Thus, $(\sum(R), \uplus, \cdot)$ is a Krasner hyperring. $\quad\square$

Corollary 7.2. *If A is a subhyperring of R and A belongs to $\sum(R)$, then A is contained in $D(R)$.*

The following example shows that not all subhyperrings of a Krasner hyperring R are in $\sum(R)$.

Example 7.1. Let $(R, +, \cdot)$ be a Krasner hyperring such that $(R, +)$ have the following table:

+	a	b	c	d
a	a	b	c	d
b	b	a	c	d
c	c	c	$\{a, b, d\}$	$\{c, d\}$
d	d	d	$\{c, d\}$	$\{a, b, c\}$

and semigroup $(R, \cdot)$ have the following operation: $x \cdot y = a$, for all $x, y \in R$. It is clear that $A = \{a, b\}$ is a subhyperring of R, but $A \notin \sum(R)$. Moreover, $D(R) = c + d + d = R \in \sum(R)$.

If R is a Krasner hyperring, we denote $\sum_{C_\alpha}(P)$ the set hyper-sums A of the elements of R such that $C_\alpha(A) = A$.

Theorem 7.17. *Let R be a Krasner hyperring and $(x_1, \ldots, x_n) \in P^n$ be such that $\sum_{i=1}^n x_i \in \sum_{C_\alpha}(R)$, then there exists $(y_1, \ldots, y_n) \in P^n$ such that $\sum_{i=1}^n x_i + \sum_{i=1}^n y_i = D(R)$.*

Proof. For $1 \leq j \leq n$, let a_j be an element of $D(R)$. Then, there exists $y_j \in R$ such that $a_j \in x_j + y_j$. Since $D(R)$ is an α-part, we have $x_j + y_j \subseteq D(R)$. Therefore,

$$\sum_{i=1}^n x_i + y_n = D(R) + \sum_{i=1}^n x_i + y_n$$

$$= \sum_{i=1}^{n-1} x_i + D(R) + x_n + y_n$$

$$= \sum_{i=1}^{n-1} x_i + D(R)$$

$$= D(R) + \sum_{i=1}^{n-1} x_i,$$

and so

$$\sum_{i=1}^{n} x_i + y_n + y_{n-1} = D(R) + \sum_{i=1}^{n-2} x_i + x_{n-1} + y_{n-1}$$

$$= D(R) + \sum_{i=1}^{n-2} x_i.$$

Continuing in the same manner, one arrives at $\sum_{i=1}^{n} x_i + \sum_{i=2}^{n} y_i = D(R) + x_1$, whence finally

$$\sum_{i=1}^{n} x_i + \sum_{i=1}^{n} y_i = D(R) + x_1 + y_1 = D(R). \qquad \square$$

Theorem 7.18. *Let R be a Krasner hyperring. If $R \backslash D(R)$ is a hyper-sum, then $D(R)$ is also a hyper-sum.*

Proof. Since $D(R)$ is an α-part, it follows that $R \backslash D(R)$ also is an α-part. Now, by using Theorem 7.17, the proof is complete. $\qquad \square$

Chapter 8

Some Special Hyperrings

8.1 Graded Hyperrings

The notion of a graded hyperring was introduced and studied by Farzalipour and Ghiasvand [33]. We present here some of their results.

All hyperrings presented in this section are Krasner commutative hyperrings with identity 1.

Definition 8.1. Let G be a monoid with identity e. A hyperring (R, G) is called a if there exists a family $\{R_g\}_{g \in G}$ of canonical subhypergroups of R indexed by the elements $g \in G$ such that $R = \oplus_{g \in G} R_g$ and $R_g R_h \subseteq R_{gh}$, for all $g, h \in G$. An element of a graded hyperring R is called *homogeneous* if it belongs to $\cup_{g \in G} R_g$.

The graded hyperring (R, G) is usually denoted by R, and the set of homogeneous elements is denoted by $h(R)$. If $x \in R_g$, for some $g \in G$, then we say that x is of degree g, and it is denoted by x_g.

Every $x \in R$ can be uniquely written as $x \in \sum_{g \in G} x_g$, with $x_g \in R_g$, such that all except finitely many x_gs are 0. Every hyperring is a G-graded hyperring by considering $R_e = R$ and $R_g = 0$, for all $g \neq e$.

Lemma 8.1. *If $R = \oplus_{g \in G} R_g$ is a graded hyperring, then R_e is a subhyperring of R, where e is the identity element of the monoid G.*

Proof. $(R_e, +)$ is a canonical subhypergroup of R. Let $x_e, y_e \in R_e$. Then, $x_e \cdot y_e \in R_e R_e \subseteq R_{e \cdot e} = R_e$; hence, $x_e \cdot y_e \in R_e$. For all $x_e, y_e, z_e \in R_e$, $x_e \cdot (y_e + z_e) = x_e \cdot y_e + x_e \cdot z_e$ and $(x_e + y_e) \cdot z_e = x_e \cdot z_e + y_e \cdot z_e$. Therefore, R_e is a subhyperring of R. $\square$

Example 8.1. Let $G = (\mathbb{Z}_2, \cdot)$ be a monoid with identity $e = 1$ (by multiplication operation) and $R = \{0, a, b, c\}$. Consider the hyperring $(R, +, \cdot)$, where the hyperoperation $+$ and the operation $\cdot$ are defined on R as follows:

+	0	a	b	c
0	0	a	b	c
a	a	$\{0, b\}$	$\{a, c\}$	b
b	b	$\{a, c\}$	$\{0, b\}$	a
c	c	b	a	0

$\cdot$	0	a	b	c
0	0	0	0	0
a	0	a	b	c
b	0	b	b	0
c	0	c	0	c

Consider $R_0 = \{0, b\}$ and $R_1 = \{0, c\}$. We have that R_0 and R_1 are canonical hypergroups of $(R, +)$, and we can write $0 \in 0 + 0, a \in b + c, b \in b + 0$ and $c \in 0 + c$ uniquely; hence, $R = R_0 \oplus R_1$. Also, $R_i R_j \subseteq R_{ij}$, for all $i, j \in \mathbb{Z}_2$, and so (R, G) is a graded hyperring.

Example 8.2. Let $R = \{0, a, b, c, d\}$. Consider the hyperring $(R, +, \cdot)$, where the hyperoperation $+$ and the operation $\cdot$ are defined on R as follows:

+	0	a	b	c	d
0	0	a	b	c	d
a	a	0	c	$\{b, d\}$	c
b	b	c	0	a	0
c	c	$\{b, d\}$	a	0	a
d	d	c	0	a	0

$\cdot$	0	a	b	c	d
0	0	0	0	0	0
a	0	a	b	c	d
b	0	b	0	b	0
c	0	c	b	a	d
d	0	d	0	d	0

$R_0 = \{0, a\}$ and $R_1 = \{0, c\}$ are subhypergroups of $(R, +)$. Moreover, $0 \in 0 + 0, a \in a + 0, b \in b + c, c \in 0 + c$, and $d \in a + c$. Hence, $R = R_0 \oplus R_1$ but $R_1 R_1 \not\subseteq R_1$ since $c \cdot c = a \notin R_1$. Thus, R is not a $\mathbb{Z}_2$-graded hyperring. On the other hand, we note that $R_0 = \{0, a\}, R_1 = \{0, c\}$ and $R_2 = \{0, b\}$ are subhypergroups of $(R, +)$; however, we have $a \in a + 0 + 0$ and $a \in 0 + c + b$, then $R \neq R_0 \oplus R_1 \oplus R_2$. So R is not a $\mathbb{Z}_3$-graded hyperring.

Example 8.3. Let $G = (\mathbb{Z}_2, \cdot)$ be a monoid with identity $e = 1$ and $R = \{0, 1, 2, 3\}$. Consider the hyperring $(R, +, \cdot)$, where the hyperoperation $+$ and the operation $\cdot$ are defined on R as follows:

+	0	1	2	3
0	0	1	2	3
1	1	$\{0,1\}$	3	$\{2,3\}$
2	2	3	0	1
3	3	$\{2,3\}$	1	$\{0,1\}$

$\cdot$	0	1	2	3
0	0	0	0	0
1	0	0	0	0
2	0	0	2	2
3	0	0	2	2

Clearly, $R_0 = \{0, 1\}$ and $R_1 = \{0, 2\}$ are subhypergroups of $(R, +)$. We have $0 \in 0 + 0, 1 \in 1 + 0, 2 \in 0 + 2$, and $3 \in 1 + 2$. Hence, $R = R_0 \oplus R_1$, and so (R, G) is a graded hyperring. Since $0\gamma 1$ and $2\gamma 3$, $\gamma(0) = \gamma(1)$ and $\gamma(2) = \gamma(3)$. Then, the quotient R/γ^* is a graded ring and $R_0 = R/\gamma^*$ and $R_1 = \gamma^*(0)$.

Definition 8.2. Let $R = \oplus_{g \in G} R_g$ be a graded hyperring. A subhyperring S of R is called a *graded subhyperring* of R if $S = \oplus_{g \in G}(S \cap R_g)$. Equivalently, S is graded if for every element $f \in S$, all the homogeneous components of f (as an element of R) are in S.

Definition 8.3. Let I be a hyperideal of a graded hyperring R. Then, I is a *graded hyperideal* if $I = \oplus_{g \in G}(I \cap R_g)$. For all $a \in I$ and for some $r_g \in h(R)$ that $a \in \sum_{g \in G} r_g$, then $r_g \in I \cap R_g$, for all $g \in G$.

Lemma 8.2. *Let X be a non-empty subset of a commutative graded hyperring R. Let $\{A_i\}_{i \in I}$ be the family of all graded hyperideals in R which contains X. Then, $\cap_{i \in I} A_i$ is also a graded hyperideal which contains X.*

Proof. We have that $\cap_{i \in I} A_i$ is a hyperideal of R containing X. Let $r \in \cap_{i \in I} A_i$; hence, $r \in \sum_{g \in G} r_g$, where $r_g \in R_g$. So, we have $r \in A_i$, for all $i \in I$. Therefore, for all $g \in G, r_g \in A_i$ since A_i is a graded hyperideal. Hence, for all $g \in G, r_g \in (\cap_{i \in I} A_i) \cap R_g$, and so $\cap_{i \in I} A_i$ is a graded hyperideal. The graded hyperideal $\cap_{i \in I} A_i$ is called a graded hyperideal generated by X and is denoted by $\langle X \rangle$. If $X = \{x_{g_1}, \ldots, x_{g_n}\}$, then $\langle X \rangle$ is said to be finitely generated by $\langle x_{g_1}, \ldots, x_{g_n} \rangle = \{t | t \in \sum_{i=1}^{i=n} r_i x_{g_i}, r_i \in R\}$. A graded hyperideal generated by a single homogeneous element $x_g \in h(R)$ is called a *principal* and denoted by $\langle x_g \rangle$. Let $R = \oplus_{g \in G} R_g$ be a graded hyperring and I be a graded hyperideal of R. Then, the quotient hyperring $(R/I, \oplus, \circ)$, where $(a + I) \circ (b + I) = ab + I$, for all $a, b \in R$ and $(a+I) \oplus (b+I) = \{t+I \mid t \in a+b\}$, for all $a, b \in R$, is also a graded hyperring with $R/I = \oplus_{g \in G}(R/I)_g$, where $(R/I)_g = (R_g + I)/I$. $\square$

Definition 8.4. If $P \neq R$ is a graded hyperideal of a graded hyperring R. Then, P is called a *graded prime hyperideal* of R if $a_g b_h \in P$ implies that $a_g \in P$ or $b_h \in P$, for $a_g, b_h \in h(R)$.

Definition 8.5. A graded hyperring $R = \oplus_{g \in G} R_g$ is a *graded hyperintegral domain* if $a_g b_h = 0$ implies that $a_g = 0$ or $b_h = 0$, for $a_g, b_h \in h(R)$.

Theorem 8.1. *Let $P \neq R$ be a graded hyperideal of a commutative graded hyperring R with identity 1. The,n P is graded prime if and only if R/P is a graded hyperintegral domain.*

Proof. Let P be a graded prime hyperideal of R. Let $(a_g + P) \circ (b_h + P) = 0 + P = P$, so $a_g b_h + P = P$. Then, $a_g b_h \in P$ because $a_g b_h + P = \cup\{a_g b_h + x \mid x \in P\}$, and for all $t \in a_g b_h + P$, there exists $y \in P$ such that $t \in a_g b_h + y$, so $a_g b_h \in t - y \subseteq P$ since $t, y \in P$, and P is a graded hyperideal. Hence, $a_g \in P$ or $b_h \in P$ since P is a graded prime hyperideal; therefore, $a_g + P = P$ or $a_h + P = P$, as needed. Conversely, let $a_g b_h \in P$ for some $a_g, b_h \in h(R)$. Hence, $a_g b_h + P = (a_g + P) \circ (b_h + P) = P$, so $a_g + P = P$ or $b_h + P = P$ since R/P is a graded hyperintegral domain. Therefore, $a_g \in P$ or $b_h \in P$. $\square$

Definition 8.6. Let R be a graded hyperring. The graded hyperideal M of R is said to be *maximal* if for every graded hyperideal J of R, $M \subseteq J \subseteq R$, it implies that $J = M$ or $J = R$.

Theorem 8.2. *Let R be a graded hyperring with identity 1. Then, every graded maximal hyperideal is graded prime.*

Proof. Let M be a graded maximal hyperideal of R. Let $a_g b_h \in M$ and $a_g \notin M$, for $a_g, b_h \in h(R)$. So, $M \subseteq \langle ag \rangle + M \subseteq R$ since M is a graded maximal hyperideal, then $R = \langle ag \rangle + M$. As $1 \in R$, we set $1 \in ra_g + x$ for some $r \in R$ and $x \in M$. Hence, $b_h \in rb_h a_g + b_h x \subseteq M$ since M is a graded hyperideal. Therefore, $b_h \in M$, as needed. $\square$

Definition 8.7. Let R be a commutative graded hyperring with identity. R is called a *graded hyperfield* if its non-zero homogeneous elements are invertible.

Theorem 8.3. *If R is a graded hyperring with identity, then R has a graded maximal hyperideal.*

Proof. The proof is similar to nongraded hyperrings. $\square$

Theorem 8.4. *Let R be a commutative graded hyperring with identity and $M \neq R$ be a graded maximal hyperideal of R. Then, M is a graded maximal hyperideal if and only if R/M is a graded hyperfield.*

Proof. Let $M \neq a_g + M$ be a homogeneous element of R/M, so $a_g \notin M$. Thus, $M \subseteq \langle a_g \rangle + M \subseteq R$; therefore, $\langle a_g \rangle + M = R$ since M is a graded maximal hyperideal. Hence, $1 \in ra_g + x$ for some $r \in R$ and $x \in M$. So, $1 + M = ra_g + x + M$; hence, $1 + M = ra_g + M = (r + M) \circ (a_g + M)$, and so $a_g + M$ is unite. Conversely, let $M \subset L \subseteq R$, where L is a graded hyperideal of R. So, there exists $x \in L$ such that $x \notin M$. Hence, we can write $x \in \sum_{g \in G} r_g$, where $r_g \in L \cap R_g$. Therefore, there exists $g \in G$ such that $r_g \notin M$ because if for all $g \in G, r_g \in M$, then $\sum_{g \in G} r_g \subseteq M$ since M is a graded hyperideal, and so $x \in M$, which is a contradiction. Hence, $r_g + M \neq 0_{R/M}$, then $(r_g + M) \circ (x + M) = r_g x + M = 1 + M$ for some $x + M \in R/M$. So, $1 - r_g x \subseteq M \subseteq L$, and since $r_g x \in L$, we have $1 \in L$ since L is a graded hyperideal. Then, $L = R$, as needed. $\square$

Let R be a graded hyperring and $S \subseteq h(R)$ be a multiplicative close subset of R. Then, the hyperring of the fractions $S^{-1}R$ is a graded hyperring, which is called the *graded hyperring of fractions*. Indeed, $S^{-1}R = \oplus_{g \in G}(S^{-1}R)_g$, where $(S^{-1}R)_g = \{r/s \mid r \in R, s \in S, \ g = (degs)^{-1}(degr)\}$.

Theorem 8.5. *A graded hyperideal $P \neq R$ in a commutative graded hyperring R with identity is a graded prime hyperideal if and only if $h(R) - P$ is a multiplicative close subset in R.*

Proof. Let P be a graded prime hyperideal of R. Assume that $x_g, y_h \in h(R) - P$, so $x_g y_h \notin P$ since P is graded prime. Therefore, $x_g y_h \in h(R) - P$, and so $h(R) - P$ is a multiplicative close subset of R. Conversely, $x_g y_h \in P$ and $x_g \notin P$, where $x_g, y_h \in h(R)$. So, $y_h \in P$

because if $y_h \notin P$, then $x_g y_h \in h(R) - P$, which is a contradiction. Hence, P is graded prime. $\qquad\square$

Definition 8.8. Let I be a graded hyperideal in a commutative graded hyperring R with identity. The *graded radical* of I (in short, $Grad(I)$) is the set of all $x \in R$ such that for each $g \in G$, there exists $n_g > 0$, with $x_{n_g} g \in I$. Note that if r is a homogeneous element of R, then $r \in Grad(I)$ if and only if $r^n \in I$ for some positive integer n.

Definition 8.9. A graded hyperideal $Q \neq R$ in a commutative graded hyperring R is said to be *graded primary* if $a_g b_h \in Q$. Then, $a_g \in Q$ or $b_h \in Grad(Q)$. for $a_g, b_h \in h(R)$.

Theorem 8.6. *If Q is a graded primary hyperideal in a commutative graded hyperring with identity, then $Grad(Q)$ is a graded prime.*

Proof. Let $a_g b_h \in Grad(Q)$ and $a_g \notin Grad(Q)$. So, there exists a positive integer n such that $(a_g b_h)^n \in Q$ and $a_g^n \notin Q$. Hence, $b_h^n \in Q$, and so $b_h \in Grad(Q)$, as required. $\qquad\square$

8.2 Hyperring Extensions of Krasner Hyperfield

The results of this section were obtained by Connes and Consani [16] in order to study the hyperring of adéle classes. Thus, hyperring theory, due to Krasner, represents a perfect framework for the study of the algebraic structure of the Adéle class space $\mathbb{H}_{\mathbb{K}}$ of a global field $\mathbb{K}$.

Let R be a ring and G be a multiplicative subgroup of the group $R^\times$ of the invertible elements of R. In the following, we recall the construction given by Krasner.

Let R be a commutative ring and $G \subset R^\times$ be a subgroup of its multiplicative group. Then, the following operations define a hyperring structure on the set R/G of orbits for the action of G on R by multiplication:

- Hyperaddition $x + y = (xG + yG)/G$, for all $x, y \in R/G$.
- Multiplication $xG \cdot yG = xyG$, for all $x, y \in R/G$.

Moreover, for any $x_i \in R/G$, one has

$$\sum x_i = \left(\sum x_i G\right)\big/G.$$

In particular, one can start with a field K and consider the hyperring $K/K^\times$. If $|K| > 2$, then $|K/K^\times| = 2$, i.e., $K/K^\times$ is the Krasner hyperfield **K**.

Definition 8.10. Let **K** be the hyperfield $(\{0,1\}, +, \cdot)$ with additive neutral element 0, satisfying the condition $1 + 1 = \{0, 1\}$, and with the usual multiplication having identity 1. Let S be the hyperfield $S = \{-1, 0, 1\}$, with the hyperaddition given by the rule of signs

$$1 + 1 = 1, \quad -1 - 1 = -1, \quad 1 - 1 = -1 + 1 = \{-1, 0, 1\}$$

and the usual multiplication also given by the rule of multiplication of signs.

In this section, we prove that a hyperring of the form R/G is a hyperring extension of K if and only if $G \cup \{0\}$ is a subfield of R.

Theorem 8.7. *In a hyperring extension R of the Krasner hyperfield K, one has $x + x = \{0, x\}$, for all $x \in R$; moreover, $a \in a + b \Leftrightarrow b \in \{0, a\}$. In particular, there is no hyperfield extension of **K** of cardinality 3 or 4.*

Proof. Since $1 + 1 = \{0, 1\}$, one gets $x + x = \{0, x\}$ using distributivity. Assume that $a \in a + b$ in R. Then, since $a + a = \{0, a\}$, one has $-a = a$ so that by the reversibility condition in the definition of a hypergroup, one has $b \in a - a = \{0, a\}$. Conversely, if $b \in \{0, a\}$, it follows that $a \in a + b$. If F is a hyperfield extension of K of cardinality 3, then F contains an element $\alpha \notin \{0, 1\}$. However, one then gets a contradiction since the subset $1 + \alpha$ cannot contain 0 or 1 or α. If F is a hyperfield extension of K of cardinality 4, then let ξ_j be the three non-zero elements of F. Then, by applying the first part of this theorem, the sum $\xi_j + \xi_k$, for $j \neq k$, is forced to be the third non-zero element ξ_l of F. This contradicts the associativity of the hyperaddition for $\sum \xi_j$. $\square$

Remark 8.1. The same proof shows that in a hyperring extension R of the hyperfield S, one has

$$a \in a + b \quad \Leftrightarrow \quad b \in \{0, a, -a\}.$$

Theorem 8.8. *Let R be a commutative ring and $G \subseteq R^\times$ be a subgroup of the multiplicative group of units in R. Assume that $G \neq \{1\}$. Then, the hyperring R/G contains K as a sub-hyperfield if and only if $\{0\} \cup G$ is a subfield of R.*

Proof. To verify whether $\mathbf{K} \subseteq R/G$, it suffices to compute $1 + 1$ in R/G. By definition, $1 + 1$ is the union of all classes under the multiplicative action of G, of the elements of the form $g_1 + g_2$, for $g_j \in G$ ($j = 1, 2$). Thus, the hyperring R/G contains K as a sub-hyperfield if and only if $G + G = \{0\} \cup G$. If this equality holds, then $\{0\} \cup G$ is stable under addition. Moreover, $0 \in G + G$, whence $g_1 = -g_2$ for some $g_j \in G$; hence, $-1 = g_1 g_2^{-1} \in G$. Thus, $\{0\} \cup G$ is an additive subgroup of R. In fact, since $R^\times$ is a group, it follows that $G \cup \{0\}$ is a subfield of R.

Conversely, let $F \subseteq R$ be a subfield, and assume that F is not reduced to the finite field $\mathbb{F}_2$. Then, the multiplicative group $G = F^\times$ fulfills $G \neq \{1\}$. Moreover, $G + G \subseteq F$ and $0 \in G + G$ as $1 - 1 = 0$. Additionally, since G contains at least two distinct elements x, y, one has $x - y \neq 0$, and thus $G + G = F$. Thus, in R/G, one has $1 + 1 = \{0, 1\}$, and thus $\mathbf{K} \subseteq R/G$. $\qquad\square$

Example 8.4. There exists a hyperfield extension of $\mathbf{K}$ of cardinality 5. Let us consider the set $H = \{0, 1, \alpha, \alpha^2, \alpha^3\}$, where $\alpha^4 = 1$. The table of hyperaddition in H is given by the following table:

0	1	α	α^2	α^3
1	$\{0,1\}$	$\{\alpha^2, \alpha^3\}$	$\{\alpha, \alpha^3\}$	$\{\alpha, \alpha^2\}$
α	$\{\alpha^2, \alpha^3\}$	$\{0, \alpha\}$	$\{1, \alpha^3\}$	$\{1, \alpha^2\}$
α^2	$\{\alpha, \alpha^3\}$	$\{1, \alpha^3\}$	$\{0, \alpha^2\}$	$\{1, \alpha\}$
α^3	$\{\alpha, \alpha^2\}$	$\{1, \alpha^2\}$	$\{1, \alpha\}$	$\{0, \alpha^3\}$

This hyperfield structure is obtained, with $\alpha = 1 + \sqrt{-1}$, as the quotient of the finite field $\mathbb{F}_9$ by the multiplicative group $\mathbb{F}_3^\times = \{1, -1\}$. It follows from the previous theorem that $F = \mathbb{F}_9/\mathbb{F}_3^\times$ is a hyperfield extension of $\mathbf{K}$.

Denote by $Spec(R)$ the set of prime ideals of a hyperring R.

Theorem 8.9. *For any hyperring R, the map*

$$\varphi : Spec(R) \to Hom(R, \mathbf{K}), \quad \varphi(p) = \varphi_p$$

$$\varphi_p(x) = 0, \quad \text{for all } x \in p, \quad \varphi_p(x) = 1, \quad \text{for all } x \notin p$$

determines a natural bijection of sets.

Proof. The map $\varphi_p : R \to \mathbf{K}$ is multiplicative since the complement of a prime ideal p in R is a multiplicative set. It is compatible with the hyperaddition, and so the map φ is well defined. To define the inverse of φ, one assigns to a homomorphism of the hyperrings $\rho \in Hom(R, \mathbf{K})$ its kernel, which is a prime ideal of R that uniquely determines ρ. □

Now, in order to describe the elements of the set $Hom(R, \mathbf{S})$, for any ring R, we give the following definition:

Definition 8.11. Let R be a ring. A symmetric cone P in R is a subset $P \subseteq R$ such that:

- $0 \notin P$, $P + P \subseteq P$, $PP \subseteq P$;
- $P^c + P^c \subseteq P^c$, where P^c is the complement of P in R;
- $a \in P$ and $ab \in P$ imply $b \in P$;
- $P - P = R$.

The following theorem shows that the notion of a symmetric cone in a hyperring is equivalent to that of an element of $Hom(R, \mathbf{S})$.

Theorem 8.10. (1) *A homomorphism from a ring R to the hyperring $\mathbf{S}$ is determined by its kernel $p \in Spec(R)$ and a total order on the field of fractions of the integral domain R/p.*

(2) *A homomorphism from a ring R to the hyperring $\mathbf{S}$ is determined by a symmetric cone of R.*

Proof. (1) Let $\rho \in Hom(R, \mathbf{S})$. The kernel of ρ is unchanged by composing ρ with the absolute value map $\pi : \mathbf{S} \to \mathbf{K}$, $\pi(x) = |x|$. Thus, $ker(\rho)$ is a prime ideal $p \subseteq R$. Moreover, the map ρ descends to the quotient R/p, which is an integral domain. Let F be the field of fractions of R/p. Let $P \subseteq F$ be the set of fractions of the form $x = a/b$, where $\rho(a) = \rho(b) \neq 0$. This subset of F is well defined since $a/b = c/d$ means that $ad = bc$, and it follows that $\rho(c) = \rho(d) \neq 0$. We have $\rho(0) = 0, \rho(1) = 1$ and $\rho(-1) = -1$ since $0 \in \rho(1) + \rho(-1)$. Thus, P is stable by addition since one can assume in the computation of $a/b + c/d$ that $\rho(a) = \rho(b) = \rho(c) = \rho(d) = 1$, so that $\rho(ad + bc) = \rho(cd) = 1$. P is also multiplicative. Moreover, for $x \in F$, $x \neq 0$, one has $x, -x \in P$, for some choice of the sign. Thus, F is an ordered field and ρ is the composition of the canonical morphism $R \to F$ with the map $F \to F/F_+^\times \sim \mathbf{S}$.

Conversely, if an order on the field of fractions of the integral domain R/p is given, one can use the natural identification $F/F_+^\times \sim \mathbf{S}$ to obtain the morphism ρ.

(2) One can check that, given a symmetric cone $P \subseteq R$, the following formula defines an element: $\rho \in Hom(R, \mathbf{S})$:

$$\rho(x) = \begin{cases} 1 & \text{if } x \in P \\ -1 & \text{if } x \in -P \\ 0 & \text{otherwise} \end{cases}$$

Not that if $\rho \in Hom(R, \mathbf{S})$, then $P = \rho^{-1}(1)$ is a symmetric cone. □

Chapter 9

Differential Krasner Hyperring

9.1 Derivations in Krasner Hyperrings

The concept of derivation on rings has been introduced by Posner [84]. Differential rings, differential fields, and differential algebras are rings, fields, and algebras, respectively, equipped with a derivation, which is a unary function that is linear and satisfies the Leibniz product rule. In Ref. [12], Chvalina and Chvalinova gave a construction of hyperstructures determined by quasi-orders defined by means of derivation operators on differential rings. Asokkumar [7] and Kamali Ardekani and Davvaz [42, 43] applied the notion of derivative to Krasner hyperrings. In this section, we present the results obtained by Asokkumar [7] and Kamali Ardekani and Davvaz [42].

Definition 9.1. Let R be a Krasner hyperring. A map $d : R \to R$ is said to be a *inclusion derivation* of R if d satisfies the following conditions:

(1) $d(x + y) \subseteq d(x) + d(y)$;
(2) $d(xy) \in d(x)y + xd(y)$, for all $x, y \in R$.

The hyperring R endowed with a derivation d is called a *weak differential hyperring*. If the map d is such that $d(x + y) = d(x) + d(y)$, for all $x, y \in R$, and satisfies condition (2). Then, d is called a *derivation* of R. In this case, the hyperring is called a *differential hyperring*. Since the hyperaddition of the hyperring R is commutative, condition (2) of the derivation is equivalent to $d(xy) \in xd(y) + d(x)y$, for all $x, y \in R$.

By the above definition for every derivation d on hyperring R, we have $d(0) = 0$ and $d(-x) = -d(x)$, for all $x \in R$.

Example 9.1. Let R be a Krasner hyperring and

$$M(R) = \left\{ \begin{pmatrix} a & b \\ 0 & 0 \end{pmatrix} \middle| a, b \in R \right\}.$$

A hyperaddition $\oplus$ is defined on $M(R)$ by

$$\begin{pmatrix} u & b \\ 0 & 0 \end{pmatrix} \oplus \begin{pmatrix} c & d \\ 0 & 0 \end{pmatrix} = \left\{ \begin{pmatrix} x & y \\ 0 & 0 \end{pmatrix} \middle| x \in a + c, y \in b + d \right\}.$$

Then, $(M(R), \oplus)$ is a canonical hypergroup. The matrix $\begin{pmatrix} 0 & 0 \\ 0 & 0 \end{pmatrix}$ is the additive identity of $M(R)$. For each matrix, $\begin{pmatrix} a & b \\ 0 & 0 \end{pmatrix}$ of $M(R)$, there exists a unique matrix $\begin{pmatrix} -a & -b \\ 0 & 0 \end{pmatrix}$ such that $\begin{pmatrix} 0 & 0 \\ 0 & 0 \end{pmatrix} \in \begin{pmatrix} a & b \\ 0 & 0 \end{pmatrix} \oplus \begin{pmatrix} -a & -b \\ 0 & 0 \end{pmatrix}$.

Now, define a multiplication on $M(R)$ by $\begin{pmatrix} a & b \\ 0 & 0 \end{pmatrix} \otimes \begin{pmatrix} c & d \\ 0 & 0 \end{pmatrix} = \begin{pmatrix} ac & ad \\ 0 & 0 \end{pmatrix}$, for all $\begin{pmatrix} a & b \\ 0 & 0 \end{pmatrix}, \begin{pmatrix} c & d \\ 0 & 0 \end{pmatrix} \in M(R)$. The multiplication is well defined and associative. Therefore, $(M(R), \otimes)$ is a semigroup. Moreover, for $\begin{pmatrix} a & b \\ 0 & 0 \end{pmatrix}, \begin{pmatrix} x & y \\ 0 & 0 \end{pmatrix}, \begin{pmatrix} p & q \\ 0 & 0 \end{pmatrix} \in M(R)$, we have

$$\begin{pmatrix} a & b \\ 0 & 0 \end{pmatrix} \otimes \left[\begin{pmatrix} x & y \\ 0 & 0 \end{pmatrix} \oplus \begin{pmatrix} p & q \\ 0 & 0 \end{pmatrix} \right]$$

$$= \left\{ \begin{pmatrix} ar & as \\ 0 & 0 \end{pmatrix} \middle| r \in x + p, \ s \in y + q \right\}$$

and

$$\left[\begin{pmatrix} a & b \\ 0 & 0 \end{pmatrix} \otimes \begin{pmatrix} x & y \\ 0 & 0 \end{pmatrix} \right] \oplus \left[\begin{pmatrix} a & b \\ 0 & 0 \end{pmatrix} \otimes \begin{pmatrix} p & q \\ 0 & 0 \end{pmatrix} \right]$$

$$= \begin{pmatrix} ax & ay \\ 0 & 0 \end{pmatrix} \oplus \begin{pmatrix} ap & aq \\ 0 & 0 \end{pmatrix}$$

$$= \left\{ \begin{pmatrix} l & m \\ 0 & 0 \end{pmatrix} \middle| l \in ax + ap, \; m \in ay + aq \right\},$$

whence it follows that the left distributivity holds in $M(R)$. Similarly, the right distributive law in $M(R)$ can be checked. Thus, $M(R)$ is a Krasner hyperring.

Now, define a function d on $M(R)$ by $d\left(\begin{pmatrix} a & b \\ 0 & 0 \end{pmatrix} \right) = \begin{pmatrix} 0 & b \\ 0 & 0 \end{pmatrix}$.
We check that d is a derivation in $M(R)$.

Indeed, for $\begin{pmatrix} a & b \\ 0 & 0 \end{pmatrix}, \begin{pmatrix} c & d \\ 0 & 0 \end{pmatrix} \in M(R)$, we have

$$d\left(\begin{pmatrix} a & b \\ 0 & 0 \end{pmatrix} \oplus \begin{pmatrix} c & d \\ 0 & 0 \end{pmatrix} \right) = d\left(\begin{pmatrix} a & b \\ 0 & 0 \end{pmatrix} \right) \oplus d\left(\begin{pmatrix} c & d \\ 0 & 0 \end{pmatrix} \right)$$

$$= \left\{ \begin{pmatrix} 0 & x \\ 0 & 0 \end{pmatrix} \middle| x \in b + d \right\}.$$

Also, we have

$$d\left(\begin{pmatrix} a & b \\ 0 & 0 \end{pmatrix} \otimes \begin{pmatrix} c & d \\ 0 & 0 \end{pmatrix} \right) = \begin{pmatrix} 0 & ad \\ 0 & 0 \end{pmatrix}$$

$$= \left[d\left(\begin{pmatrix} a & b \\ 0 & 0 \end{pmatrix} \right) \otimes \begin{pmatrix} c & d \\ 0 & 0 \end{pmatrix} \right]$$

$$\oplus \left[\begin{pmatrix} a & b \\ 0 & 0 \end{pmatrix} \otimes d\left(\begin{pmatrix} c & d \\ 0 & 0 \end{pmatrix} \right) \right].$$

Example 9.2. Let $\mathbb{Q}^+ = \{x \in \mathbb{Q} \mid x \geq 0\}$, where $\mathbb{Q}$ is the set of rational numbers. The binary hyperoperation $+$ and the binary operation $\cdot$ are defined as follows:

$$x + x = \{y \in \mathbb{Q}^+ \mid y \leq x\}, \quad \text{for all } x \in \mathbb{Q}^+,$$
$$x + y = \max\{x, y\}, \quad \text{for all } x, y \in \mathbb{Q}^+, \; x \neq y,$$
$$x \cdot y = xy, \quad \text{for all } x, y \in \mathbb{Q}^+.$$

Then, $(\mathbb{Q}^+, +, \cdot)$ is a Krasner hyperring. The function $d : \mathbb{Q}^+ \to \mathbb{Q}^+$ defined by $d(x) = x$, for all $x \in \mathbb{Q}^+$, is a derivation since for all

$x, y \in \mathbb{Q}^+$,

$d(x) + d(y) = d(x + y)$,

$d(x \cdot y) = xy \in \{t \in \mathbb{Q}^+ \mid t \le xy\} = xy + xy = d(x) \cdot y + x \cdot d(y)$.

Example 9.3. Let $(G, \cdot, e)$ be a finite group with m elements, $m > 3$, and define a hyperaddition and a multiplication on $H = G \cup \{0\}$ by

$$x + 0 = 0 + x = \{x\}, \quad \text{for all } x \in H,$$

$$x + x = \{x, 0\}, \quad \text{for all } x \in G,$$

$$x + y = y + x = H \setminus \{x, y\}, \quad \text{for all } x, y \in G, \ x \ne y,$$

$$x \odot 0 = 0 \odot x = 0, \quad \text{for all } x \in H,$$

$$x \odot y = x \cdot y, \quad \text{for all } x, y \in G.$$

Then, $(H, +, \odot)$ is a Krasner hyperring. The function $d : H \to H$ defined by $d(x) = x$, for all $x \in H$, is a derivation since

$$d(x) + d(y) = d(x + y), \quad \text{for all } x, y \in H,$$

$$d(x \odot 0) = d(0) = 0 \in \{0\} = 0 + 0 = d(x) \odot 0 + x \odot d(0),$$
$$\text{for all } x \in H,$$

$$d(x \odot y) = d(x \cdot y) = x \cdot y \in \{x \cdot y, 0\} = x \cdot y + x \cdot y$$
$$= d(x) \odot y + x \odot d(y), \quad \text{for all } x, y \in G.$$

Example 9.4. Consider Example 9.3, and let $(G, \cdot, e)$ be an abelian group which has no elements of order 2. Then, the function $d_1 : H \to H$ defined by

$$d_1(x) = \begin{cases} 0 & x = 0 \\ x^{-1} & \text{for all } x \in G \end{cases}$$

is a derivation function since

$$d_1(x + 0) = d_1(0 + x) = \{d_1(x)\} = \{x^{-1}\} = d_1(x) + d_1(0)$$
$$= d_1(0) + d_1(x), \quad \text{for all } x \in H;$$

$$d_1(x + x) = \{d_1(x), d_1(0)\} = \{x^{-1}, 0\} = x^{-1} + x^{-1} = d_1(x) + d_1(x),$$
$$\text{for all } x \in G;$$

$$d_1(x + y) = H \setminus \{x^{-1}, y^{-1}\} = x^{-1} + y^{-1} = d_1(x) + d_1(y),$$
$$\text{for all } x, y \in G, x \ne y.$$

Hence, the first condition of the definition of derivation is valid. Also, we have

$$d_1(x \odot 0) = d_1(0) = 0 \in \{0\} = x^{-1} \odot 0 + x \odot 0$$
$$= d_1(x) \odot 0 + x \odot d_1(0), \quad \text{for all } x \in H;$$
$$d_1(x) \odot y + x \odot d_1(y) = x^{-1} \odot y + x \odot y^{-1} = x^{-1} \cdot y + x \cdot y^{-1},$$
$$\text{for all } x, y \in G.$$

By the above relations, in order to prove the second condition of the definition of derivation, it is enough to show that $d(xy) = x^{-1}y^{-1} \in x^{-1} \cdot y + x \cdot y^{-1}$, for all $x, y \in G$. We have $x^{-1} \cdot y + x \cdot y^{-1} = H \setminus \{x^{-1} \cdot y, x \cdot y^{-1}\}$ since G has no elements of order 2. If $x^{-1} \cdot y^{-1} = x^{-1} \cdot y$, then $y = y^{-1}$, and if $x^{-1} \cdot y^{-1} = x \cdot y^{-1}$, then $x = x^{-1}$. Hence, $x^{-1}y^{-1} \notin \{x^{-1} \cdot y, x \cdot y^{-1}\}$ since G has no elements of order 2. Therefore, $x^{-1}y^{-1} \in H \setminus \{x^{-1} \cdot y, x \cdot y^{-1}\} = x^{-1} \cdot y + x \cdot y^{-1}$.

The following example shows that the identity function is not always a derivation.

Example 9.5. Let $(G, \cdot, e)$ be a group. Define a hyperaddition and a multiplication on $H = G \cup \{0\}$ as follows:

$$x + 0 = 0 + x = \{x\}, \quad \text{for all } x \in H,$$
$$x + x = H \setminus \{x\}, \quad \text{for all } x \in G,$$
$$x + y = y + x = \{x, y\}, \quad \text{for all } x, y \in G, \ x \neq y,$$
$$x \odot 0 = 0 \odot x = 0, \quad \text{for all } x \in H,$$
$$x \odot y = x \cdot y, \quad \text{for all } x, y \in G.$$

Then, $(H, +, \odot)$ is a Krasner hyperring. The function $d : H \to H$ defined by $d(x) = x$, for all $x \in H$, is not a derivation since $d(x) \odot y + x \odot d(y) = x \cdot y + x \cdot y = H - \{x \cdot y\}$, for all $x, y \in G$. So, $d(x \odot y) = x \cdot y \notin d(x) \odot y + x \odot d(y)$.

Definition 9.2. Let d be a nontrivial weak derivation (resp., derivation) in a Krasner hyperring R. A hyperideal I of R is said to be a *weak differential hyperideal* (resp., *differential hyperideal*) of R if $d(I) \subseteq I$.

Let S be a non-empty subset of a Krasner hyperring R. The set

$$Ann(S) = \{x \in R \mid xS = 0\}$$

is called the *annihilator* of S in R.

Example 9.6. Consider the Krasner hyperring $R = \{0, a, b, c\}$ with the hyperaddition and the multiplication defined as follows:

$\oplus$	0	a	b	c	$\odot$	0	a	b	c
0	0	a	b	c	0	0	0	0	0
a	a	$\{0,b\}$	$\{a,c\}$	b	a	0	a	b	c
b	b	$\{a,c\}$	$\{0,b\}$	a	b	0	b	b	0
c	c	b	a	0	c	0	c	0	c

The map $d : R \to R$ defined by $d(0) = 0, d(a) = b, d(b) = b, d(c) = 0$ is a derivation in R, and $Ann(0, c) = \{0, b\}$ is a hyperideal of R.

Since $d(Ann(0, c)) = d(\{0, b\}) = \{0, b\} = (Ann(0, c))$, we see that $Ann(0, c) = \{0, b\}$ is a differential hyperideal of R.

Let H be a canonical hypergroup and N be a canonical subhypergroup of H. For any two elements $a, b \in H$, we define $a \sim b$ if $a \in b + N$. Then, $\sim$ is an equivalence relation on H. We denote the equivalence class of an element $x \in H$ by $\bar{x}$. We have $\bar{x} = x + N$. We denote the quotient set $\{\bar{x} \mid x \in H\}$ by H/N. If we define $\bar{x} \oplus \bar{y} = \{\bar{z} \mid z \in x + y\}$, for all $\bar{x}, \bar{y} \in H/N$, then H/N is a canonical hypergroup.

Let R be a Krasner hyperring and I be a hyperideal of R. Since I is a canonical subhypergroup of R, $R/I = \{\bar{x} \mid x \in R\}$ is a canonical hypergroup under the hyperaddition defined above. Now, we define $\bar{x} \otimes \bar{y} = \overline{xy} = xy + I$, for all $\bar{x}, \bar{y} \in R/I$, then R/I is a Krasner hyperring.

Theorem 9.1. *Let R be a differential hyperring. Then, for all differential hyperideal I of R, the factor hyperring R/I is a differential hyperring.*

Proof. Let d be a derivation of R. Let us define a map $D : R/I \to R/I$ by $D(a + I) = d(a) + I$, for all $a + I \in R/I$. If $a, b \in R$ such that $a + I = b + I$, then $a \in b + I$. Since $d(I) \subseteq I$, we obtain $d(a) \in d(b) + I$, whence $d(a) + I = d(b) + I$. Hence, $D(a + I) = D(b + I)$. Therefore, D is a well-defined map.

Let $r + I, s + I \in R/I$. Now, $D((r + I) + (s + I)) = D(\{x + I \mid x \in r + s\}) = \{d(x) + I \mid x \in r + s\}$. Further, $D(r + I) + D(s + I) = (d(r) + I) + (d(s) + I) = \{x + I \mid x \in d(r) + d(s)\} = \{x + I \mid x \in d(r + s)\} = \{d(y) + I \mid y \in r + s\}$. Since d is a derivation of R, we get $D((r + I) + (s + I)) = D(r + I) + D(s + I)$. Also, we have $D((r + I)(s + I)) = D(rs + I) = d(rs) + I$. However, $D(r + I)(s + I) + (r + I)D(s + I) = (d(r) + I)(s + I) + (r + I)(d(s) + I) = \{x + I \mid x \in d(r)s + rd(s)\}$. Since $d(rs) \in d(r)s + rd(s)$, we get $d(rs) + I \in \{x + I \mid x \in d(r)s + rd(s)\}$. Hence, $D((r + I)(s + I)) \in D(r + I)(s + I) + (r + I)D(s + I)$. Thus, D is a derivation of R/I. $\qquad\square$

Theorem 9.2. *Let R be a differential hyperring and I a differential hyperideal of R. Then, there exists a one-to-one correspondence between the set of all differential hyperideals of R containing I and the the set of all differential hyperideals of R/I.*

Proof. Let A be the set of all differential hyperideals of R containing I, and let F be the set of all differential hyperideals of R/I. Define a map $f : A \to F$ by $f(J) = J/I$, where J is a differential hyperideal of R containing I. Since J is a differential hyperideal of R, it follows that J/I is a differential hyperideal of R/I. Therefore, the map f is well defined. Let J, K be two differential hyperideals of R containing I such that $f(J) = f(K)$. Then, $J/I = K/I$. Now, for $x \in J$, we have $x + I \in J/I = K/I$. Therefore, $x + I = y + I$, for some $y in K$. Hence, $x \in y + I \subseteq K$, and $J \subseteq K$. Similarly, we can prove that $K \subseteq J$, and hence $K = J$. Therefore, the function f is one to one. Clearly, the map f is onto. Hence, f is a bijective map. $\qquad\square$

Lemma 9.1. *Let d be a derivation on a Krasner hyperring R. For all $x, y \in R$, define $x^0 y = y$ and $d^0(x) = x$. Then, for all $n \in \mathbb{N}$ and $x, y \in R$:*

(1) *If R is commutative, then $d(x^n) \in n(x^{n-1}.d(x))$.*
(2) *$d^n(xy) \in \sum_{i=0}^{n} \binom{n}{i} d^{n-i}(x)d^i(y)$, where d^n denotes the derivation of order n.*

Proof. (1) The proof follows easily by induction.
(2) It is trivial that the statement is valid for $n = 1$. Now, let the statement be valid for $n = k - 1$ (induction hypothesis).

We have

$$d^k(xy) = d(d^{k-1}(xy)) \in d\left(\sum_{i=0}^{k-1} \binom{k-1}{i} d^{k-i-1}(x)d^i(y)\right)$$

$$\subseteq \sum_{i=0}^{k-1} \binom{k-1}{i}\left(d^{k-i}(x)d^i(y) + d^{k-i-1}(x)d^{i+1}(y)\right)$$

$$= \sum_{i=0}^{k} \binom{k}{i} d^{k-i}(x)d^i(y).$$

□

Lemma 9.2. *Let R be a Krasner hyperring and $[x, y]$ denote the set $xy - yx$, for all $x, y \in R$. Then, for all $x, y, z \in R$, we have:*

(1) $[x + y, z] = [x, z] + [y, z]$;
(2) $[xy, z] \subseteq x[y, z] + [x, z]y$;
(3) *if* $x \in Z(R)$, *then* $[xy, z] = x[y, z]$;
(4) *if* d *is a derivation of* R, *then* $d[x, y] \subseteq [d(x), y] + [x, d(y)]$.

Proof. For $x, y, z \in R$,

(1) $[x+y, z] = (x+y)z - z(x+y) = xz - zx + yz - zy = [x, z] + [y, z]$.
(2) $\quad [xy, z] = xyz - zxy \subseteq xyz - xzy + xzy - zxy = x(yz - zy)$
$\qquad\qquad + (xz - zx)y$
$\qquad = x[y, z] + [x, z]y$.

(3) If $x \in Z(R)$, then we have

$$[xy, z] = xyz - zxy = xyz - xzy = x(yz - zy) = x[y, z].$$

(4) Suppose that d is a derivation of R. Then,

$$d[x, y] = d(xy - yx) = d(xy) - d(yx)$$
$$\subseteq d(x)y + xd(y) - d(y)x - yd(x)$$
$$= d(x)y - yd(x) + xd(y) - d(y)x = [d(x), y] + [x, d(y)].$$

□

Theorem 9.3. *Let d be a derivation on a Krasner hyperring R and n be the smallest natural number such that $d^n(R) = 0$. Then, for all $y \in R$, $d(y) = 0$ or there is $0 < k < n$ such that $0 \in n(d^{n-1}(x_0)d^k(y))$, where $0 \neq x_0 \in R$ is a fixed element.*

Proof. Suppose that n is the smallest natural number such that $d^n(R) = 0$. Then, $d^{n-1}(R) \neq 0$. So, there is $0 \neq x_0 \in R$ such that $d^{n-1}(x_0) \neq 0$. Let $d(y) \neq 0$, where $y \in R$. Then, there is $0 < k < n$ such that $d^k(y) \neq 0$ and $d^{k+1}(y) = 0$. By Lemma 9.1, we have

$$0 = d^n(x_0 d^{k-1}(y)) \in \sum_{i=0}^{n} \binom{n}{i} d^{n-i}(x_0) d^{k+i-1}(y)$$

$$= d^n(x_0) d^{k-1}(y) + n(d^{n-1}(x_0) d^k(y))$$

$$+ \sum_{i=0}^{n-2} \binom{n}{i+2} (d^{n-i-2}(x_0) d^{k+i+1}(y))$$

$$= n(d^{n-1}(x_0) d^k(y)). \qquad \square$$

Theorem 9.4. *Let d be a good homomorphism and derivation on a Krasner hyperring R. Then, for all $x, y \in R$,*

$$d(x)yd(x) \in (d(x)yx + xd(yx) - xd(yx))$$

$$\cap (xyd(x) + d(xy)x - d(xy)x).$$

Proof. We have, for all $x, y \in R$,

$$d(x)d(y) = d(xy) \in d(x)y + xd(y). \qquad (9.1)$$

Replace y by yx, in (9.1),

$$d(xy)d(x) = d(x)d(y)d(x) = d(x)d(yx) = d(xyx) \in d(x)yx + xd(yx).$$

On the other hand, $d(xy)d(x) \in d(x)yd(x) + xd(y)d(x) = d(x)yd(x) + xd(yx)$. So, $d(x)yd(x) \in d(x)yx + xd(yx) - xd(yx)$. Now, we replace x by yx in $(*)$,

$$d(y)d(xy) = d(y)d(x)d(y) = d(yx)d(y) = d(yxy) \in d(yx)y + yxd(y).$$

On the other hand, $d(y)d(xy) \in d(y)d(x)y + d(y)xd(y) = d(yx)y + d(y)xd(y)$. So, $d(y)xd(y) \in yxd(y) + d(yx)y - d(yx)y$. By changing the role of x and y, we have $d(x)yd(x) \in xyd(x) + d(xy)x - d(xy)x$. This completes the proof. $\qquad \square$

9.2   Derivation of Prime Hyperrings

In this section, we study the concept of derivation on prime hyperrings. These results were obtained by Kamali Ardekani and Davvaz [42].

Definition 9.3. A Krasner hyperring R is called a *prime* if $xRy = 0$ implies that either $x = 0$ or $y = 0$. Also, R is called a *semiprime* if $xRx = 0$ implies that $x = 0$. Obviously, every prime hyperring is a semiprime hyperring; however, the converse is not always true.

Example 9.7. Every hyperdomain is prime.

Example 9.8. All of the Krasner hyperrings in Examples 9.2, 9.15, and 9.16 are prime and semiprime hyperrings.

Example 9.9. Let M be the Krasner hyperring defined in Example 9.1. The hyperring $(M, \oplus, \odot)$ is not a semiprime hyperring since for all $x, y \in R$ and $0 \neq b \in R$, we have

$$\begin{pmatrix} 0 & b \\ 0 & 0 \end{pmatrix} \odot \begin{pmatrix} x & y \\ 0 & 0 \end{pmatrix} \odot \begin{pmatrix} 0 & b \\ 0 & 0 \end{pmatrix} = \bar{0}, \text{ but } \begin{pmatrix} 0 & b \\ 0 & 0 \end{pmatrix} \neq \bar{0}.$$

Put $M' = \left\{ \begin{pmatrix} a & a \\ 0 & 0 \end{pmatrix} \middle| a \in R \right\}$. Then, $(M', \oplus, \odot)$ is a prime (resp., semiprime) hyperring if and only if R is a prime (resp., semiprime) hyperring.

The following example shows that a semiprime hyperring is not a prime hyperring, in general.

Example 9.10. Let $R = \{e, a, b, c, d, f\}$. Consider the following tables:

+	e	a	b	c	d	f
e	e	a	b	c	d	f
a	a	a	$\{e,a,b\}$	d	d	$\{c,d,f\}$
b	b	$\{e,a,b\}$	b	f	$\{c,d,f\}$	f
c	c	d	f	e	a	b
d	d	d	$\{c,d,f\}$	a	a	$\{e,a,b\}$
f	f	$\{c,d,f\}$	f	b	$\{e,a,b\}$	b

·	e	a	b	c	d	f
e	e	e	e	e	e	e
a	e	a	b	e	a	b
b	e	a	b	e	a	b
c	e	e	e	c	c	c
d	e	a	b	c	d	f
f	e	a	b	c	d	f

It is easy to check that $(R, +, \cdot)$ is a semiprime hyperring. But $(R, +, \cdot)$ is not a prime hyperring since $aRc = e$ and $a, c \neq e$.

Example 9.11. Let $R = \{e, a, b, c\}$. Consider the following tables:

+	e	a	b	c
e	e	a	b	c
a	a	$\{e, a\}$	c	$\{b, c\}$
b	b	c	$\{e, b\}$	$\{a, c\}$
c	c	$\{b, c\}$	$\{a, c\}$	R

·	e	a	b	c
e	e	e	e	e
a	e	e	e	e
b	e	a	b	c
c	e	a	b	c

Then $(R, +, \cdot)$ is a Krasner hyperring, while R is not semiprime since $aRa = e$ but $a \neq e$.

Lemma 9.3. *Let I be a non-zero hyperideal on a prime hyperring R. Then, for all $x, y \in R$:*

(1) *if $Ix = 0$ or $xI = 0$, then $x = 0$;*
(2) *if $xIy = 0$, then $x = 0$ or $y = 0$;*
(3) *if $x \in Z(R)$ and $xy = 0$, then $x = 0$ or $y = 0$;*
(4) *if $x \in R$ such that $[I, x] = 0$, then $x \in Z(R)$;*
(5) *if $x \in Z(R)$ and $xy \in Z(R)$, then $x = 0$ or $y \in Z(R)$.*

Proof. (1) Suppose that $Ix = 0$. Then, $uRx \subseteq Ix = \{0\}$, for all $u \in I$. So, $x = 0$ since R is prime and $I \neq 0$. In the case of $xI = 0$, the proof is similar.

(2) Suppose that $xIy = 0$, then $xIRy \subseteq xIy = \{0\}$. Thus, $xIRy = 0$. Hence, $xI = 0$ or $y = 0$ since R is prime. So, by (1), $x = 0$ or $y = 0$.

(3) Suppose that $x \in Z(R)$ and $xy = 0$. Then, for all $r \in R$, $0 = r0 = rxy = xry$. Therefore, $xRy = 0$, which implies that $x = 0$ or $y = 0$ since R is prime.

(4) By Lemma 9.17 (2), we have $0 = [utr, x] \subseteq ut[r, x] + [ut, x]r = ut[r, x]$, for all $u \in I$ and $t, r \in R$. Therefore, for all $s \in [r, x]$, we

have $uts = 0$, which means that $uRs = 0$. Hence, $s = 0$ since R is prime and $I \neq 0$. This shows that $x \in Z(R)$.

(5) Suppose that $xy \in Z(R)$. Then, $0 \in [xy, r]$, for all $r \in R$. Therefore, $0 \in [xy, r] = xyr - rxy = xyr - xry = x[y, r]$. So, $0 = t0 \in tx[y, r] = xt[y, r]$, for all $t \in R$. This implies that $0 \in xR[y, r]$. Hence, $x = 0$ or $0 \in [y, r]$, for all $r \in R$, since R is prime. Then, $x = 0$ or $y \in Z(R)$. $\square$

Lemma 9.4. *Let d be a derivation on a prime hyperring R and I be a non-zero hyperideal on R. Then, for all $x \in R$:*

(1) *if $d(I) = 0$, then $d = 0$;*
(2) *if $d(I)x = 0$ or $xd(I) = 0$, then $x = 0$ or $d = 0$;*
(3) *if $d(R)x = 0$ or $xd(R) = 0$, then $x = 0$ or $d = 0$.*

Proof. (1) For all $u \in I$ and $x \in R$, we have $0 = d(ux) \in d(u)x + ud(x) = ud(x)$. Therefore, $Id(x) = 0$, which implies that $d = 0$ by Lemma 9.3(1).

(2) Suppose that $d(I)x = 0$. Then, $0 = d(yu)x \in d(y)ux + yd(u)x = d(y)ux$, for all $u \in I$ and $y \in R$. Therefore, $d(y)Ix = 0$, which implies that $d = 0$ or $x = 0$ by Lemma 9.3(2). In the case of $xd(I) = 0$, the proof is similar.

(3) In (2), substitute R with I. $\square$

Definition 9.4. Let R be a Krasner hyperring and d be a derivation on R. Then, $x \in R$ is called a *constant element associated with d* if $d(x) = 0$. We denote by $C_d(R)$ the set of all the constant elements of R associated with the derivation d. It is trivial that $C_d(R)$ is a subhyperring of R.

Theorem 9.5. *Let d be a derivation on a prime hyperring R such that $d(R) \subseteq Z(R)$. Also, let there be a constant element $c \in R$ associated with d such that $c \notin Z(R)$. Then, $d = 0$.*

Proof. There is $x_0 \in R$ such that $cx_0 \neq x_0c$ since $c \notin Z(R)$. We have $d(xc) \in d(x)c + xd(c) = d(x)c$, for all $x \in R$. So, $d(x)c = d(xc) \in Z(R)$. Therefore, $d(x)cx_0 = x_0d(x)c = d(x)x_0c$. This means that $0 \in d(x)[c, x_0]$. Then, there is $t \in [c, x_0]$ such that $d(x)t = 0$. So, $d(x) = 0$ or $t = 0$ by Lemma 9.3(3). If $t = 0$, then $0 \in [c, x_0] = cx_0 - x_0c$, which is a contradiction. So, $d(x) = 0$, for all $x \in R$. $\square$

Lemma 9.5. *Let H and K be canonical subhypergroups of the canonical hypergroup $(G, +, 0)$. Then, $H \cup K$ is a canonical subhypergroup of G if and only if $H \subseteq K$ or $K \subseteq H$.*

Proof. If $H \subseteq K$ or $K \subseteq H$, then it is clear that $H \cup K$ is a canonical subhypergroup of G. Now, suppose that $H \cup K$ is a canonical subhypergroup of G and $H \nsubseteq K$ and $K \nsubseteq H$. Then, there are $a, b \in H \cup K$ such that $a \in H \setminus K$ and $b \in K \setminus H$. Also, we have $a + b \subseteq H \cup K$ since $H \cup K$ is a canonical subhypergroup. Now, one of the following cases two happens:

Case 1: $(a + b) \cap H \neq \emptyset$, then there exists $x \in (a + b) \cap H$. So, $b \in x - a \subseteq H$, which is a contradiction.

Case 2: $(a + b) \cap K \neq \emptyset$, in which case there exists $y \in (a + b) \cap K$. So, $a \in y - b \subseteq K$, which is a contradiction. $\square$

Theorem 9.6. *Let d be a non-zero derivation on a prime hyperring R and I be a non-zero hyperideal on R. Then:*

(1) *if $I \subseteq Z(R)$, then R is commutative;*
(2) *if $0 \in [u, R]Id(u)$, for all $u \in I$, then R is commutative.*

Proof. (1) We have $rsu = rus = (ru)s = s(ru) = sru$, for all $r, s \in R$ and $u \in I$. So, $0 \in rsu - sru = [r, s]u$. Therefore, $0 \in [r, s]$, for all $r, s \in R$ by Lemma 9.3 (1).

(2) Since $0 \in [u, r]Id(u)$, for $u \in I$ and $r \in R$, $0 \in [u, r]$ or $d(u) = 0$ by Lemma 9.3 (2). Put $A = \{u \in I \mid d(u) = 0\}$ and $B = \{u \in I \mid u \in Z(R)\}$. It is clear that A and B are canonical subhypergroups of I and $I = A \cup B$. So, $I = A$ or $I = B$ by Lemma 9.5. If $I = A$, i.e., is $d(I) = 0$, then $d = 0$ by Lemma 9.4(1), which is a contradiction. Therefore, $I = B$, i.e., $I \subseteq Z(R)$. This implies that R is commutative by (1). $\square$

Definition 9.5. A Krasner hyperring R is called *n-torsion-free*, where $n \in \mathbb{N}$, if $0 \in nx = \underbrace{x + x + \cdots + x}_{n}$, where $x \in R$, implies that $x = 0$.

Example 9.12. In Example 9.10, R is a 3-torsion-free hyperring, whereas R is not a 2-torsion-free hyperring since $e \in 2c$ but $c \neq e$.

Theorem 9.7. *Let I be a non-zero hyperideal of the 2-torsion-free hyperring R. Then:*

(1) *if d is a derivation of R such that $d^2(I) = 0$, then $d = 0$;*

(2) *if d_1 and d_2 are derivations of R such that $d_1 d_2(I) = 0$, then $d_1 = 0$ or $d_2 = 0$.*

Proof. (1) By Lemma 9.1, we have for all $u, v \in I$,

$$0 = d^2(uv) \in d^2(u)v + 2d(u)d(v) + ud^2(v) = 2d(u)d(v).$$

So, $d(u)d(v) = 0$ since R is a 2-torsion-free hyperring. Therefore, $d = 0$ by Lemma 9.4(1) and (2).

(2) We have for all $u, v \in I$,

$$0 = d_1 d_2(uv) \in d_1(d_2(u)v + ud_2(v))$$
$$\subseteq d_1 d_2(u)v + d_2(u)d_1(v) + d_1(u)d_2(v) + ud_1 d_2(v)$$
$$= d_2(u)d_1(v) + d_1(u)d_2(v).$$

Replacing u by $d_2(u)$ in the above equation, we get

$$0 \in d_2^2(u)d_1(v) + d_1 d_2(u)d_2(v) = d_2^2(u)d_1(v),$$

i.e., $d_2^2(u)d_1(v) = 0$. So, $d_1 = 0$ or $d_2^2(I) = 0$ by Lemma 9.4(1) and (2). Therefore, $d_1 = 0$ or $d_2 = 0$ by (1). $\square$

In the following lemma and theorem, R will be a Krasner hyperring such that its center, i.e., $Z(R)$, is a ring.

Example 9.13. In Example 9.11, the center of the hyperring R is a ring since $Z(R) = \{0\}$. It is clear that in Example 9.9, the center of M' is a ring if and only if the center of R is a ring. In Example 9.10, $Z(R) = \{e, c\} \cong \mathbb{Z}_2$. So, $Z(R)$ is a ring.

Lemma 9.6. *Let R be a Krasner hyperring such that its center, i.e., $Z(R)$, is a ring. Also, let d be a derivation on R. Then, $d(x) \in Z(R)$, for all $x \in Z(R)$.*

Proof. Suppose that $x \in Z(R)$. Then, $d(xr) = d(rx)$, for all $r \in R$. So,

$$\in 0d(xr) - d(rx) \subseteq d(x)r + xd(r) - d(r)x - rd(x)$$
$$= d(x)r + xd(r) - xd(r) - rd(x)$$
$$= d(x)r + (x - x)d(r) - rd(x) = d(x)r - rd(x).$$

Therefore, $d(x)r = rd(x)$, for all $r \in R$. $\square$

Theorem 9.8. *Let R be a prime hyperring such that its center, i.e., $Z(R)$, is a ring. Also, let I be a non-zero hyperideal of R. Then, in all the following cases, R is commutative:*

(1) *if d is a derivation such that $d^2 \neq 0$ and $d(R) \subseteq Z(R)$;*
(2) *if R is a 2-torsion-free hyperring and d is a non-zero derivation such that $d(I) \subseteq Z(R)$;*
(3) *if for all subset A of R, $0 \in 6A$ implies that $0 \in A$ and d is a non-zero derivation such that $d(I) \subseteq I$ and $d^2(I) \subseteq Z(R)$;*
(4) *if for all subset A of R, $0 \in 6A$ implies that $0 \in A$ and d_1, d_2 are non-zero derivations such that $d_2(I) \subseteq I$, $d_1 d_2(I) \subseteq Z(R)$ and $d_1 d_2^2(I) = 0$.*

Proof. (1) Suppose that $d(R) \subseteq Z(R)$. Then, $[d(x), y] = 0$, for all $x, y \in R$. Replace x by xz, where $z \in R$. Hence, $0 = [d(xz), y] \subseteq [d(x)z, y] + [xd(z), y] = d(x)[z, y] + d(z)[x, y]$, by Lemma 9.17(3). Replacing z by $d(z)$, we get $0 \in d(x)[d(z), y] + d^2(z)[x, y] = d^2(z)[x, y]$. So, $d^2(z) = 0$ or $0 \in [x, y]$ by Lemma 9.3(3). Hence, R is commutative since $d^2 \neq 0$.

(2) If for all $x \in Z(R)$, we have $d(x) = 0$. Then, $d(Z(R)) = 0$ and so $d^2(I) = 0$. So, $d = 0$ by Theorem 9.7(1), and this is a contradiction. Hence, there is $x_0 \in Z(R)$ such that $d(x_0) \neq 0$. By Lemmas 9.17(3) and 9.7, we have for all $u \in I$ and $y \in R$,

$$0 = [d(ux_0), y] \subseteq [d(u)x_0 + ud(x_0), y]$$
$$= d(u)[x_0, y] + d(x_0)[u, y] = d(x_0)[u, y].$$

That is, $0 \in d(x_0)[u, y]$. So, there is $t \in [u, y]$ such that $0 = d(x_0)t$. Therefore, by Lemmas 9.3(3) and 9.7, we get $t = 0$ since $d(x_0) \neq 0$. This means that $I \subseteq Z(R)$. So, R is commutative by Theorem 9.18(1).

(3) Suppose that $u \in I$, then by Lemmas 9.17(3) and 9.7, we have for all $y \in R$,

$$0 \in [d^2(d(u)d(u)), y]$$
$$\subseteq [d(d^2(u)d(u) + d(u)d^2(u)), y]$$
$$= 2[d(d^2(u)d(u)), y] \subseteq 2[d^3(u)d(u) + d^2(u)d^2(u), y]$$
$$= 2d^3(u)[d(u), y] + d^2(u)[d^2(u), y] = 2d^3(u)[d(u), y].$$

Hence, $0 \in d^3(u)[d(u), y]$ by hypothesis. So, $d^3(u) = 0$ or $0 \in [d(u), y]$ by Lemmas 9.3(3) and 9.7. Therefore, $d^3(u) = 0$ or $d(u) \in Z(R)$.

Suppose that $d^3(u) = 0$. Then, by Lemmas 9.1 and 9.17(3), we have for all $y \in R$,

$$0 = [d^2(ud(u)), y] \subseteq [d^2(u)d(u) + 2d(u)d^2(u) + ud^3(u), y]$$

$$= 3[d(u)d^2(u), y] = 3d^2(u)[d(u), y].$$

Hence, $0 \in d^2(u)[d(u), y]$ by hypothesis. So, $d^2(u) = 0$ or $0 \in [d(u), y]$ by Lemma 9.3(3). Therefore, $d^2(u) = 0$ or $d(u) \in Z(R)$, for all $u \in I$. Put $A = \{u \in I \mid d(u) \in Z(R)\}$ and $B = \{u \in I \mid d^2(u) = 0\}$. It is clear that A and B are canonical subhypergroups of I and $I = A \cup B$. So, $I = A$ or $I = B$ by Lemma 9.5. If $I = B$, i.e., $d^2(I) = 0$, then $d = 0$ by Theorem 9.7(1), and this is a contradiction. So, $I = A$, i.e., $d(I) \subseteq Z(R)$. Now, (2) completes the proof.

(4) By Lemma 9.17(3), for all $u \in I$ and $x \in R$,

$$0 = [d_1 d_2(d_2(u)d_2(u)), x] \subseteq [d_1(d_2^2(u)d_2(u) + d_2(u)d_2^2(u)), x]$$

$$\subseteq 2[d_2^2(u)d_1 d_2(u), x] = 2d_1 d_2(u)[d_2^2(u), x].$$

Hence, $0 \in d_1 d_2(u)[d_2^2(u), x]$ by hypothesis. So, $d_1 d_2(u) = 0$ or $d_2^2(u) \in Z(R)$ by Lemma 9.3(3), for all $u \in I$. Put $A = \{u \in I \mid d_2^2(u) \in Z(R)\}$ and $B = \{u \in I \mid d_1 d_2(u) = 0\}$. It is clear that A and B are canonical subhypergroups of I and $I = A \cup B$. So, $I = A$ or $I = B$ by Lemma 9.5. If $I = B$, i.e., $d_1 d_2(I) = 0$, then $d_1 = 0$ or $d_2 = 0$ by Theorem 9.7(2), which a contradiction. So, $I = A$, i.e., $d_2^2(I) \subseteq Z(R)$. Now, (3) completes the proof. $\square$

Definition 9.6. A Krasner hyperring R is called *differentiable* if there is at least a derivation on R. A Krasner hyperring R with all of derivations is called a *differential hyperring*. A hyperfield R is called a *differential hyperfield* if R is a differential hyperring. A subhyperring H of the differential hyperring R is called a *differential subhyperring* if for all derivations d of R, we have $d(h) \in H$, for all $h \in H$. A hyperideal I of the differential hyperring R is called a *differential hyperideal* if I is differential subhyperring of R.

Example 9.14. For every differential hyperring R, $\langle 0 \rangle_R$ is a differential hyperideal.

A differential hyperideal $I (\neq R)$ of a differential hyperring R is called a *prime* if for all $x, y \in R$, $xy \in I$ implies that $x \in I$ or $y \in I$. The intersection of all differential prime hyperideals of R that contains differential hyperideal I is called a *radical I* and denoted by $Rad(I)$. If the differential hyperring R does not have any differential prime hyperideal containing I, we define $Rad(I) = R$. A differential hyperideal I is called a *differential radical hyperideal* if $Rad(I) = I$.

We usually use the perfix Δ instead of saying that R is differential, and Δ is the set of all derivations on R. Also, if R is a differential hyperring, i.e., R is a Δ-hyperring, then we usually use the notion Δ-hyperideal instead of saying that I is a differential hyperideal of R.

Let R and S be Δ_1 and Δ_2-hyperrings, respectively. By a *differential good homomorphism* of R onto S, we mean a good homomorphism φ such that $d_2\varphi(x) = \varphi d_1(x)$, for all $x \in R$, $d_1 \in \Delta_1$ and $d_2 \in \Delta_2$.

Theorem 9.9. *Let R and S be Δ_1 and Δ_2-hyperrings, respectively. Also, let $\varphi : R \to S$ be a differential good homomorphism. Then:*

(1) *$ker\varphi$ is a Δ_1-hyperideal;*
(2) *if I is a Δ_2-hyperideal of S, then $\varphi^{-1}(I)$ is a Δ_1-hyperideal of R.*

Proof. It is trivial that $ker\varphi$ is a hyperideal of R. For all $d_1 \in \Delta_1$, $d_2 \in \Delta_2$ and $x \in ker\varphi$, we have $\varphi d_1(x) = d_2\varphi(x) = d_2(0) = 0$. So, $d_1(x) \in ker\varphi$.

The proof of Part (2) is similar. □

Theorem 9.10. *Let $(R, +, \cdot)$ be a Δ-hyperring and I and J be Δ-hyperideals of R. Then, $IJ = \{x \mid x \in \sum_{i=1}^n a_i b_i, a_i \in I, b_i \in J, n \in \mathbb{N}\}$ is also a Δ-hyperideal of R.*

Proof. It is easy to see that IJ is a hyperideal. If $x \in IJ$, then $x \in \sum_{i=1}^n a_i b_i$, for some $a_i \in I$, $b_i \in J$ and $n \in \mathbb{N}$. So, for all $d \in \Delta$, we have $d(x) \in d(\sum_{i=1}^n a_i b_i) = \sum_{i=1}^n d(a_i b_i) = \sum_{i=1}^n d(a_i)b_i + a_i d(b_i) \subseteq IJ$. □

Theorem 9.11. *Let R be a Δ-hyperring and P be a Δ-hyperideal of R. Then, $J = \{a \in R \mid ra \in P, \text{ forall } r \in R\}$ is a Δ-hyperideal of R.*

Proof. It is easy to check that $P \subseteq J$ and J is a hyperideal of R. We prove that J is a differential. Suppose that $a \in J$. Then, $ra \in P$, for all $r \in R$. So, $d(ra) \in d(P) \subseteq P$. On the other hand, $d(ra) \in d(r)a + rd(a)$. Therefore, $rd(a) \in -d(r)a + d(ra) \subseteq P$, for all $r \in R$. Hence, $rd(a) \in P$, for all $r \in R$, and this implies that $d(a) \in J$. So, J is a Δ-hyperideal. $\qquad\square$

Let $(R_1, +_1, \cdot_1)$ and $(R_2, +_2, \cdot_2)$ be Δ_1 and Δ_2-hyperrings, respectively. Then, $(R_1 \times R_2, +, \cdot)$ is a hyperring, where for all $(a, b), (c, d) \in R_1 \times R_2$ hyeroperation $+$ and operation $\cdot$ are defined as $(a, b) + (c, d) = \{(x, y) \mid x \in a +_1 c, y \in b +_2 d\}$ and $(a, b) \cdot (c, d) = (a \cdot_1 c, b \cdot_2 d)$. For all $d_1 \in \Delta_1$ and $d_2 \in \Delta_2$, we define the function $d_1 \times d_2 : R_1 \times R_2 \to R_1 \times R_2$ as $(d_1 \times d_2)(x, y) = (d_1(x), d_2(y))$, for all $(x, y) \in R_1 \times R_2$. Then, $d_1 \times d_2$ is a derivation on $R_1 \times R_2$.

Theorem 9.12. *Let I be a Δ-radical hyperideal of the commutative Δ-hyperring R. Then, $(I : r) = \{x \in R \mid xr \in I\}$, for all $r \in R$, is also a Δ-radical hyperideal.*

Proof. Let $x, y \in (I : r)$. Then, $(x - y)r = xr - yr \subseteq I$. So, $x - y \subseteq (I : r)$. Now, suppose that $x \in (I : r)$ and $t \in R$. Then, $xtr = xrt \in It = I$. So, $xt \in (I : r)$. It shows that $(I : r)$ is a hyperideal. Let $x \in (I : r)$ and d is a derivation of R, then $d(x)rd(xr) \in d(x)rd(x)r + d(x)rxd(r)$. So, $(d(x)r)^2 \in d(x)rd(xr) - d(x)rxd(r) \subseteq I$. Therefore, $d(x)r \in Rad(I) = I$, which means that $d(x) \in (I : r)$. So, I is a Δ-hyperideal. Obviously, $(I : r) \subseteq Rad((I : r))$. Let $x \in Rad((I : r))$. Then, there is $n \in \mathbb{N}$ such that $x^n \in (I : r)$. Therefore, $x^n r \in I$. So, we have $(xr)^n = x^n r^n = r^{n-1}(x^n r) \in r^{n-1} I = I$ since R is commutative. Hence, $xr \in Rad(I) = I$, which means $x \in (I : r)$. So, $(I : r)$ is a Δ-radical hyperideal. $\qquad\square$

Theorem 9.13. *Let d be a derivation on Δ-hyperfield R such that $d(1) = 0$, where 1 is the unit element of R. Then, $C_d(R)$ is also a Δ-hyperfield.*

Proof. Let $x, y \in C_d(R)$, then $d(x + y) = d(x) + d(y) = 0$. So, $x + y \subseteq C_d(R)$. Also, we have $d(xy) \in d(x)y + xd(y) = 0$, which means that $xy \in C_d(R)$. Now, suppose that $0 \neq x \in C_d(R)$. Then, $d(x) = 0$. Since R is a hyperfield, there is $y \in R$ such that $xy = 1$. We have $0 = d(1) = d(xy) \in d(x)y + xd(y) = xd(y)$, i.e., $xd(y) = 0$. So, $d(y) = 0$ since R is a Δ-hyperfield and $x \neq 0$. This shows that $y \in C_d(R)$. $\qquad\square$

9.3 (θ, σ)-Derivations of Krasner Hyperrings

In this section, we introduce the notion of θ-derivation and (θ, σ)-derivation of Krasner hyperrings. Then, we present some results in this respect. These results were obtained by Kamali Ardekani and Davvaz [43].

Definition 9.7. Let $(R, +, \cdot)$ be a Krasner hyperring and θ, σ be good homomorphisms on R. The function $d : R \to R$ is called (θ, σ)-*derivation* if for all $x, y \in R$:

(1) $d(x + y) = d(x) + d(y)$;
(2) $d(x \cdot y) \in d(x) \cdot \theta(y) + \sigma(x) \cdot d(y)$.

If $\theta = \sigma$, then d is called θ-*derivation*.

By the above definition, for every (θ, σ)-derivation d of the Krasner hyperring R, we have $d(0) = 0$ and $d(-x) = -d(x)$, for all $x \in R$, since $d(0) = d(0 \cdot 0) \in d(0) \cdot \theta(0) + \sigma(0) \cdot d(0) = \{0\}$ and $0 = d(0) \in d(x - x) = d(x) + d(-x)$, for all $x \in R$. Thus, $d(-x) = -d(x)$, for all $x \in R$. Now, we consider some examples.

Example 9.15. Let $(G, \cdot, e)$ be a finite abelian group with m elements, $m > 3$, that has no elements of order 2, and define a hyper-addition and a multiplication on $H = G \cup \{0\}$ by

$$x + 0 = 0 + x = \{x\}, \quad \text{for all } x \in H,$$
$$x + x = \{x, 0\}, \quad \text{for all } x \in G,$$
$$x + y = y + x = H - \{x, y\}, \quad \text{for all } x, y \in G, \ x \neq y,$$
$$x \odot 0 = 0 \odot x = 0, \quad \text{for all } x \in H,$$
$$x \odot y = x \cdot y, \quad \text{for all } x, y \in G.$$

Then, $(H, +, \odot)$ is a Krasner hyperring. Moreover, let

$$\theta(x) = \begin{cases} 0 & \text{if } x = 0, \\ x^{-1} & \text{for all } x \in G. \end{cases}$$

Then, θ is a good homomorphism on H since

$$\theta(x+0) = \theta(0+x) = \theta(x) = \{x^{-1}\} = \theta(x) + \theta(0) = \theta(0) + \theta(x),$$

for all $x \in G$,

$$\theta(x+x) = \{\theta(x), \theta(0)\} = \{x^{-1}, 0\} = x^{-1} + x^{-1} = \theta(x) + \theta(x),$$

for all $x \in G$,

$$\theta(x+y) = H - \{x^{-1}, y^{-1}\} = x^{-1} + y^{-1} = \theta(x) + \theta(y),$$

for all $x, y \in G$, $x \neq y$,

$$\theta(x \odot 0) = \theta(0) = 0 = \theta(x) \odot \theta(0), \quad \text{for all } x \in H,$$

$$\theta(x \odot y) = \theta(x \cdot y) = x^{-1} \cdot y^{-1} = \theta(x) \odot \theta(y), \quad \text{for all } x, y \in G.$$

The function $d : H \to H$, defined by $d(x) = x$, for all $x \in H$, is a θ-derivation since the first condition of the definition of θ-derivation is valid. Also, we have

$$d(x \odot 0) = d(0) = 0 \in \{0\}$$

$$= x \odot 0 + x^{-1} \odot 0$$

$$= d(x) \odot \theta(0) + \theta(x) \odot d(0), \quad \text{for all } x \in H \text{ and}$$

$d(x) \odot \theta(y) + \theta(x) \odot d(y) = x \odot y^{-1} + x^{-1} \odot y = x \cdot y^{-1} + x^{-1} \cdot y$, for all $x, y \in G$. Based on the above relations, to prove the second condition of the definition of derivation, it is sufficient to show that $d(x \odot y) = x \cdot y \in x \cdot y^{-1} + x^{-1} \cdot y$, for all $x, y \in G$. We have $x \cdot y^{-1} + x^{-1} \cdot y = H - \{x \cdot y^{-1}, x^{-1} \cdot y\}$ since G has no elements of order 2. If $x \cdot y = x \cdot y^{-1}$, then $y = y^{-1}$, and if $x \cdot y = x^{-1} \cdot y$, then $x = x^{-1}$. Hence, $x \cdot y \notin \{x \cdot y^{-1}, x^{-1} \cdot y\}$ since G has no elements of order 2. Therefore, $x \cdot y \in H - \{x \cdot y^{-1}, x^{-1} \cdot y\}$.

Also, θ is a θ-derivation since the first condition of the definition of θ-derivation is valid. Also, we have

$$\theta(x \odot 0) = \theta(0) = 0 \in \{0\} = x^{-1} \odot 0 + x^{-1} \odot 0$$

$$= \theta(x) \odot \theta(0) + \theta(x) \odot \theta(0), \quad \text{for all } x \in H,$$

$$\theta(x \odot y) = x^{-1} \cdot y^{-1} \in \theta(x) \odot \theta(y) + \theta(x) \odot \theta(y)$$

$$= x^{-1} \odot y^{-1} + x^{-1} \odot y^{-1}$$

$$= \{x^{-1} \cdot y^{-1}, 0\}, \quad \text{for all } x, y \in G.$$

θ is a d-derivation since the first condition of the definition of (d, θ)-derivation is valid, and we have

$$\theta(x \odot 0) = \theta(0) = 0 \in \{0\} = x^{-1} \odot 0 + x \odot 0$$
$$= \theta(x) \odot d(0) + d(x) \odot \theta(0), \quad \text{for all } x \in H,$$
$$\theta(x \odot y) = x^{-1} \cdot y^{-1} \in \theta(x) \odot d(y) + d(x) \odot \theta(y)$$
$$= x^{-1} \odot y + x \odot y^{-1}$$
$$= x^{-1} \cdot y + x \cdot y^{-1} = H - \{x^{-1} \cdot y, x \cdot y^{-1}\},$$
$$\text{for all } x, y \in G.$$

d is nota (d, θ)-derivation since for all $x, y \in G$,

$$d(x \odot y) = x \cdot y \notin d(x) \odot d(y) + \theta(x) \odot d(y) = x \odot y + x^{-1} \odot y$$
$$= x \cdot y + x^{-1} \cdot y = H - \{x \cdot y, x^{-1} \cdot y\}.$$

Similarly, one can show that θ is not (d, θ)-derivation.

Example 9.16. Let $(G, \cdot, e)$ be an abelian group that has no elements of order 2. Define a hyperaddition and a multiplication on $H = G \cup \{0\}$ by

$$x + 0 = 0 + x = x, \quad \text{for all } x \in H,$$
$$x + x = H - \{x\}, \quad \text{for all } x \in G,$$
$$x + y = y + x = \{x, y\}, \quad \text{for all } x, y \in G, \ x \neq y,$$
$$x \odot 0 = 0 \odot x = 0, \quad \text{for all } x \in H,$$
$$x \odot y = x \cdot y, \quad \text{for all } x, y \in G.$$

Then, $(H, +, \odot)$ is a Krasner hyperring. Let $d, \theta, \sigma : H \to H$ be defined by $d(x) = \sigma(x) = x$, for all $x \in H$ and

$$\theta(x) = \begin{cases} 0 & D & \text{if } x = 0 \\ x^{-1}D & \text{for all } x \in G. \end{cases}$$

Then, θ and σ are good homomorphisms and d is s (θ,σ)-derivation since

$$d(x \odot 0) = d(0) = 0 \in \{0\} = x \odot 0 + x \odot 0$$
$$= d(x) \odot \theta(0) + \sigma(x) \odot d(0), \quad \text{for all } x \in H,$$
$$d(x \odot y) = x \cdot y \in d(x) \odot \theta(y) + \sigma(x) \odot d(y) = x \odot y^{-1} + x \odot y$$
$$= x \cdot y^{-1} + x \cdot y = \{x \cdot y^{-1}, x \cdot y\}, \quad \text{for all } x,y \in G.$$

Similarly, d is a (σ,θ)-derivation. However, d is not a θ-derivation since $d(x \odot y) = x \cdot y \notin d(x) \odot \theta(y) + \theta(x) \odot d(y) = x \cdot y^{-1} + x^{-1} \cdot y = \{x \cdot y^{-1}, x^{-1} \cdot y\}$, for all $x,y \in G$.

Example 9.17. Let R be a Krasner hyperring and $M = \left\{ \begin{pmatrix} x & y \\ 0 & 0 \end{pmatrix} \middle| x,y \in R \right\}$. Also, let the functions $d,\theta,\sigma : M \to M$ be defined by $d\left(\begin{pmatrix} x & y \\ 0 & 0 \end{pmatrix} \right) = \begin{pmatrix} 0 & y \\ 0 & 0 \end{pmatrix}$, $\theta\left(\begin{pmatrix} x & y \\ 0 & 0 \end{pmatrix} \right) = \begin{pmatrix} x & 0 \\ 0 & 0 \end{pmatrix}$ and $\sigma\left(\begin{pmatrix} x & y \\ 0 & 0 \end{pmatrix} \right) = \bar{0}$. Then, it is easy to check that M is a Krasner hyperring, θ and σ are good homomorphisms and d is a (σ,θ)-derivation.

Theorem 9.14. *Let $(R,+,\cdot)$ be a Krasner hyperring and $\theta,\sigma \in Hom(R,+,\cdot)$. For every $n \in \mathbb{N}$ and $x,y \in R$, let d be a (θ,σ)-derivation on R.*

(1) If R is commutative, then $d(x^n) \in \left(\sum_{i=0}^{n-1} \theta(x)^{n-1-i} \cdot \sigma(x)^i \right) \cdot d(x)$.

(2) If $d\theta = \theta d$ and $d\sigma = \sigma d$, then

$$d^n(x \cdot y) \in \sum_{i=0}^{n} \binom{n}{i} d^{n-i}\sigma^i(x) \cdot d^i\theta^{n-i}(y),$$

where d^n shows (θ,σ)-derivation of order n.

Proof. (1) We prove the statement by induction on n. Obviously, the statement is valid in the case of $n = 1$. Suppose that the statement is valid in the case pf $n = k$ (the hypothesis of the induction).

We have

$$d(x^{k+1}) \in d(x^k) \cdot \theta(x) + \sigma(x)^k \cdot d(x)$$

$$\subseteq \left(\sum_{i=0}^{k-1} \theta(x)^{k-1-i} \cdot \sigma(x)^i \right) \cdot d(x) \cdot \theta(x) + \sigma(x)^k \cdot d(x)$$

$$= \left(\sum_{i=0}^{k-1} \theta(x)^{k-i} \cdot \sigma(x)^i \right) \cdot d(x) + \sigma(x)^k \cdot d(x)$$

$$= \left(\sum_{i=0}^{k} \theta(x)^{k-i} \cdot \sigma(x)^i \right) \cdot d(x).$$

(2) Suppose that d is a (θ, σ)-derivation on R. We prove the statement by induction on n. It is clear that the statement is valid in the case of $n = 1$. Suppose that the statement is valid in the case of $n = k$ (the hypothesis of the induction). We have

$$d^{k+1}(x \cdot y) = d\big(d^k(x \cdot y)\big) \in d \left(\sum_{i=0}^{k} \binom{k}{i} d^{k-i} \sigma^i(x) \cdot d^i \theta^{k-i}(y) \right)$$

$$\subseteq \sum_{i=0}^{k} \binom{k}{i} \Big(d^{k+1-i} \sigma^i(x) \cdot d^i \theta^{k+1-i}(y)$$

$$+ d^{k-i} \sigma^{i+1}(x) \cdot d^{i+1} \theta^{k-i}(y) \Big)$$

$$= \sum_{i=0}^{k+1} \binom{k+1}{i} d^{k+1-i} \sigma^i(x) \cdot d^i \theta^{k+1-i}(y).$$

This completes the proof. □

Theorem 9.15. *Let θ, σ be good homomorphisms and d be a (θ, σ)-derivation on a Krasner hyperring R. Also, let $d\theta = \theta d$, $d\sigma = \sigma d$, and n be the smallest natural number such that $d^n(R) = 0$. Then, for all $y \in R$, $d(y) = 0$, or there is $0 < k < n$ such that $0 \in n\sigma d^{n-1}(x_0) \cdot \theta^{n-1} d^k(y)$, where $0 \neq x_0 \in R$ is a fixed element.*

Proof. Suppose that n is the smallest natural number such that $d^n(R) = 0$. Then, $d^{n-1}(R) \neq 0$. Hence, there is $0 \neq x_0 \in R$ such that

$d^{n-1}(x_0) \neq 0$. If $d(y) \neq 0$, $y \in R$, then there is $1 < k < n$ such that $d^k(y) \neq 0$ and $d^{k+1}(y) = 0$. By Theorem 9.1 and hypothesis, we have

$$0 = d^n\left(x_0 \cdot d^{k-1}(y)\right) \in \sum_{i=0}^{n} \binom{n}{i} d^{n-i}\sigma^i(x_0) \cdot d^{k+i-1}\theta^{n-i}(y)$$

$$= d^n(x_0) \cdot d^{k-1}\theta^n(y) + nd^{n-1}\sigma(x_0) \cdot d^k\theta^{n-1}(y)$$

$$+ \sum_{i=0}^{n-2} \binom{n}{i+2} d^{n-i-2}\sigma^{i+2}(x_0) \cdot d^{k+i+1}\theta^{n-i-2}(y)$$

$$= n\sigma d^{n-1}(x_0) \cdot \theta^{n-1}d^k(y).$$

This completes the proof.   □

Theorem 9.16. *Let θ, σ, d be good homomorphisms and d be also a (θ, σ)-derivation on a Krasner hyperring R. Then, for all $x, y \in R$:*

(1) $d(x) \cdot \theta(y) \cdot d(x) \in d(x) \cdot \theta(y \cdot x) + \sigma(x) \cdot d(y \cdot x) - \sigma(x) \cdot d(y \cdot x);$
(2) $d(x) \cdot \sigma(y) \cdot d(x) \in \sigma(x \cdot y) \cdot d(x) + d(x \cdot y) \cdot \theta(x) - d(x \cdot y) \cdot \theta(x).$

Proof. (1) We have, for all $x, y \in R$,

$$d(x) \cdot d(y) = d(x \cdot y) \in d(x) \cdot \theta(y) + \sigma(x) \cdot d(y). \qquad (9.2)$$

Replacing y by $y \cdot x$ in (9.2), we obtain

$$d(x \cdot y) \cdot d(x) = d(x) \cdot d(y) \cdot d(x) = d(x) \cdot d(y \cdot x)$$

$$= d(x \cdot y \cdot x) \in d(x) \cdot \theta(y \cdot x) + \sigma(x) \cdot d(y \cdot x).$$

On the other hand, $d(x \cdot y) \cdot d(x) \in d(x) \cdot \theta(y) \cdot d(x) + \sigma(x) \cdot d(y) \cdot d(x) = d(x) \cdot \theta(y) \cdot d(x) + \sigma(x) \cdot d(y \cdot x)$. So, $d(x) \cdot \theta(y) \cdot d(x) \in d(x) \cdot \theta(y \cdot x) + \sigma(x) \cdot d(y \cdot x) - \sigma(x) \cdot d(y \cdot x)$.

(2) Replacing x by $y \cdot x$ in (9.2), we obtain

$$d(y) \cdot d(x \cdot y) = d(y) \cdot d(x) \cdot d(y) = d(y \cdot x) \cdot d(y)$$

$$= d(y \cdot x \cdot y) \in d(y \cdot x) \cdot \theta(y) + \sigma(y \cdot x) \cdot d(y).$$

On the other hand, $d(y) \cdot d(x \cdot y) \in d(y) \cdot d(x) \cdot \theta(y) + d(y) \cdot \sigma(x) \cdot d(y) = d(y \cdot x) \cdot \theta(y) + d(y) \cdot \sigma(x) \cdot d(y)$. So, $d(y) \cdot \sigma(x) \cdot d(y) \in \sigma(y \cdot x) \cdot d(y) + d(y \cdot x) \cdot \theta(y) - d(y \cdot x) \cdot \theta(y)$. By interchanging the roles of x and y, we have $d(x) \cdot \sigma(y) \cdot d(x) \in \sigma(x \cdot y) \cdot d(x) + d(x \cdot y) \cdot \theta(x) - d(x \cdot y) \cdot \theta(x)$.
This completes the proof.   □

Theorem 9.17. *Let* $(R, +, \cdot)$ *be a Krasner hyperring, and the nota-tion* $[x, y]$ *denotes the set* $x \cdot y - y \cdot x$, *for all* $x, y \in R$. *Then, for all* $x, y, z \in R$:

(1) *we have* $[x \cdot y, z] \subseteq x \cdot [y, z] + [x, z] \cdot y$;
(2) *if* d *is a* θ-*derivation of Krasner hyperring* R, *then* $d[x, y] \subseteq [d(x), \theta(y)] + [\theta(x), d(y)]$.

Proof. We prove (2). Suppose that d is a θ-derivation of the Krasner hyperring R. Then,

$$d[x, y] = d(x \cdot y - y \cdot x) = d(x \cdot y) - d(y \cdot x)$$
$$\subseteq d(x) \cdot \theta(y) + \theta(x) \cdot d(y) - d(y) \cdot \theta(x) - \theta(y) \cdot d(x)$$
$$= d(x) \cdot \theta(y) - \theta(y) \cdot d(x) + \theta(x) \cdot d(y) - d(y) \cdot \theta(x)$$
$$= [d(x), \theta(y)] + [\theta(x), d(y)].$$

$\square$

Theorem 9.18. *Let* θ *be a good homomorphism,* σ *be an epimorphism,* d *be a non-zero* (θ, σ)-*derivation on a prime Krasner hyperring* $(R, +, \cdot)$, *and* I *be a non-zero hyperideal on* R. *Then:*

(1) *if* $I \subseteq Z$, *then* R *is commutative;*
(2) *if* $0 \in [u, R] \cdot I \cdot d(u)$, *for all* $u \in I$, *then* R *is commutative.*

Proof. We prove (2). Since $0 \in [u, r] \cdot I \cdot d(u)$, for all $u \in I$ and $r \in R$, $0 \in [u, r]$ or $d(u) = 0$. Put $A = \{u \in I \mid d(u) = 0\}$ and $B = \{u \in I \mid u \in Z\}$. It is clear that $I = A \cup B$. Thus, $I = A$ or $I = B$ by Lemma 9.5 since A and B are canonical subhypergroups of I. If $I = A$, i.e., $d(I) = 0$, then $d = 0$ by Lemma 9.4 (1), which is a contradiction. Therefore, $I = B$, i.e., $I \subseteq Z$. This implies that R is commutative by (1). $\square$

Note. If R is a Krasner hyperring and $x \in Z$, then $[x \cdot y, z] = x \cdot [y, z]$, for all $y, z \in R$.

In the following lemma and theorem, R will be a Krasner hyper-ring such that its center Z is a ring.

Lemma 9.7. *Let* θ *be an epimorphism and* d *be a* θ-*derivation on a Krasner hyperring* $(R, +, \cdot)$. *Then,* $d(x) \in Z$, *for all* $x \in Z$.

Proof. Suppose that $x \in Z$, then $\theta(x) \in Z$ and $d(x \cdot r) = d(r \cdot x)$, for all $r \in R$. So,

$$0 \in d(x \cdot r) - d(r \cdot x) \subseteq d(x) \cdot \theta(r) + \theta(x) \cdot d(r)$$
$$- d(r) \cdot \theta(x) - \theta(r) \cdot d(x)$$
$$= d(x) \cdot \theta(r) + \theta(x) \cdot d(r) - \theta(x) \cdot d(r) - \theta(r) \cdot d(x)$$
$$= d(x) \cdot \theta(r) + \theta(x - x) \cdot d(r) - \theta(r) \cdot d(x)$$
$$= d(x) \cdot \theta(r) - \theta(r) \cdot d(x).$$

Therefore, $d(x) \cdot \theta(r) = \theta(r) \cdot d(x)$, for all $r \in R$. Then, $d(x) \cdot s = s \cdot d(x)$, for all $s \in R$, since θ is onto. Thus, $d(x) \in Z$. $\square$

Theorem 9.19. *Let I be a non-zero hyperideal and θ be an epimorphism of a prime hyperring R.*

(1) *If σ is a good homomorphism and d is a (θ, σ)-derivation on R such that $d^2 \neq 0$ and $d(R) \subseteq Z$, then $\sigma(R)$ is commutative, and in particular, if σ is onto, then R is commutative.*

(2) *If R is a 2-torsion-free hyperring and d is a non-zero θ-derivation such that $d\theta = \theta d$ and $d(I) \subseteq Z$, then R is commutative.*

(3) *If for all subsets $\emptyset \neq A$ of R, $0 \in 6A$ implies that $0 \in A$, θ is one to one and d is a non-zero θ-derivation such that $\theta d = d\theta$ and $d^2(I) \subseteq Z$, then R is commutative.*

(4) *If for all subsets $\emptyset \neq A$ of R, $0 \in 6A$ implies that $0 \in A$, θ is one to one and d_1, d_2 are non-zero θ-derivations such that $d_1\theta = \theta d_1$, $d_2\theta = \theta d_2$, $d_2(I) \subseteq I$, $d_1 d_2(I) \subseteq Z$, $\theta(I) \subseteq I$, and $d_1 d_2^2(I) = 0$, then R is commutative.*

Proof. (1) Suppose that $d(R) \subseteq Z$, then $[d(x), y] = 0$, for all $x, y \in R$. Replace x by $x \cdot z$, where $z \in R$, then $0 = [d(x \cdot z), y] \subseteq [d(x) \cdot \theta(z), y] + [\sigma(x) \cdot d(z), y] = d(x) \cdot [\theta(z), y] + d(z) \cdot [\sigma(x), y]$. Replacing z by $d(z)$, we get $0 \in d(x) \cdot [\theta(d(z)), y] + d^2(z) \cdot [\sigma(x), y] = d^2(z)[\sigma(x), y]$. So, $d^2(z) = 0$ or $0 \in [\sigma(x), y]$. Hence, $0 \in [\sigma(x), y]$, since $d^2 \neq 0$.

(2) By Lemma 9.7, we have for all $u \in I$, $x \in Z$ and $y \in R$,

$$0 = [d(u \cdot x), y] \subseteq [d(u) \cdot \theta(x) + \theta(u) \cdot d(x), y]$$
$$= d(u) \cdot [\theta(x), y] + d(x) \cdot [\theta(u), y] = d(x) \cdot [\theta(u), y].$$

That is, $0 \in d(x) \cdot [\theta(u), y]$. So, there is $t \in [\theta(u), y]$ such that $0 = d(x) \cdot t$. Therefore, $d(x) = 0$ or $t = 0$ by Lemma 9.7. If $d(x) = 0$, for all $x \in Z$, i.e., $d(Z) = 0$, then $d^2(I) = 0$. So, $d = 0$ by Theorem 9.7(1), which is a contradiction. Thus, $t = 0$, i.e., $\theta(I) \subseteq Z$. So, R is commutative by Theorem 9.18(1).

(3) By Lemma 9.7, we have for all $u \in I$ and $y \in R$,

$$0 = [d^2(d(u) \cdot d(u)), y] \subseteq [d(d^2(u) \cdot \theta d(u) + \theta d(u) \cdot d^2(u)), y]$$

$$= 2[d(d^2(u) \cdot \theta d(u)), y] \subseteq 2[d^3(u) \cdot \theta^2 d(u) + \theta d^2(u) \cdot \theta d^2(u), y]$$

$$= 2d^3(u) \cdot [\theta^2 d(u), y] + \theta d^2(u)[\theta d^2(u), y] = 2d^3(u) \cdot [\theta^2 d(u), y].$$

Hence, $0 \in d^3(u)[\theta^2 d(u), y]$ by hypothesis. Thus, $d^3(u) = 0$ or $0 \in [\theta^2 d(u), y]$. Therefore, $d^3(u) = 0$ or $\theta^2 d(u) \in Z$. Suppose that $d^3(u) = 0$. Then, by Theorem 9.1, we have

$$0 = [d^2(\theta(u) \cdot d(u)), y]$$

$$\subseteq [d^2\theta(u) \cdot \theta^2 d(u) + 2d\theta^2(u) \cdot d^2\theta(u) + \theta^3(u) \cdot d^3(u), y]$$

$$= 3[d\theta^2(u) \cdot d^2\theta(u), y] = 3d^2\theta(u)[d\theta^2(u), y].$$

Hence, $0 \in d^2\theta(u)[d\theta^2(u), y]$ by hypothesis. Thus, $d^2\theta(u) = \theta d^2(u) = 0$ or $0 \in [d\theta^2(u), y]$. Therefore, $\theta d^2(u) = 0$ or $d\theta^2(u) = \theta^2 d(u) \in Z$. Put $A = \{u \in I \mid d\theta^2(u) \in Z\}$ and $B = \{u \in I \mid \theta d^2(u) = 0\}$. It is clear that $I = A \cup B$. Since A and B are canonical sub-hypergroups of I, $I = A$ or $I = B$ by Lemma 9.5. If $I = B$, i.e., $\theta d^2(u) = 0$, then $d^2(u) = 0$. So, $d = 0$ by Theorem 9.7(1), which is a contradiction. Thus, $I = A$, i.e., $d\theta^2(I) \subseteq Z$. Similar to (2), we can prove that R is commutative.

(4) For all $u \in I$ and $x \in R$,

$$0 = [d_1 d_2(d_2(u) \cdot d_2(u)), x] \subseteq [d_1(d_2^2(u) \cdot \theta d_2(u) + \theta d_2(u) \cdot d_2^2(u)), x]$$

$$\subseteq 2[\theta d_2^2(u) \cdot \theta d_1 d_2(u), x] = 2\theta d_1 d_2(u)[\theta d_2^2(u), x].$$

Hence, $0 \in \theta d_1 d_2(u)[\theta d_2^2(u), x]$ by hypothesis. So, $\theta d_1 d_2(u) = 0$ or $\theta d_2^2(u) \in Z$. Put $A = \{u \in I \mid \theta d_2^2(u) \in Z\}$ and $B = \{u \in I \mid \theta d_1 d_2(u) = 0\}$. It is clear that $I = A \cup B$. Since A and B are canonical subhypergroups of I, $I = A$ or $I = B$ by Lemma 9.5. If $I = B$, i.e., $\theta d_1 d_2(I) = 0$, then $d_1 = 0$ or $d_2 = 0$ by Theorem 9.7(2), which is a contradiction. Thus, $I = A$, i.e., $\theta d_2^2(I) \subseteq Z$. Now, (3) completes the proof. $\qquad\square$

9.4   Semi-derivations in Krasner Hyperrings

The results of this section are about semi-derivations in Krasner hyperrings and were obtained by Yilmaz and Yazarli [104].

Definition 9.8. Let R be a Krasner hyperring. A map $f : R \to R$ is said to be a semi-derivation associated with a function $g : R \to R$ if for all $r, s \in R$:

(1) $f(r + s) \subseteq f(r) + f(s)$;
(2) $f(rs) \in f(r)g(s) + rf(s) = f(r)s + g(r)f(s)$;
(3) $f(g(r)) = g(f(r))$.

The Krasner hyperring R endowed with a semi-derivation f is called an *f-semi-differential hyperring*. If the map f satisfies the conditions (2), (3), and $f(r + s) = f(r) + f(s)$, for all $r, s \in R$, then f is called a *strong semi-derivation* of R and the hyperring is called *strongly f-semi-differential hyperring*. Every derivation is a semi-derivation.

Example 9.18. Let $M(R)$ be the a Krasner hyperring defined in Example 9.1. Now, define a function f on $M(R)$ by $d\left(\begin{pmatrix} a & b \\ 0 & 0 \end{pmatrix} \right) = \begin{pmatrix} a & 0 \\ 0 & 0 \end{pmatrix}$ and a function g on $M(R)$ by $d\left(\begin{pmatrix} a & b \\ 0 & 0 \end{pmatrix} \right) = \begin{pmatrix} 0 & b \\ 0 & 0 \end{pmatrix}$. It is easy to check that f is a strong semi-derivation in $M(R)$.

Example 9.19. Let $R = \{r, s, t, w\}$ be a set with the hyperaddition $\oplus$ and the multiplication $\odot$ defined as follows:

$\oplus$	r	s	t	w
r	r	s	t	w
s	s	R	$\{s,t,w\}$	$\{s,t,w\}$
t	t	$\{s,t,w\}$	R	$\{s,t,w\}$
w	w	$\{s,t,w\}$	$\{s,t,w\}$	R

$\odot$	r	s	t	w
r	r	r	r	r
s	r	s	t	w
t	r	t	w	s
w	r	w	s	t

Then, $(R, \oplus, \odot)$ is a Krasner hyperring. We define a map $f :$ $R \to R$ by $f(r) = r$, $f(s) = t$, $f(t) = w$ and $f(w) = s$. Here, f is a strong semi-derivation of R, where $g : R \to R$ is defined by $g(r) = r, g(s) = w, g(t) = s, g(w) = t$.

Definition 9.9. Let R be a Krasner hyperring and $P(R)$ be the set of non-empty subsets of R. A map $f : R \to P(R)$ is said to be a *hypersemi-derivation associated with the function* $g : R \to R$ if it satisfies, for all $r, s \in R$:

(1) $f(r + s) \subseteq f(r) + f(s)$;
(2) $f(rs) \subseteq f(r)g(s) + rf(s) = f(r)s + g(r)f(s)$;
(3) $f(g(r)) = g(f(r))$.

If the map f satisfies $f(r + s) = f(r) + f(s)$, for all $r, s \in R$, and the conditions (2) and (3), then f is called a *strong hypersemi-derivation* of R.

Example 9.20. Consider the Krasner hyperring given in the above example. We define $f : R \to P(R)$ by $f(r) = r, f(s) = \{s, t\}, f(t) = \{t, w\}, f(w) = \{s, w\}$ and $g : R \to R$ by $g(r) = r, g(s) = t, g(t) = w, g(w) = s$. Then, f is a hypersemi-derivation associated with the function g.

Lemma 9.8. *Let R be a prime hyperring and $f : R \to R$ be a semi-derivation associated with a surjective function g and $r \in R$. If $rf(x) = 0$ (or $f(x)r = 0$) for all $x \in R$, then either $r = 0$ or $f = 0$.*

Proof. Let $s, t \in R$ and suppose $rf(x) = 0$, for all $x \in R$. Then, we get $0 = rf(st) \in r(f(s)g(t) + sf(t)) = rf(s)g(t) + rsf(t) = rsf(t)$. By the primeness of R, we have $r = 0$ or $f(t) = 0$, for all $t \in R$, i.e., $f = 0$. When $f(x)r = 0$, the proof is similar. $\square$

Lemma 9.9. *Let R be a prime hyperring and C be the extended centroid of R. Suppose that f_1 and f_2 are semi-derivations of R such that $f_1(r)sf_2(t) = f_2(r)sf_1(t)$, for all $r, s, t \in R$ and $f_1 \neq 0$. Then, there exists $\lambda \in C$ such that $f_2(r) = \lambda f_1(r)$, for all $r \in R$.*

Proof. Given $r \in R$, we have $f_1(r)sf_2(r) = f_2(r)sf1(r)$, for all $s \in R$. If $f_1(r) \neq 0$, then we have $f_2(r) = \lambda(r)f_1(r)$, for some $\lambda(r)$ in C. If $r, t \in R$ such that $f_1(r) \neq 0$ and $f_1(t) \neq$, then we can write $0 \in \lambda(t)f_1(r)sf_1(t) - \lambda(t)f_1(r)sf_1(t) = \lambda(t)_1(r)sf_1(t) - f_1(r)sf_2(t) = \lambda(t)f_1(r)sf_1(t) - f_2(r)sf_1(t) = \lambda(t)f_1(r)sf_1(t) - \lambda(r)f_1(r)sf_1(t)$.

Thus, we get $0 \in \lambda((t) - \lambda(r))f_1(r)sf_1(t)$. Hence, there exists $\alpha \in \lambda(t) - \lambda(r)$ such that $0 = f_1(r)sf_1(t)$. Since R is a prime hyperring, we conclude based on the previous lemma that $\alpha = 0$. It means that $\lambda(r) = \lambda(t)$, for all $r, t \in R$. For $f_1(r) \neq 0$, the proof is completed. Let $f_1(r) = 0$. Since $f_1 \neq$ and R is prime, we get $f_2(r) = 0$, for all $r \in R$. Therefore, we obtain $f_2(r) = \lambda f_1(r)$, for all $r \in R$. $\square$

Example 9.21. Let R be a Krasner hyperring. We define a map $f : R \to P(R)$ by $f(r) = r - g(r)$, for each $r \in R$, where g is a strong endomorphism. Clearly, the map f is well defined. Let $x \in f(r + s)$. Then, $x \in f(t)$, for some $t \in r + s$, and so $x \in t - g(t) \subseteq r + s - g(r + s) = r - g(r) + s - g(s) = f(r) + f(s)$. Thus, $f(r+s) \subseteq f(r) + f(s)$, for all $r, s \in R$. Now, $f(rs) = rs - g(rs) = rs - g(r)g(s) + 0 \subseteq rs - rg(s) + rg(s) - g(r)g(s) = r(s - g(s)) + (r - g(r))g(s) = rf(s) + f(r)g(s)$. Then, we get $f(rs) \subseteq f(r)g(s) + rf(s)$. Similarly, $f(rs) \subseteq f(r)s + g(r)f(s)$, for all $r, s \in R$. If $x \in f(g(r))$, then $x \in g(r) - g(g(r)) = g(r - g(r)) = g(f(r))$. Also, if $x \in g(f(r))$, then $x \in g(r - g(r)) = g(r) - g(g(r)) = f(g(r))$. Hence, for all $r \in R$, we have $f(g(r)) = g(f(r))$. Therefore, f is a hypersemi-derivation. In the following theorem, we give the structure of a semi-derivation in a prime hyperring.

Theorem 9.20. *Let R be a prime hyperring and f be a semi-derivation of R associated with the strong endomorphism $g : R \to R$. Then, one of the following holds:*

(1) *f is an ordinary derivation;*
(2) *f is a hypersemi-derivation, and there exists a $\lambda \in C$ such that $f(r) = \lambda(r - g(r))$, for all $r \in R$, where C is the extended centroid of R.*

Proof. We can write $0 \in f(r)(s - g(s)) - (r - g(r))f(s)$, for all $r, s \in R$. In particular, $0 \in f(r)(st - g(st)) - (r - g(r))f(st)$, for all $r, s, t \in R$. Here, $f(r)(st - g(st)) \subseteq f(r)st - f(r)g(s)g(t) \subseteq f(r)(s - g(s))g(t) + f(r)s(t - g(t))$ and $(r - g(r))f(st) \subseteq (r - g(r))f(s)g(t) + (r - g(r))sf(t)$, for all $r, s, t \in R$. Thus, we have $0 \in f(r)(s - g(s))g(t) + f(r)s(t - g(t)) - (r - g(r))f(s)g(t) - (r - g(r))sf(t) = [f(r)(s - g(s)) - (r - g(r))f(s)]g(t) + f(r)s(t - g(t)) - (r - g(r))sf(t)$, and so we get $0 \in f(r)s(t - g(t)) - (r - g(r))sf(t)$. If $g = 1$, the identity map on R, f is a derivation. Hence, we suppose that the set $1 - g \neq 0$. Hence, we obtain that $f(r)s(1 - g)(t) = (1 - g)(r)sf(t)$, for all $r, s, t \in R$. The desired result is obtained using Lemma 9.9. $\square$

9.5   Symmetric Bi-derivation in Krasner Hyperrings

In this section, we present symmetric bi-derivation and strong symmetric bi-derivation in Krasner hyperrings. These results were derived by Durna [32].

Definition 9.10. Let R be a Krasner hyperring. A mapping $D :$ $R \times R \to R$ is called a *symmetric* if $D(x,y) = D(y,x)$, for all $x,y \in R$.

Definition 9.11. Let R be a Krasner hyperring. A map $D :$ $R \times R \to R$ is said to be a *symmetric bi-derivation* of R if D satisfies:

(1) $D(x + z, y) \subseteq D(x, y) + D(z, y)$;
(2) $D(xz, y) \in D(x, y)z + xD(z, y)$, for all $x, y, z \in R$.

The Krasner hyperring R equipped with a symmetric bi-derivation D is called a *D-differential hyperring*.
If the map D is such that $D(x + z, y) = D(x, y) + D(z, y)$, for all $x, y, z \in R$, and satisfies condition (2), then D is called a *strong symmetric bi-derivation* of R. In this case, the hyperring is called a *strongly D-differential hyperring*.

Theorem 9.21. *Let R be a Krasner hyperring, $D : R \times R \to R$ be a symmetric bi-derivation of R, and 1 be the identity element of R. Then:*

(1) $D(a, 0) = 0$, *for all $a \in R$;*
(2) $D(-a, b) = -D(a, b)$, *for all $a, b \in R$;*
(3) $D(1, a) \in D(1, a) + D(1, a)$, *for all $a \in R$.*

Proof.   (1) $D(a, 0) = D(a, 0 \cdot 0) \in D(a, 0)0 + 0D(a, 0) = 0 + 0 = 0$, and so $D(a, 0) = 0$.
(2) For all $a, b \in R$, $0 = D(a, 0) = D(a, b - b) \subseteq D(a, b) + D(a, -b)$. That is, $D(a, b) \in 0 - D(a, -b)$. Hence, $D(a, b) = -D(a, -b)$. Therefore, $-D(a, b) = -(-D(a, -b)) = D(a, -b)$.
(3) $D(1, a) = D(1 \cdot 1, a) \in D(1, a) \cdot 1 + 1 \cdot D(1, a) = D(1, a) + D(1, a)$. Hence, $D(1, a) \in D(1, a) + D(1, a)$.   □

Example 9.22. Let R be a commutative Krasner hyperring and

$$M(R) = \left\{ \begin{pmatrix} a & 0 \\ b & 0 \end{pmatrix} \middle| a, b \in R \right\}$$

be a collection of 2×2 matrices over R. A hyperaddition $\oplus$ is defined on $M(R)$ by $\begin{pmatrix} a & 0 \\ b & 0 \end{pmatrix} \oplus \begin{pmatrix} c & 0 \\ d & 0 \end{pmatrix} = \left\{ \begin{pmatrix} x & 0 \\ y & 0 \end{pmatrix} \middle| x \in a + c, y \in b + d \right\}$, for all $\begin{pmatrix} a & 0 \\ b & 0 \end{pmatrix}, \begin{pmatrix} c & 0 \\ d & 0 \end{pmatrix} \in M(R)$. Then, $(M(R), \oplus)$ is a canonical hypergroup. The multiplication $\otimes$ is defined on $M(R)$ by $\begin{pmatrix} a & 0 \\ b & 0 \end{pmatrix} \otimes \begin{pmatrix} c & 0 \\ d & 0 \end{pmatrix} = \begin{pmatrix} ac & 0 \\ bc & 0 \end{pmatrix}$, for all $\begin{pmatrix} a & 0 \\ b & 0 \end{pmatrix}, \begin{pmatrix} c & 0 \\ d & 0 \end{pmatrix} \in M(R)$. The multiplication $\otimes$ is well defined and associative. Therefore, $(M(R), \otimes)$ is a semigroup; moreover, $M(R)$ is a Krasner hyperring.

Define a function D on $M(R)$ by $D\left(\begin{pmatrix} a & 0 \\ b & 0 \end{pmatrix}, \begin{pmatrix} c & 0 \\ d & 0 \end{pmatrix} \right) = \begin{pmatrix} 0 & 0 \\ ac & 0 \end{pmatrix}$. This map is well defined and symmetric. Let us show that D is a symmetric bi-derivation.

For all $\begin{pmatrix} a & 0 \\ b & 0 \end{pmatrix}, \begin{pmatrix} c & 0 \\ d & 0 \end{pmatrix}, \begin{pmatrix} e & 0 \\ f & 0 \end{pmatrix} \in M(R)$, $D\left(\begin{pmatrix} a & 0 \\ b & 0 \end{pmatrix} \oplus \begin{pmatrix} e & 0 \\ f & 0 \end{pmatrix}, \begin{pmatrix} c & 0 \\ d & 0 \end{pmatrix} \right) = D\left\{ \begin{pmatrix} r & 0 \\ s & 0 \end{pmatrix} \middle| r \in a + e, s \in b + f \right\},$ $\begin{pmatrix} c & 0 \\ d & 0 \end{pmatrix} \right\} = \left\{ \begin{pmatrix} 0 & 0 \\ rc & 0 \end{pmatrix} \middle| r \in a + e \right\}$ and $D\left(\begin{pmatrix} a & 0 \\ b & 0 \end{pmatrix}, \begin{pmatrix} c & 0 \\ d & 0 \end{pmatrix} \right) \oplus D\left(\begin{pmatrix} e & 0 \\ f & 0 \end{pmatrix}, \begin{pmatrix} c & 0 \\ d & 0 \end{pmatrix} \right) = \begin{pmatrix} 0 & 0 \\ ac & 0 \end{pmatrix} \oplus \begin{pmatrix} 0 & 0 \\ ec & 0 \end{pmatrix} = \left\{ \begin{pmatrix} 0 & 0 \\ l & 0 \end{pmatrix} \middle| l \in ac + ec \right\}$. On the other hand, $D\left(\begin{pmatrix} a & 0 \\ b & 0 \end{pmatrix} \otimes \begin{pmatrix} e & 0 \\ f & 0 \end{pmatrix}, \begin{pmatrix} c & 0 \\ d & 0 \end{pmatrix} \right) = D\left(\begin{pmatrix} ae & 0 \\ be & 0 \end{pmatrix}, \begin{pmatrix} c & 0 \\ d & 0 \end{pmatrix} \right) = \begin{pmatrix} 0 & 0 \\ (ae)c & 0 \end{pmatrix}$ and $\left\{ D\left(\begin{pmatrix} a & 0 \\ b & 0 \end{pmatrix}, \begin{pmatrix} c & 0 \\ d & 0 \end{pmatrix} \right) \otimes \begin{pmatrix} e & 0 \\ f & 0 \end{pmatrix} \right\} \oplus \left\{ \begin{pmatrix} a & 0 \\ b & 0 \end{pmatrix} \otimes D\left(\begin{pmatrix} e & 0 \\ f & 0 \end{pmatrix}, \begin{pmatrix} c & 0 \\ d & 0 \end{pmatrix} \right) \right\} = \left\{ \begin{pmatrix} 0 & 0 \\ ac & 0 \end{pmatrix} \otimes \begin{pmatrix} e & 0 \\ f & 0 \end{pmatrix} \right\} \oplus \left\{ \begin{pmatrix} a & 0 \\ b & 0 \end{pmatrix} \otimes \begin{pmatrix} 0 & 0 \\ ec & 0 \end{pmatrix} \right\} = \begin{pmatrix} 0 & 0 \\ (ac)e & 0 \end{pmatrix} \oplus \begin{pmatrix} 0 & 0 \\ 0 & 0 \end{pmatrix} = \begin{pmatrix} 0 & 0 \\ (ac)e & 0 \end{pmatrix}.$

Definition 9.12. Let R be a Krasner hyperring and $D : R \times R \to R$ be a symmetric map. A mapping $d : R \to R$ defined by $d(x) = D(x, x)$ is called the trace of D.

We recall that a Krasner hyperring R is said to be 2-torsion-free if $0 \in x + x$, for $x \in R$, implies that $x = 0$. Also, a Krasner hyperring R is said to be a prime hyperring if $aRb = 0$, for $a, b \in R$, implies either $a = 0$ or $b = 0$.

If $D : R \times R \to R$ is a symmetric mapping which is also bi-additive, it means that it is hyperadditive in both arguments. Then, the trace of D satisfies the conditions $d(x + y) = D(x + y, x + y) \subseteq d(x) + D(x, y) + D(x, y) + d(y)$ and $d(0) = D(0, 0) = 0$. If D is a strong symmetric bi-derivation, we have $d(x + y) = d(x) + D(x, y) + D(x, y) + d(y)$. Also, $d(-x) = -d(x)$. Indeed, $0 = d(0) = d(x + (-x)) \subseteq d(x) + D(x, -x) + D(x, -x) + d(-x) = -d(x) + d(-x)$, whence $d(-x) \in 0 - (-d(x))$. Hence, $d(-x) = d(x)$.

Theorem 9.22. *Let R be a Krasner hyperring, D be a symmetric bi-derivation of R, and a be a fixed element of R. Then, $S = \{x \in R \mid D(x, a) = 0\}$ is a subhyperring of R.*

Proof. Since $D(0, a) = 0$, it follows that S is non-empty. Let $x, y \in S$. Then, $D(x, a) = 0$ and $D(y, a) = 0$. Now, $D(x + y, a) \subseteq D(x, a) + D(y, a) = 0$. For all $x \in S, D(-x, a) = -D(x, a) = 0$ and $D(xy, a) \in D(x, a)y + xD(y, a) = 0$. Thus, for all $x, y \in S, x + y \subseteq S, -x \in S, xy \in S$. So, S is a subhyperring of R. $\square$

Theorem 9.23. *Let D be a symmetric bi-derivation of a prime hyperring R and $a \in R$ such that $aD(x, y) = 0$ (or $D(x, y)a = 0$), for all $x, y \in R$. Then, either $a = 0$ or $D = 0$.*

Proof. Let $x, y, z \in R$. Suppose that $aD(x, y) = 0$, for all $x, y \in R$, then $0 = aD(xz, y) \in axD(z, y) + aD(x, y)z = axD(z, y)$. Thus, $axD(z, y) = 0$. Since R is a prime hyperring, $a = 0$ or $D(z, y) = 0$. If $a \neq 0$, then $D(z, y) = 0$. Hence, $D = 0$. Suppose that $D(x, y)a = 0$, for all $x, y \in R$. Then, $0 = D(xz, y)a \in xD(z, y)a + D(x, y)za = D(x, y)za$. Thus, $D(x, y)za = 0$. Since R is a prime hyperring, $a = 0$ or $D(x, y) = 0$. If $a \neq 0$, then $D(x, y) = 0$. Therefore, $D = 0$. $\square$

Theorem 9.24. *Let D be a strong symmetric bi-derivation with trace d of the 2-torsion-free prime hyperring R and $a \in R$ such that*

$ad(x) = 0$ (*or* $d(x)a = 0$), *for all* $x \in R$. *Then, either* $a = 0$ *or* $D = 0$.

Proof. Suppose that $ad(x) = 0$, for all $x \in R$. Then, for all $y \in R$, $0 = ad(x + y) = ad(x) + aD(x,y) + aD(x,y) + ad(y) = aD(x,y) + aD(x,y)$. For all $z \in R, 0 = aD(xz,y) + aD(xz,y) \in aD(x,y)z + axD(z,y) + aD(x,y)z + axD(z,y) = axD(z,y) + axD(z,y)$. Since R is 2-torsion-free, we obtain that $axD(z,y) = 0$, for all $x, y, z \in R$. Since R is a prime hyperring, $a = 0$ or $D(z,y) = 0$. If $a \neq 0$, then $D(z,y) = 0$. Hence $D = 0$. □

Recall that a Krasner hyperring R is said to be a reduced hyperring if it has no nilpotent elements. In other words, if $x^n = 0$ for all $x \in R$ and a natural number n, then $x = 0$.

Theorem 9.25. *Let D be a strong symmetric bi-derivation of 2-torsion-free reduced hyperring R. If $D(D(x,y),y) = 0$ for all $x, y \in R$, then $D = 0$.*

Proof. Let $D(D(x,y),y) = 0$, for all $x, y \in R$. Replacing x by $xz, z \in R$, we get $0 = D(D(xz,y),y) \in D(D(x,y)z + xD(z,y),y) = D(D(x,y)z,y) + D(xD(z,y),y) \in D(x,y)D(z,y) + D(D(x,y),y)z + xD(D(z,y),y) + D(x,y)D(z,y)$. We obtain $0 \in D(x,y)D(z,y) + D(x,y)D(z,y)$. Since R is a 2-torsion-free hyperring, we have $D(x,y)D(z,y) = 0$, for all $x, y, z \in R$. If we take x instead of z, we get $(D(x,y))^2 = 0$, for all $x, y \in R$. Since R is a reduced hyperring, we have $D(x,y) = 0$, for all $x, y \in R$. Hence, $D = 0$. □

Definition 9.13. Let D be a nontrivial symmetric bi-derivation (resp., a strong symmetric bi-derivation) of a Krasner hyperring R. A hyperideal I of R is said to be a *D-differential* (resp., *strongly D-differential*) *hyperideal* of R if $D(I,I) \subseteq I$.

Theorem 9.26. *Let R be a reduced hyperring.*

(1) *If S is a non-empty subset of R, then $Ann(S)$ is a hyperideal of R.*
(2) *If S_1 and S_2 are subsets of R such that $S_1 \subseteq S_2$, then $Ann(S_2) \subseteq Ann(S_1)$.*

Proof. The proof of (1) is obvious. Let $x \in Ann(S_2)$. Then, $S_2x = 0$. That is, $s_2x = 0$, for all $s_2 \in S_2$. This means that x annihilates all elements of S_2. In particular, x annihilates all elements of S_1. Therefore, $x \in Ann(S_1)$. This completes the proof of (2). $\square$

Corollary 9.1. *Let R be a reduced hyperring and I be a D-differential hyperideal of R, then $Ann(I) \subseteq Ann(D(I,I))$.*

Proof. Since I is a D-differential hyperideal of R, we have $D(I,I) \subseteq I$. It follows that $Ann(I) \subseteq Ann(D(I,I))$. $\square$

Theorem 9.27. *Let D be a symmetric bi-derivation of a reduced hyperring R. Then, for any subset S of R, $D(Ann(S), Ann(S)) \subseteq Ann(S)$.*

Proof. If $x,y \in Ann(S)$, $Sx = 0$ and $Sy = 0$. For $s \in S$, $0 = D(sx,y) \in D(s,y)x + sD(x,y)$. Multiplying by s on the right, we obtain $0 \in D(s,y)xs + sD(x,y)s$. Hence, we have $sD(x,y)s = 0$. Thus, $(sD(x,y))^2 = 0$. Since R is a reduced hyperring, we obtain $sD(x,y) = 0$. Hence, $D(x,y) \in Ann(S)$. This means that $D(Ann(S), Ann(S)) \subseteq Ann(S)$. $\square$

Example 9.23. Consider the reduced hyperring $R = \{0, a, b, c\}$ with the hyperaddition $+$ and the multiplication $\cdot$ defined as follows:

+	0	a	b	c		$\cdot$	0	a	b	c
0	0	a	b	c		0	0	0	0	0
a	a	$\{0,b\}$	$\{a,c\}$	b		a	0	a	b	c
b	b	$\{a,c\}$	$\{0,b\}$	a		b	0	b	b	0
c	c	b	a	0		c	0	c	0	c

The map $D : R \times R \to R$, defined by $D(a,a) = D(b,b) = D(a,b) = D(b,a) = b$ and $D(x,y) = 0$ if x,y in the other cases, is a symmetric bi-derivation of R.

$Ann(0,b) = \{0,c\}$ is a hyperideal of R. Since $D(Ann(0,b), Ann(0,b)) = D(\{0,c\}, \{0,c\}) = \{0\} \subseteq Ann(0,b)$, it follows that $Ann(0,b)$ is a D-differential hyperideal of R.

9.6   Permuting Tri-derivations in Krasner Hyperrings

In this section, we study permuting tri-derivations and strong permuting tri-derivations of hyperrings. These notions are introduced and analyzed by Unal [79].

Definition 9.14. Let R be a Krasner hyperring. A map $D : R \times R \times R \to R$ is called a *permuting map* if it satisfies the condition $D(x, y, z) = D(x, z, y) = D(y, x, z) = D(y, z, x) = D(z, x, y) = D(z, y, x)$, for all $x, y, z \in R$.

Definition 9.15. Let R be a Krasner hyperring. A map $D : R \times R \times R \to R$ is said to be a *permuting tri-derivation* of R if D satisfies the following conditions:

(1) $D(x + w, y, z) \subseteq D(x, y, z) + D(w, y, z)$;
(2) $D(xw, y, z) \in D(x, y, z)w + xD(w, y, z)$, for all $x, y, z, w \in R$.

The Krasner hyperring R equipped with a permuting tri-derivation D is called a *D-differential hyperring*.

If the map D is such that $D(x + w, y, z) = D(x, y, z) + D(w, y, z)$, for all $x, y, z, w \in R$, and satisfies condition (ii), then D is called a *strong permuting tri-derivation* of R. In this case, the Krasner hyperring R is called a *strongly D-differential hyperring*.

Theorem 9.28. *Let R be a Krasner hyperring and $D : R \times R \times R \to R$ be a permuting tri-derivation of R. Then:*

(1) $D(a, b, 0) = 0$, for all $a, b \in R$;
(2) $D(-a, b, c) = -D(a, b, c)$, for all $a, b, c \in R$;
(3) $D(1, a, b) \in D(1, a, b) + D(1, a, b)$; for all $a, b \in R$, where 1 is the identity element of R.

Proof.   (1) $D(a, b, 0) = D(a, b, 0 \cdot 0) \in D(a, b, 0) \cdot 0 + 0 \cdot D(a, b, 0)$, and so $D(a, b, 0) = 0$.

(2) For every $a, b, c \in R, 0 = D(a, b, 0) = D(a, b, c - c) \subseteq D(a, b, c) + D(a, b, -c)$. Hence, $D(a, b, c) \in 0 - D(a, b, -c)$, whence $D(a, b, c) = -D(a, b, -c)$. Therefore, we obtain $-D(a, b, c) = -(-D(a, b, -c)) = D(a, b, -c)$.

(3) $D(1, a, b) = D(1 \cdot 1, a, b) \in D(1, a, b) \cdot 1 + 1 \cdot D(1, a, b) = D(1, a, b) + D(1, a, b)$, for all $a, b \in R$. Therefore, $D(1, a, b) \in D(1, a, b) + D(1, a, b)$. $\qquad\square$

Example 9.24. Consider the Krasner hyperring $R = \{0, a, b\}$ with the hyperaddition and multiplication defined as follows:

+	0	a	b
0	0	a	b
a	a	$\{a, b\}$	R
b	b	R	$\{a, b\}$

·	0	a	b
0	0	0	0
a	0	b	a
b	0	a	b

Define a map $D : R \times R \times R \to R$ by $D(0, 0, 0) = 0, D(a, 0, 0) = D(0, a, 0) = D(0, 0, a) = D(b, 0, 0) = D(0, b, 0) = D(0, 0, b) = 0$, $D(a, a, 0) = D(a, 0, a) = D(0, a, a) = b, D(b, b, 0) = D(b, 0, b) = D(0, b, b) = a, D(a, b, 0) = D(a, 0, b) = D(0, a, b) = D(b, a, 0) = D(b, 0, a) = D(0, b, a) = a, D(b, b, a) = D(b, a, b) = D(a, b, b) = a, D(a, a, b) = D(a, b, a) = D(b, a, a) = b, D(a, a, a) = D(b, b, b) = a$. Then, D is a strong permuting tri-derivation of R.

Example 9.25. Let R be a commutative hyperring, and recall the Krasner hyperring $M(R) = \left\{ \begin{pmatrix} 0 & a \\ 0 & b \end{pmatrix} \middle| a, b \in R \right\}$, where the hyperaddition $\oplus$ is defined on $M(R)$ by

$$\begin{pmatrix} 0 & a \\ 0 & b \end{pmatrix} \oplus \begin{pmatrix} 0 & c \\ 0 & d \end{pmatrix} = \left\{ \begin{pmatrix} 0 & x \\ 0 & y \end{pmatrix} \middle| x \in a + c, \ y \in b + d \right\}$$

and the multiplication on $M(R)$ is defined by $\begin{pmatrix} 0 & a \\ 0 & b \end{pmatrix} \otimes \begin{pmatrix} 0 & c \\ 0 & d \end{pmatrix} = \begin{pmatrix} 0 & ad \\ 0 & bd \end{pmatrix}$, for all $\begin{pmatrix} 0 & a \\ 0 & b \end{pmatrix}, \begin{pmatrix} 0 & c \\ 0 & d \end{pmatrix} \in M(R)$.

Now, define a function D on $M(R)$ by

$$D\left(\begin{pmatrix} 0 & a \\ 0 & b \end{pmatrix}, \begin{pmatrix} 0 & c \\ 0 & d \end{pmatrix}, \begin{pmatrix} 0 & e \\ 0 & f \end{pmatrix} \right) = \begin{pmatrix} 0 & bdf \\ 0 & 0 \end{pmatrix}.$$

Then, D is a permuting tri-derivation.

Indeed, for all $\begin{pmatrix} 0 & a \\ 0 & b \end{pmatrix}, \begin{pmatrix} 0 & c \\ 0 & d \end{pmatrix}, \begin{pmatrix} 0 & e \\ 0 & f \end{pmatrix}, \begin{pmatrix} 0 & g \\ 0 & h \end{pmatrix} \in M(R)$,

$$D\left(\begin{pmatrix} 0 & a \\ 0 & b \end{pmatrix} \oplus \begin{pmatrix} 0 & g \\ 0 & h \end{pmatrix}, \begin{pmatrix} 0 & c \\ 0 & d \end{pmatrix}, \begin{pmatrix} 0 & e \\ 0 & f \end{pmatrix} \right) = \left\{ \begin{pmatrix} 0 & sdf \\ 0 & 0 \end{pmatrix} \middle| \right.$$

$s \in b + h \Big\}$, and $D\left(\begin{pmatrix} 0 & a \\ 0 & b \end{pmatrix}, \begin{pmatrix} 0 & c \\ 0 & d \end{pmatrix}, \begin{pmatrix} 0 & e \\ 0 & f \end{pmatrix} \right) \oplus$

$$D\left(\begin{pmatrix} 0 & g \\ 0 & h \end{pmatrix}, \begin{pmatrix} 0 & c \\ 0 & d \end{pmatrix}, \begin{pmatrix} 0 & e \\ 0 & f \end{pmatrix} \right) = \left\{ \begin{pmatrix} 0 & 0 \\ 0 & l \end{pmatrix} \middle| l \in bdf + hdf \right\}.$$

We also have $D\left(\begin{pmatrix} 0 & a \\ 0 & b \end{pmatrix} \otimes \begin{pmatrix} 0 & g \\ 0 & h \end{pmatrix}, \begin{pmatrix} 0 & c \\ 0 & d \end{pmatrix}, \begin{pmatrix} 0 & e \\ 0 & f \end{pmatrix} \right) =$

$\begin{pmatrix} 0 & bhdf \\ 0 & 0 \end{pmatrix} = \left\{ D\left(\begin{pmatrix} 0 & a \\ 0 & b \end{pmatrix}, \begin{pmatrix} 0 & c \\ 0 & d \end{pmatrix}, \begin{pmatrix} 0 & e \\ 0 & f \end{pmatrix} \right) \otimes \begin{pmatrix} 0 & g \\ 0 & h \end{pmatrix} \right\} \oplus$

$\left\{ \begin{pmatrix} 0 & a \\ 0 & b \end{pmatrix} \otimes D\left(\begin{pmatrix} 0 & g \\ 0 & h \end{pmatrix}, \begin{pmatrix} 0 & c \\ 0 & d \end{pmatrix}, \begin{pmatrix} 0 & e \\ 0 & f \end{pmatrix} \right) \right\}.$

Hence, D is a strong permuting tri-derivation.

Definition 9.16. Let R be a Krasner hyperring and $D : R \times R \times R \to R$ be a permuting tri-derivation. A mapping $d : R \to R$ defined by $d(x) = D(x, x, x)$ is called the *trace* of D.

If $D : R \times R \times R \to R$ is a permuting tri-mapping, then the trace of D satisfies the following relation: $d(x + y) = D(x + y, x + y, x + y) \subseteq d(x) + D(x, x, y) + D(x, y, x) + D(x, y, y) + D(y, x, x) + D(y, x, y) + D(y, y, x) + d(y)$ and $d(0) = D(0, 0, 0)$. If D is a strong permuting tri-derivation, then $d(x + y) = d(x) + D(x, x, y) + D(x, y, x) + D(x, y, y) + D(y, x, x) + D(y, x, y) + D(y, y, x) + d(y)$. Since $0 = d(0) = d(x + (-x)) \subseteq dx + D(x, x, -x) + D(x, -x, x) + D(x, -x, -x) + D(-x, x, x) + D(-x, x, -x) + D(-x, -x, x) + d(-x) = -d(x) + d(-x)$, we have $d(-x) \in 0 - (-d(x))$. Therefore, we obtain $d(-x) = d(x)$.

Theorem 9.29. *Let R be a hyperring, D be a permuting tri-derivation of R, and a, b be fixed elements of R. Then, $S = \{x \in R \mid D(x, a, b) = 0\}$ is a subhyperring of R.*

Proof. S is non-empty since $D(0, a, b) = 0$. Hence, $D(x, a, b) = 0$ and $D(y, a, b) = 0$, for $x, y \in S$. Hence, $D(x + y, a, b) \subseteq D(x, a, b) + D(y, a, b)$. For all $x \in S$, $D(-x, a, b) = -D(x, a, b) = 0$. Also, $D(xy, a, b) \in D(x, a, b)y + xD(y, a, b) = 0$. Thus, for all $x, y \in S, x + y \subseteq S, -x \in S, xy \in S$. Therefore, S is a subhyperring of R. $\square$

Theorem 9.30. *Let D be a permuting tri-derivation of a prime hyperring R and $a \in R$ such that $aD(x, y, z) = 0$ (or $D(x, y, z)a = 0$), for all $x, y, z \in R$. Then, either $a = 0$ or $D = 0$.*

Proof. Assume that $aD(x, y, z) = 0$, for all $x, y, z \in R$, then we have $0 = aD(xt, y, z) \in axD(t, y, z) + aD(x, y, z)t = axD(t, y, z)$. Hence, $axD(t, y, z) = 0$. Since R is a prime hyperring, we obtain $a = 0$ or $D(t, y, z) = 0$. If $a \neq 0$, then $D(t, y, z) = 0$. Thus,

$D = 0$. Suppose that $D(x, y, z)a = 0$, for all $x, y, z \in R$, then $0 = D(xt, y, z)a \in xD(t, y, z)a + D(x, y, z)ta = D(x, y, z)ta$. Hence, $D(x, y, z)ta = 0$. Since R is a prime hyperring, we obtain $a = 0$ or $D(x, y, z) = 0$. If $a \neq 0$, then we have $D(x, y, z) = 0$. Therefore, $D = 0$. □

Theorem 9.31. *Let R be a prime hyperring with $\operatorname{char} R \neq 2, 3$ and D be a strong permuting tri-derivation with trace d of R and $a \in R$ such that $ad(x) = 0$ (or $d(x)a = 0$), for all $x \in R$. Then, either $a = 0$ or $D = 0$.*

Proof. Assume that $ad(x) = 0$, for all $x \in R$. Replacing x by $x + y$, we obtain $0 = ad(x + y) = ad(x) + 3aD(x, x, y) + 3aD(x, y, y) + ad(y)$. Since $\operatorname{char} R \neq 3$, we obtain

$$aD(x, x, y) + aD(x, y, y) = 0.$$

Writing $-x$ for x in the above formula, we have

$$aD(x, y, y) = 0.$$

Replacing x by xy in the above equality, we obtain $axd(y) = 0$. Since R is a prime hyperring, we obtain $a = 0$ or $d(y) = 0$, for all $x \in R$. Therefore, $a = 0$ or $D = 0$. □

Theorem 9.32. *Let D be a permuting tri-derivation of a 2-torsion-free reduced hyperring R. If $D(D(x, y, z), y, z) = 0$, for all $x, y, z \in R$, then $D = 0$.*

Proof. Suppose that $D(D(x, y, z), y, z) = 0$, for all $x, y, z \in R$. Replacing x by $xt, t \in R$, we obtain

$$
\begin{aligned}
0 = D(D(xt, y, z), y, z) &\in D(D(x, y, z)t + xD(t, y, z), y, z) \\
&\in D(D(x, y, z)t, y, z) + D(xD(t, y, z), y, z) \\
&\in D(D(x, y, z)t, y, z) + D(xD(t, y, z), y, z) \\
&\in D(x, y, z)D(t, y, z) + D(D(x, y, z), y, z)t \\
&\quad + D(x, y, z)D(t, y, z) + xD(D(t, y, z), y, z) \\
&\in D(x, y, z)D(t, y, z) + D(x, y, z)D(t, y, z).
\end{aligned}
$$

Since R is a 2-torsion-free hyperring, we get $D(x, y, z)D(t, y, z) = 0$. If we take x instead of t, we have $D(x, y, z)^2 = 0$, for all $x, y, z \in R$. Since R is a reduced hyperring, we have $D(x, y, z) = 0$, for all $x, y, z \in R$. Hence, we obtain $D = 0$. □

Chapter 10

Ordered Krasner Hyperrings

Omidi and Davvaz [74, 75] introduced the concept of Krasner hyperring, $(R, +, \cdot)$, besides a binary relation $\leq$, where $\leq$ is a partial order that satisfies the following conditions: (1) If $a \leq b$, then $a + c \leq b + c$, meaning that for any $x \in a+c$, there exists $y \in b+c$ such that $x \leq y$. The case $c + a \leq c + b$ is defined similarly. (2) If $a \leq b$ and $0 \leq c$, then $a \cdot c \leq b \cdot c$ and $c \cdot a \leq c \cdot b$. This structure is called an ordered Krasner hyperring.

In this chapter, we study the notion of ordered Krasner hyperring. The main references are Refs. [74–78].

10.1 Basic Results in Ordered Krasner Hyperrings

A *partially ordered ring* is a ring $(R, +, \cdot)$, together with a compatible partial order, i.e., a partial order $\leq$ on the underlying set R that is compatible with the ring operations in the sense that it satisfies: (1) for all $a, b, c \in R$, $a \leq b$ implies that $a + c \leq b + c$; (2) for all $a, b \in R$, $0 \leq a$, and $0 \leq b$, we have $0 \leq a \cdot b$. An ordered ring, also called a *totally ordered ring*, is a partially ordered ring $(R, \leq)$ where $\leq$ is additionally a total order. An element $a \in R$ such that $0 \leq a$ is called *positive*. If P is the set of positive elements of a partially ordered ring, then $P + P \subseteq P$ and $P \cdot P \subseteq P$. Furthermore, $P \cap (-P) = \{0\}$. If R is an ordered ring, then the set $\{x : x \in R, x \geq 0\}$ is called the *positive cone*. The positive cone of an ordered ring completely defines the order $x \leq y$ if and only if $y - x \in P$. An *ordered field* is an ordered ring which is also a field. It is easy to see that if $a, b, c \in R$ with $a \leq b$

and $0 \leq c$, then $a \cdot c \leq b \cdot c$. Note that every ring is an ordered ring with the trivial order.

Definition 10.1. An algebraic hyperstructure $(R, +, \cdot, \leq)$ is called an *ordered Krasner hyperring* if $(R, +, \cdot)$ is a Krasner hyperring with a partial order relation $\leq$, such that for all a, b and c in R:

(1) If $a \leq b$, then $a + c \leq b + c$, meaning that for any $x \in a + c$, there exists $y \in b + c$ such that $x \leq y$. The case $c + a \leq c + b$ is defined similarly.
(2) If $a \leq b$ and $0 \leq c$, then $a \cdot c \leq b \cdot c$ and $c \cdot a \leq c \cdot b$.

An element $a \in R$ is called a *positive* if $0 \leq a$. The set of all positive elements of R is called the *positive cone* of R and is denoted by $P = R^+$. The element $x \in R$ is called a *negative* if $x \leq 0$. The set of all negative elements of R is called the *negative cone* of R and is denoted by R^-.

Note that the concept of ordered Krasner hyperrings is a generalization of the concept of ordered rings. Indeed, every ordered ring is an ordered Krasner hyperring.

Definition 10.2. An *ordered Krasner hyperfield* is an ordered Krasner hyperring, where R is also a Krasner hyperfield.

Definition 10.3. An *ordered hyperdomain* is an ordered Krasner hyperring R, where R is a commutative hyperring with unit element and $a \cdot b = 0$ implies that $a = 0$ or $b = 0$, for all $a, b \in R$.

Example 10.1. Let $(R, +, \cdot)$ be a Krasner hyperring. Define a partial order on R by

$$x \leq_R y \quad \text{if and only if} \quad x = y \quad \text{for all } x, y \in R.$$

Then, $(R, +, \cdot, \leq_R)$ forms an ordered Krasner hyperring.

Example 10.2. Let $R = \{a, b, c\}$ be a set with the hyperoperation $\oplus$ and the binary operation $\odot$ defined as follows:

$\oplus$	a	b	c
a	a	b	c
b	b	b	$\{a,b,c\}$
c	c	$\{a,b,c\}$	c

$\odot$	a	b	c
a	a	a	a
b	a	b	c
c	a	b	c

Then, $(R, \oplus, \odot)$ is a Krasner hyperring. We have that $(R, \oplus, \odot, \leq)$ is an ordered Krasner hyperring where the order relation $\leq$ is defined by

$$\leq := \{(a,a), (b,b), (c,c), (a,b), (a,c)\}.$$

The covering relation and the figure of R are given by

$$\prec = \{(a,b), (a,c)\}.$$

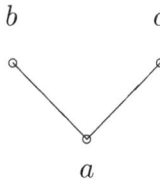

Example 10.3. Define the hyperoperation $\oplus$ on the unit interval $[0,1]$ by

$$x \oplus y = \begin{cases} \{\max\{x,y\}\}, & x \neq y \\ [0,x], & x = y \end{cases}$$

Then, $([0,1], \oplus, \cdot)$ is a Krasner hyperring where $\cdot$ is the usual multiplication. Consider $([0,1], \oplus, \cdot)$ as a poset with natural ordering, and define $x \cdot y$ to be the set of all upper bounds of $\{x,y\}$. Thus, $([0,1], \oplus, \cdot)$ is an ordered Krasner hyperring.

Example 10.4. Define the hyperoperation $\boxplus$ on the set $\{0,1\}$ by

$$0 \boxplus 0 = \{0\}, \ 1 \boxplus 0 = 0 \boxplus 1 = \{1\}, \ 1 \boxplus 1 = \{0,1\}.$$

Then, $(\{0,1\}, \boxplus, \cdot)$ is a Krasner hyperring, where $\cdot$ is the usual multiplication. Consider $(\{0,1\}, \boxplus, \cdot)$ as a poset with natural ordering. Thus, $(\{0,1\}, \boxplus, \cdot)$ is an ordered Krasner hyperring.

Let $(R_1, +_1, \cdot_1, \leq_1)$ and $(R_2, +_2, \cdot_2, \leq_2)$ be two ordered Krasner hyperrings. Then, the direct product of ordered Krasner hyperrings R_1 and R_2 is an ordered Krasner hyperring where for all (s_1, s_2) and (t_1, t_2) in $R_1 \times R_2$, we define:

(1) $(s_1, s_2) + (t_1, t_2) = \{(x,y) : x \in s_1 +_1 t_1, y \in s_2 +_2 t_2\}$,
(2) $(s_1, s_2) \cdot (t_1, t_2) = (s_1 \cdot_1 t_1, s_2 \cdot_2 t_2)$,
(3) $(s_1, s_2) \leq (t_1, t_2)$ if and only if $s_1 \leq_1 t_1$ and $s_2 \leq_2 t_2$.

Example 10.5. A *preorder* on an arbitrary non-empty set X is a binary relation on X which is reflexive and transitive. An antisymmetric preorder is said to be an order. Let ρ be a preorder relation on a Krasner hyperring $(R, +, \cdot)$. We say that ρ is *stable* if for every $a, b, x \in R$, $a\rho b$ implies $a + x\rho b + x$, $x \cdot a\rho x \cdot b$, and $a \cdot x\rho b \cdot x$. Let ρ be a stable preorder on a Krasner hyperring R. We construct an ordered Krasner hyperring $R/\rho = (R/\sim_\rho, \oplus, \odot, \leq)$. We define a binary relation $\sim_\rho$ on R as follows: For every $a, b \in R$, $a \sim_\rho b \Leftrightarrow a\rho b$ and $b\rho a$. Then, $\sim_\rho$ is a congruence relation on R. It can be shown easily that $\sim_\rho$ is an equivalence relation on R. Let $x, y, u, v \in R$ such that $x \sim_\rho y$ and $u \sim_\rho v$. Then, $x\rho y$, $y\rho x$, $u\rho v$, and $v\rho u$. Since ρ is a stable preorder on R, it follows that $x \cdot u\rho x \cdot v$ and $x \cdot v\rho y \cdot v$. Since ρ is transitive, it follows that $x \cdot u\rho y \cdot v$. Similarly, we obtain $y \cdot v\rho x \cdot u$. Thus, we have $x \cdot u \sim_\rho y \cdot v$. Similarly, we get $x + u \sim_\rho y + v$. Hence, $\sim_\rho$ is a congruence relation on R. We write $[a]$ for the congruence class containing a specified element a. Now, let $R/\sim_\rho = \{[a] : a \in R\}$ be the set of equivalence classes. The congruence $\sim_\rho$ determines a Krasner hyperring $(R/\sim_\rho, \oplus, \odot)$, with the hyperoperation $\oplus$ and the binary operation $\odot$ defined as follows:

$$[a] \oplus [b] = \{[z] : z \in a + b\},$$
$$[a] \odot [b] = [a \cdot b].$$

The definition of a congruence ensures that $\oplus$ and $\odot$ are well defined. We define an order relation $\leq$ on $R/\sim_\rho$ as follows: For any $a, b \in R$, $[a] \leq [b] \Leftrightarrow a\rho b$. It is easy to see that $(R/\sim_\rho, \oplus, \odot, \leq)$ is an ordered Krasner hyperring.

Example 10.6. Let $(\mathbb{Q}^+, +, \cdot)$ be a canonical hypergroup defined on $\mathbb{Q}^+ = \{x \in \mathbb{Q} \mid x \geq 0\}$, for all x, $x + x = \{y \mid y \leq x\}$, if $x \neq y$, $x + y = \max\{x, y\}$, where max is intended with respect to the natural order. If one considers in $\mathbb{Q}^+$ as product the ordinary multiplication $\cdot$, then $(\mathbb{Q}^+, +, \cdot)$ has the structure of a Krasner hyperring. We define $\leq$ as the natural ordering on $\mathbb{Q}^+$; then, $(\mathbb{Q}^+, +, \cdot, \leq)$ is an ordered Krasner hyperring.

Definition 10.4. A *homomorphism* from an ordered Krasner hyperring $(R_1, +_1, \cdot_1, \leq_1)$ onto an ordered Krasner hyperring $(R_2, +_2, \cdot_2, \leq_2)$ is a function $\varphi : R_1 \to R_2$ such that we have:

(1) $\varphi(a +_1 b) \subseteq \varphi(a) +_2 \varphi(b)$;
(2) $\varphi(a \cdot_1 b) = \varphi(a) \cdot_2 \varphi(b)$;
(3) if $a \leq_1 b$, then $\varphi(a) \leq_2 \varphi(b)$.

Also, φ is called a good (strong) homomorphism if in condition (1), the equality is valid. An *isomorphism* from $(R_1, +_1, \cdot_1, \leq_1)$ into $(R_2, +_2, \cdot_2, \leq_2)$ is a bijective good homomorphism from $(R_1, +_1, \cdot_1, \leq_1)$ onto $(R_2, +_2, \cdot_2, \leq_2)$. The kernel of φ, $\ker\varphi$, is defined by $\ker\varphi = \{x \in R_1 \mid \varphi(x) = 0_2\}$, where 0_2 is the zero of $(R_2, +_2, \cdot_2)$. In fact, $\ker\varphi$ may be empty.

Example 10.7. Let $([0,1], \oplus, \cdot)$ be an ordered Krasner hyperring defined as in Example 10.3 and $(\{0,1\}, \boxplus, \cdot)$ be an ordered Krasner hyperring defined as in Example 10.4. Define $\varphi : [0,1] \longrightarrow \{0,1\}$ by $\varphi(0) = 0$ and $\varphi(x) = 1$, for all $x \in (0,1]$. Then, we have

$$\varphi(x \oplus y) = \begin{cases} \varphi(\{0\}) = \{0\} = 0 \boxplus 0 = \varphi(x) \boxplus \varphi(y), & x = y = 0 \\ \varphi(\{x\}) = \{1\} = 1 \boxplus 0 = \varphi(x) \boxplus \varphi(y), & x > y = 0 \\ \varphi(\{x\}) = \{1\} \subsetneqq \{0,1\} = 1 \boxplus 1 = \varphi(x) \boxplus \varphi(y), & x > y > 0 \\ \varphi([0,x]) = \{0,1\} = 1 \boxplus 1 = \varphi(x) \boxplus \varphi(y), & x = y > 0 \end{cases}$$

and

$$\varphi(x \cdot y) = \varphi(x) \cdot \varphi(y) = \begin{cases} 0, & x = 0 \quad \text{or} \quad y = 0 \\ 1, & x \neq 0 \quad \text{and} \quad y \neq 0 \end{cases}$$

and

$$\text{if } x \leq_{[0,1]} y, \quad \text{then } \varphi(x) \leq_{\{0,1\}} \varphi(y).$$

It follows that φ is a homomorphism from $([0,1], \oplus, \cdot)$ onto $(\{0,1\}, \boxplus, \cdot)$, which is not a good (strong) homomorphism.

Example 10.8. Let $([0,1], \oplus, \cdot)$ be an ordered Krasner hyperring defined as in Example 10.3. Define $\varphi : [0,1] \longrightarrow [0,1]$ by $\varphi(x) = 1$, for all $x \in [0,1]$. Since $1 \oplus 1 = [0,1]$ and $1 \cdot 1 = 1$ and $x \leq_{[0,1]} y$ implies $\varphi(x) \leq_{[0,1]} \varphi(y)$, it follows that φ is a homomorphism from the ordered Krasner hyperring $([0,1], \oplus, \cdot)$ onto itself. Note that $\ker\varphi = \emptyset$. Thus, the kernel of an ordered Krasner hyperring homomorphism may be empty.

Let $(R_1, +_1, \cdot_1)$ and $(R_2, +_2, \cdot_2)$ be two Krasner hyperrings. Then, $(R_1 \times R_2, +, \cdot)$ is a Krasner hyperring, where the hyperoperation $+$ and operation $\cdot$ defined as follows:

$$(x_1, x_2) + (y_1, y_2) = \{(a, b) : a \in x_1 +_1 y_1, b \in x_2 +_2 y_2\},$$

$$(x_1, x_2) \cdot (y_1, y_2) = (x_1 \cdot_1 y_1, x_2 \cdot_2 y_2).$$

The lexicographical order defined on $R_1 \times R_2$ is as follows: $(x_1, x_2) \leq (y_1, y_2)$ if and only if $x_1 \leq_1 y_1$ or $x_1 = y_1$ and $x_2 \leq_2 y_2$. In the following, we prove that $(R_1 \times R_2, +, \cdot, \leq)$ is an ordered Krasner hyperring and is called the direct product of ordered Krasner hyperrings $(R_1, +_1, \cdot_1, \leq_1)$ and $(R_2, +_2, \cdot_2, \leq_2)$.

Theorem 10.1. *Let* $(R_1, +_1, \cdot_1, \leq_1)$ *and* $(R_2, +_2, \cdot_2, \leq_2)$ *be two ordered Krasner hyperrings. Then,* $(R_1 \times R_2, +, \cdot, \leq)$ *is an ordered Krasner hyperring.*

Proof. Suppose that $(x_1, x_2) \leq (y_1, y_2)$, for $(x_1, x_2), (y_1, y_2) \in R_1 \times R_2$, and $(t_1, t_2) \in (a_1, a_2) + (x_1, x_2)$, for $(a_1, a_2) \in R_1 \times R_2$. Then, $t_1 \in a_1 +_1 x_1$ and $t_2 \in a_2 +_2 x_2$. Since $(x_1, x_2) \leq (y_1, y_2)$, we have two cases:

(1) $x_1 \leq y_1$. Then, $t_1 \in a_1 +_1 x_1 \leq_1 a_1 +_1 y_1$, so there exists $s_1 \in a_1 +_1 y_1$ such that $t_1 \leq_1 s_1$. Now, if $s_2 \in a_2 +_2 y_2$, then $(t_1, t_2) \leq (s_1, s_2) \in (a_1, a_2) + (y_1, y_2)$.
(2) $x_1 = y_1$ and $x_2 \leq_2 y_2$. Then, $t_2 \in a_2 +_2 x_2 \leq_2 a_2 +_2 y_2$, so there exists $s_2 \in a_2 +_2 y_2$ such that $t_2 \leq_2 s_2$. Thus, $(t_1, t_2) \leq (t_1, s_2) \in (a_1, a_2) + (y_1, y_2)$.

Suppose that $(x_1, x_2) \leq (y_1, y_2)$ for $(x_1, x_2), (y_1, y_2) \in R_1 \times R_2$. We have two cases:

(1) $x_1 \leq y_1$. Then, $a_1 \cdot_1 x_1 \leq_1 a_1 \cdot_1 y_1$.
(2) $x_1 = y_1$ and $x_2 \leq_2 y_2$. Then, $a_2 \cdot_2 x_2 \leq_2 a_2 \cdot_2 y_2$. Thus, $(a_1, a_2) \cdot (x_1, x_2) \leq (a_1, a_2) \cdot (y_1, y_2)$, for $(a_1, a_2) \in R_1 \times R_2$.

Therefore, $(R_1 \times R_2, +, \cdot, \leq)$ is an ordered Krasner hyperring. $\square$

Example 10.9. Let $(R, +, \cdot, \leq)$ be an ordered ring and S be a nonempty subset of R. Then, $(R, +_s, \cdot_s, \leq)$ is an ordered Krasner hyperring, where $+_s, \cdot_s$ is the S-hyperoperation defined as $x +_s y = xSy$ and $x \cdot_s y = x \cdot y$, for every $x, y \in R$. Since $x \leq y$ for $x, y \in R$. Then, for every $a \in R$, we should prove that $a +_s x \leq a +_s y$. If $t \in a +_s x = aSx$, then there exists $z \in S$ such that $t = azx$. Now, since $(R, +, \cdot, \leq)$ is an ordered ring, we obtain $azx \leq azy \in aSy = a +_s y$. Thus, we have $a +_s x \leq a +_s y$. Therefore, $(R, +_s, \cdot, \leq)$ is an ordered Krasner hyperring.

Definition 10.5. Let $(R, +, \cdot, \leq)$ be an ordered Krasner hyperring. A subset I of R is called a *hyperideal* of R if it satisfies the following conditions:

(1) $(I, +)$ is a canonical subhypergroup of $(R, +)$;
(2) $x \cdot y \in I$ and $y \cdot x \in I$, for all $x \in I$ and $y \in R$;
(3) when $x \in I$ and $y \in R$ such that $y \leq x$, it implies that $y \in I$.

Theorem 10.2. *Let φ be a homomorphism from an ordered Krasner hyperring R into an ordered Krasner hyperring T such that $\ker\varphi \neq \emptyset$. Then, $\ker\varphi$ is a hyperideal of R.*

Proof. Let $r \in \ker\varphi$. Then, $\varphi(r) = 0$, so $\{\varphi(0)\} = \varphi(0) + 0 = \varphi(0) + \varphi(r) \supseteq \varphi(0 + r) = \varphi(\{r\}) = \{\varphi(r)\}$. This implies that $\varphi(0) = \varphi(r) = 0$. Thus, we have $0 \in \ker\varphi$. Let $x, y \in \ker\varphi$. Then, $\varphi(x) = 0 = \varphi(y)$. By Definition 10.4, $\varphi(x + y) \subseteq \varphi(x) + \varphi(y) = 0 + 0 = \{0\}$, it follows that $x + y \subseteq \ker\varphi$. Now, let $x \in R$. Since $0 \in x - x$, it follows that $\varphi(0) \in \varphi(x - x) \subseteq \varphi(x) + \varphi(-x)$. So, $0 \in \varphi(x) + \varphi(-x)$. Thus, $\varphi(-x)$ is the inverse of $\varphi(x)$ in the canonical hypergroup $(T, +)$. Therefore, $\varphi(-x) = -\varphi(x)$. So, $\varphi(-x) = -0 = 0$. Hence, $-x \in \ker\varphi$. Let $z \in R$. Then, $\varphi(zx) = \varphi(z)\varphi(x) = \varphi(z)0 = 0$, $\varphi(xz) = \varphi(x)\varphi(z) = 0\varphi(z) = 0$. Thus, we have $zx, xz \in \ker\varphi$. Now, let $x \in \ker\varphi$, $r \in R$, $r \leq x$. Since $x \in \ker\varphi$, it follows that $\varphi(x) = 0$. By Definition 10.4, $\varphi(r) \leq \varphi(x) = 0$. Thus, we have $\varphi(r) = 0$. Therefore, $r \in \ker\varphi$. This proves that $\ker\varphi$ is a hyperideal of R, as desired. $\square$

Definition 10.6. Suppose that $(R, +, \cdot, \leq)$ is an ordered Krasner hyperring and $A \subseteq R$ is a subhyperring. Then, A is a *convex* if for all $a \in A$ and all $r \in R$, the inequality $-a \leq r \leq a$ implies $r \in A$.

Example 10.10. Let $([0, 1], \oplus, \cdot)$ be an ordered Krasner hyperring defined as in Example 10.3. We can see that 0 is the zero of $([0, 1], \oplus, \cdot)$, and for every $x \in [0, 1]$, x is the inverse of x in $([0, 1], \oplus)$. The set $A = \{[0, a] | a \in [0, 1]\} \cup \{[0, a) | a \in (0, 1]\}$ comprises all canonical subhypergroups of the canonical hypergroup $([0, 1], \oplus)$. We can see that A is the set of all subhyperrings of R. Now, let $a \in A$, $r \in R$,

$-a \leq r \leq a$. Then, we have $a \leq r \leq a$, which implies $r = a \in A$. Therefore, A is convex.

Remark 10.1. A relation ρ^* is the *transitive closure* of a binary relation ρ if:

(1) ρ^* is transitive;
(2) $\rho \subseteq \rho^*$;
(3) for any relation ρ', if $\rho \subseteq \rho'$ and ρ' is transitive, then $\rho^* \subseteq \rho'$, i.e., ρ^* is the smallest relation that satisfies (1) and (2).

Definition 10.7. Let $(R, +, \cdot, \leq)$ be an ordered Krasner hyperring. A relation ρ on R is called *pseudo-order* if the following conditions hold:

(1) $\leq \subseteq \rho$;
(2) $a\rho b$ and $b\rho c$ imply $a\rho c$;
(3) $a\rho b$ implies $a + c\overline{\overline{\rho}}b + c$ and $c + a\overline{\overline{\rho}}c + b$, for all $c \in R$;
(4) $a\rho b$ implies $a \cdot c\overline{\overline{\rho}}b \cdot c$ and $c \cdot a\overline{\overline{\rho}}c \cdot b$, for all $c \in R$.

Theorem 10.3. *Let $(R, +, \cdot, \leq)$ be an ordered Krasner hyperring and ρ be a pseudo-order on R. Then, there exists a strongly regular relation ρ^* on R such that R/ρ^* is an ordered ring.*

Proof. Suppose that ρ^* is the relation on R defined as follows:

$$\rho^* = \{(a, b) \in R \times R \mid a\rho b \text{ and } b\rho a\}.$$

First, we show that ρ^* is a strongly regular relation on $(R, +)$ and $(R, \cdot)$. Clearly, $(a, a) \in \leq \subseteq \rho$, so $a\rho^*a$. If $(a, b) \in \rho^*$, then $a\rho b$ and $b\rho a$. Hence, $(b, a) \in \rho^*$. If $(a, b) \in \rho^*$ and $(b, c) \in \rho^*$, then $a\rho b$, $b\rho a$, $b\rho c$ and $c\rho b$. Hence, $a\rho c$ and $c\rho a$, which imply that $(a, c) \in \rho^*$. Thus, ρ^* is an equivalence relation. Now, let $a\rho^*b$ and $c \in R$. Then, $a\rho b$ and $b\rho a$. Since ρ is pseudo-order on R, by conditions (3) and (4) of Definition 10.7, we conclude that

$$a + c\overline{\overline{\rho}}b + c, c + a\overline{\overline{\rho}}c + b,$$
$$b + c\overline{\overline{\rho}}a + c, c + b\overline{\overline{\rho}}c + a,$$
$$a \cdot c\overline{\overline{\rho}}b \cdot c, c \cdot a\overline{\overline{\rho}}c \cdot b,$$
$$b \cdot c\overline{\overline{\rho}}a \cdot c, c \cdot b\overline{\overline{\rho}}c \cdot a.$$

Hence, for every $x \in a + c$ and $y \in b + c$, we have $x\rho y$ and $y\rho x$, which imply that $x\rho^* y$. So, $a + c \overline{\overline{\rho^*}} b + c$. Similarly, we obtain $c + a \overline{\overline{\rho^*}} c + b$. Thus, ρ^* is a strongly regular relation on $(R, +)$. Clearly, ρ^* is a strongly regular relation on $(R, \cdot)$. Hence, R/ρ^* with the following operations is a ring:

$$\rho^*(x) \oplus \rho^*(y) = \rho^*(z), \quad \text{for all } z \in \rho^*(x) + \rho^*(y);$$

$$\rho^*(x) \odot \rho^*(y) = \rho^*(x \cdot y).$$

Now, we define a relation $\preceq$ on R/ρ^* as follows:

$$\preceq := \{(\rho^*(x), \rho^*(y)) \in R/\rho^* \times R/\rho^* \mid \exists a \in \rho^*(x),$$

$$\exists b \in \rho^*(y) \quad \text{such that } (a, b) \in \rho\}.$$

We show that

$$\rho^*(x) \preceq \rho^*(y) \quad \Leftrightarrow \quad x\rho y.$$

Let $\rho^*(x) \preceq \rho^*(y)$. We show that for every $a \in \rho^*(x)$ and $b \in \rho^*(y)$, $a\rho b$. Since $\rho^*(x) \preceq \rho^*(y)$, there exist $x' \in \rho^*(x)$ and $y' \in \rho^*(y)$ such that $x'\rho y'$. Since $a \in \rho^*(x)$ and $x' \in \rho^*(x)$, we obtain $a\rho^* x'$, and so $a\rho x'$ and $x'\rho a$. Since $b \in \rho^*(y)$ and $y' \in \rho^*(y)$, we obtain $b\rho^* y'$, and so $b\rho y'$ and $y'\rho b$. Now, we have $a\rho x'$, $x'\rho y'$ and $y'\rho b$, which imply that $a\rho b$. Since $x \in \rho^*(x)$ and $y \in \rho^*(y)$, we conclude that $x\rho y$. Conversely, let $x\rho y$. Since $x \in \rho^*(x)$ and $y \in \rho^*(y)$, we have $\rho^*(x) \preceq \rho^*(y)$.

Finally, we prove that $(R/\rho^*, \oplus, \odot, \preceq)$ is an ordered ring. Suppose that $\rho^*(x) \in R/\rho^*$, where $x \in R$. Then, $(x, x) \in \leq \subseteq \rho$. Hence, $\rho^*(x) \preceq \rho^*(x)$. Let $\rho^*(x) \preceq \rho^*(y)$ and $\rho^*(y) \preceq \rho^*(x)$. Then, $x\rho y$ and $y\rho x$. Thus, $x\rho^* y$, which means that $\rho^*(x) = \rho^*(y)$. Now, let $\rho^*(x) \preceq \rho^*(y)$ and $\rho^*(y) \preceq \rho^*(z)$. Then, $x\rho y$ and $y\rho z$. So, $x\rho z$. This implies that $\rho^*(x) \preceq \rho^*(z)$.

Now, let $\rho^*(x) \preceq \rho^*(y)$ and $\rho^*(z) \in R/\rho^*$. Then, $x\rho y$ and $z \in R$. By conditions (3) and (4) of Definition 10.7, we have $x + z \overline{\overline{\rho}} y + z$, $z + x \overline{\overline{\rho}} z + y$, $x \cdot z \overline{\overline{\rho}} y \cdot z$ and $z \cdot x \overline{\overline{\rho}} z \cdot y$. So, for all $a \in x + z$ and $b \in y + z$, we have $a\rho b$. This implies that $\rho^*(a) \preceq \rho^*(b)$. Hence, $\rho^*(x) \oplus \rho^*(z) \preceq \rho^*(y) \oplus \rho^*(z)$. Similarly, we get $\rho^*(a) \preceq \rho^*(b)$. Hence, $\rho^*(z) \oplus \rho^*(x) \preceq \rho^*(z) \oplus \rho^*(y)$. Also, we have $\rho^*(a) \preceq \rho^*(b)$. Hence, $\rho^*(x) \odot \rho^*(z) \preceq \rho^*(y) \odot \rho^*(z)$ and $\rho^*(z) \odot \rho^*(x) \preceq \rho^*(z) \odot \rho^*(y)$. $\square$

Theorem 10.4. *Let $(R, +, \cdot, \leq)$ be an ordered Krasner hyperring and ρ be a pseudo-order on R. Let $\mathcal{X} = \{\theta \mid \theta$ be a pseudo-order on R such that $\rho \subseteq \theta\}$. Let $\mathcal{Y}$ be the set of all pseudo-orders on R/ρ^*. Then, $\operatorname{card}(\mathcal{X}) = \operatorname{card}(\mathcal{Y})$.*

Proof. For $\theta \in \mathcal{X}$, we define a relation θ' on R/ρ^* as follows:

$$\theta' := \{(\rho^*(x), \rho^*(y)) \in R/\rho^* \times R/\rho^* \mid \exists a \in \rho^*(x),$$

$$\exists b \in \rho^*(y) \text{ such that } (a, b) \in \theta\}.$$

First, we show that

$$(\rho^*(x), \rho^*(y)) \in \theta' \Leftrightarrow (x, y) \in \theta.$$

Let $(\rho^*(x), \rho^*(y)) \in \theta'$. We show that for every $a \in \rho^*(x)$ and $b \in \rho^*(y)$, $(a, b) \in \theta$. Since $(\rho^*(x), \rho^*(y)) \in \theta'$, there exist $x' \in \rho^*(x)$ and $y' \in \rho^*(y)$ such that $(x', y') \in \theta$. Since $a\rho^* x$ and $x\rho^* x'$, we have $a\rho^* x'$. So, $a\rho x'$. Since $\rho \subseteq \theta$, it follows that $a\theta x'$. Similarly, we obtain $y'\theta b$. Now, we have $a\theta x'$, $x'\theta y'$ and $y'\theta b$. Thus, we have $a\theta b$. Since $x \in \rho^*(x)$ and $y \in \rho^*(y)$, we conclude that $(x, y) \in \theta$. Conversely, let $(x, y) \in \theta$. Since $x \in \rho^*(x)$ and $y \in \rho^*(y)$, we obtain $(\rho^*(x), \rho^*(y)) \in \theta'$.

Now, let $(\rho^*(x), \rho^*(y)) \in \preceq$. Then, by Theorem 10.3, $(x, y) \in \rho \subseteq \theta$. This implies that $(\rho^*(x), \rho^*(y)) \in \theta'$. Hence, $\preceq \subseteq \theta'$. Now, suppose that $(\rho^*(x), \rho^*(y)) \in \theta'$ and $(\rho^*(y), \rho^*(z)) \in \theta'$. Then, $(x, y) \in \theta$ and $(y, z) \in \theta$, which imply that $(x, z) \in \theta$. Thus, we have $(\rho^*(x), \rho^*(z)) \in \theta'$. Also, if $(\rho^*(x), \rho^*(y)) \in \theta'$ and $\rho^*(z) \in R/\rho^*$, then $(x, y) \in \theta$ and $z \in R$. Thus, $x + z\overline{\overline{\theta}}y + z$, $z + x\overline{\overline{\theta}}z + y$, $x \cdot z\overline{\overline{\theta}}y \cdot z$ and $z \cdot x\overline{\overline{\theta}}z \cdot y$. So, for all $a \in x+z$ and $a' = x \cdot z$ and for all $b \in y+z$ and $b' = y \cdot z$, we have $a\theta b$ and $a'\theta b'$. This implies that $\theta'(\rho^*(a)) = \theta'(\rho^*(b))$ and $\theta'(\rho^*(a')) = \theta'(\rho^*(b'))$, and so $\theta'(\rho^*(x) \oplus \rho^*(z)) = \theta'(\rho^*(y) \oplus \rho^*(z))$ and $\theta'(\rho^*(x) \odot \rho^*(z)) = \theta'(\rho^*(y) \odot \rho^*(z))$. Thus, $(\rho^*(x) \oplus \rho^*(z))\theta'(\rho^*(y) \oplus \rho^*(z))$ and $(\rho^*(x) \odot \rho^*(z))\theta'(\rho^*(y) \odot \rho^*(z))$. Similarly, we obtain $(\rho^*(z) \oplus \rho^*(x))\theta'(\rho^*(z) \oplus \rho^*(y))$ and $(\rho^*(z) \odot \rho^*(x))\theta'(\rho^*(z) \odot \rho^*(y))$. Therefore, if $\theta \in \mathcal{X}$, then θ' is a pseudo-order on R/ρ^*.

Now, we define the map $\psi : \mathcal{X} \to \mathcal{Y}$ by $\psi(\theta) = \theta'$. Let $\theta_1, \theta_2 \in \mathcal{X}$ and $\theta_1 = \theta_2$. Suppose that $(\rho^*(x)\,\rho^*(y)) \in \theta'_1$ is an arbitrary element. Then, $(x, y) \in \theta_1$, and so $(x, y) \in \theta_2$. This implies that $(\rho^*(x)\,\rho^*(y)) \in \theta'_2$. Thus, we have $\theta'_1 \subseteq \theta'_2$. Similarly, we obtain $\theta'_2 \subseteq \theta'_1$. Therefore, ψ is well defined.

Let $\theta_1, \theta_2 \in \mathcal{X}$ and $\theta_1' = \theta_2'$. Suppose that $(x, y) \in \theta_1$ is an arbitrary element. Then, $(\rho^*(x)\,\rho^*(y)) \in \theta_1'$, and so $(\rho^*(x)\,\rho^*(y)) \in \theta_2'$. This implies that $(x, y) \in \theta_2$. Thus, we have $\theta_1 \subseteq \theta_2$. Similarly, we obtain $\theta_2 \subseteq \theta_1$. Therefore, ψ is one to one.

Finally, we prove that ψ is onto. Consider $\Sigma \in \mathcal{Y}$. We define a relation θ on R as follows:

$$\theta = \{(x, y) \mid (\rho^*(x), \rho^*(y)) \in \Sigma\}.$$

We show that θ is a pseudo-order on R and $\rho \subseteq \theta$. Suppose that $(x, y) \in \rho$. Then, $(\rho^*(x), \rho^*(y)) \in \preceq \subseteq \Sigma$, and so $(x, y) \in \theta$. If $(x, y) \in \leq$, then $(x, y) \in \rho \subseteq \theta$. Hence, $\leq \subseteq \theta$. Let $(x, y) \in \theta$ and $(y, z) \in \theta$. Then, $(\rho^*(x), \rho^*(y)) \in \Sigma$ and $(\rho^*(y), \rho^*(z)) \in \Sigma$. So, $(\rho^*(x), \rho^*(z)) \in \Sigma$. This implies that $(x, z) \in \theta$.

Now, let $(x, y) \in \theta$ and $z \in R$. Then, $(\rho^*(x), \rho^*(y)) \in \Sigma$ and $\rho^*(z) \in R/\rho^*$. Thus, $(\rho^*(x) \oplus \rho^*(z), \rho^*(y) \oplus \rho^*(z)) \in \Sigma$ and $(\rho^*(x) \odot \rho^*(z), \rho^*(y) \odot \rho^*(z)) \in \Sigma$. Therefore, for all $a \in x + z$ and $a' = x \cdot z$ and for all $b \in y + z$ and $b' = y \cdot z$, we have $(\rho^*(a), \rho^*(b)) \in \Sigma$ and $(\rho^*(a'), \rho^*(b')) \in \Sigma$. This means that $(a, b) \in \theta$ and $(a', b') \in \theta$. Therefore, $x + z \overline{\overline{\theta}} y + z$ and $x \cdot z \overline{\overline{\theta}} y \cdot z$. Similarly, we obtain $z + x \overline{\overline{\theta}} z + y$ and $z \cdot x \overline{\overline{\theta}} z \cdot y$. Now, obviously we have $\theta' = \Sigma$. $\square$

Remark 10.2. In Theorem 10.4, it is easy to see that $\theta_1 \subseteq \theta_2$ if and only if $\theta_1' \subseteq \theta_2'$.

Remark 10.3. If $(R, +, \cdot, \leq_R)$ and $(T, \uplus, \diamond, \leq_T)$ are two ordered rings and $\varphi : R \to T$ is a homomorphism, we denote by k the pseudo-order on R defined by $k = \{(a, b) | \varphi(a) \leq_T \varphi(b)\}$. Then, we have $\ker\varphi = k^*$.

Corollary 10.1. *Let* $(R, +, \cdot, \leq_R)$ *and* $(T, \uplus, \diamond, \leq_T)$ *be two ordered rings and* $\varphi : R \to T$ *is a homomorphism. Then,* $S/\ker\varphi \cong \operatorname{Im}\varphi$.

Let $(R, +, \cdot, \leq_R)$ be an ordered Krasner hyperring and ρ, θ be pseudo-orders on R such that $\rho \subseteq \theta$. We define a relation θ/ρ on R/ρ^* as follows:

$$\theta/\rho := \{(\rho^*(a), \rho^*(b)) \in R/\rho^* \times R/\rho^* \mid \exists x \in \rho^*(a),$$

$$\exists y \in \rho^*(b) \quad \text{such that } (x, y) \in \theta\}.$$

Then, we can see that

$$(\rho^*(a), \rho^*(b)) \in \theta/\rho \quad \Leftrightarrow \quad (a, b) \in \theta.$$

Theorem 10.5. *Let $(R, +, \cdot, \leq_R)$ be an ordered Krasner hyperring and ρ, θ be pseudo-orders on R such that $\rho \subseteq \theta$. Then:*

(1) θ/ρ is a pseudo-order on R/ρ^*.
(2) $(R/\rho^*)/(\theta/\rho)^* \cong R/\rho^*$.

Proof. (1) If $(\rho^*(a), \rho^*(b)) \in \preceq_\rho$, then $(a, b) \in \rho$. So, $(a, b) \in \theta$, which implies that $(\rho^*(a), \rho^*(b)) \in \theta/\rho$. Thus, $\preceq_\rho \subseteq \theta/\rho$. Let $(\rho^*(a), \rho^*(b)) \in \theta/\rho$ and $(\rho^*(b), \rho^*(c)) \in \theta/\rho$. Then, $(a, b) \in \theta$ and $(b, c) \in \theta$. Hence, $(a, c) \in \theta$, and so $(\rho^*(a), \rho^*(c)) \in \theta/\rho$. Now, let $(\rho^*(a), \rho^*(b)) \in \theta/\rho$ and $\rho^*(c) \in R/\rho^*$. Then, $(a, b) \in \theta$. Since θ is a pseudo-order on R, we obtain $a + c \bar{\bar{\theta}} b + c$, $c + a \bar{\bar{\theta}} c + b$, $a \cdot c \bar{\bar{\theta}} b \cdot c$, and $c \cdot a \bar{\bar{\theta}} c \cdot b$. Hence, for all $x \in a + c$ and $x' = a \cdot c$ and for all $y \in b + c$ and $y' = b \cdot c$, we have $(x, y) \in \theta$ and $(x', y') \in \theta$. This implies that $(\rho^*(x), \rho^*(y)) \in \theta/\rho$ and $(\rho^*(x'), \rho^*(y')) \in \theta/\rho$. Since ρ^* is a strongly regular relation on R, $\rho^*(x) = \rho^*(a) \oplus \rho^*(c)$, $\rho^*(y) = \rho^*(b) \oplus \rho^*(c)$, $\rho^*(x') = \rho^*(a) \odot \rho^*(c)$, and $\rho^*(y') = \rho^*(b) \odot \rho^*(c)$. So, we obtain $(\rho^*(a) \oplus \rho^*(c), \rho^*(b) \oplus \rho^*(c)) \in \theta/\rho$ and $(\rho^*(a) \odot \rho^*(c), \rho^*(b) \odot \rho^*(c)) \in \theta/\rho$. Similarly, we obtain $(\rho^*(c) \oplus \rho^*(a), \rho^*(c) \oplus \rho^*(b)) \in \theta/\rho$ and $(\rho^*(c) \odot \rho^*(a), \rho^*(c) \odot \rho^*(b)) \in \theta/\rho$. Therefore, θ/ρ is a pseudo-order on R/ρ^*.

(2) We define the map $\psi : R/\rho^* \to R/\theta^*$ by $\psi(\rho^*(a)) = \theta^*(a)$. If $\rho^*(a) = \rho^*(b)$, then $(a, b) \in \rho^*$. Hence, by the definition of ρ^*, $(a, b) \in \rho \subseteq \theta$, and $(b, a) \in \rho \subseteq \theta$. This implies that $(a, b) \in \theta^*$, and so $\theta^*(a) = \theta^*(b)$. Thus, θ is well defined. For all $\rho^*(x), \rho^*(y) \in R/\rho^*$, we have

$$\rho^*(x) \oplus \rho^*(y) = \rho^*(z), \quad \text{for all } z \in x + y;$$
$$\theta^*(x) \uplus \theta^*(y) = \theta^*(z), \quad \text{for all } z \in x + y;$$
$$\rho^*(x) \odot \rho^*(y) = \rho^*(x \cdot y);$$
$$\theta^*(x) \otimes \theta^*(y) = \theta^*(x \cdot y).$$

Thus,

$$\begin{aligned}
\psi(\rho^*(x) \oplus \rho^*(y)) &= \psi(\rho^*(z)), \quad \text{for all } z \in x + y \\
&= \theta^*(z), \quad \text{for all } z \in x + y \\
&= \theta^*(x) \boxplus \theta^*(y) \\
&= \psi(\rho^*(x)) \boxplus \psi(\rho^*(y)),
\end{aligned}$$

and

$$\psi(\rho^*(x) \odot \rho^*(y)) = \psi(\rho^*(z)), \quad \text{for } z = x \cdot y$$
$$= \theta^*(z), \quad \text{for } z = x \cdot y$$
$$= \theta^*(x) \otimes \theta^*(y)$$
$$= \psi(\rho^*(x)) \otimes \psi(\rho^*(y)),$$

and if $\rho^*(x) \preceq_\rho \rho^*(y)$, then $(x,y) \in \rho$. So, $(x,y) \in \theta$, and this implies that $\theta^*(x) \preceq_\theta \theta^*(y)$. Therefore, ψ is a homomorphism. It is easy to see that ψ is onto since

$$Im\psi = \{\psi(\rho^*(x)) \mid x \in R\} = \{\theta^*(x) \mid x \in R\} = R/\theta^*.$$

So, by Corollary 10.1, we obtain

$$(R/\rho^*)/ker\psi \cong Im\psi = R/\theta^*.$$

Suppose that

$$k := \{(\rho^*(x), \rho^*(y)) \mid \psi(\rho^*(x)) \preceq_\theta \psi(\rho^*(y))$$

Then,

$$(\rho^*(x), \rho^*(y)) \in k \quad \Leftrightarrow \quad \psi(\rho^*(x)) \preceq_\theta \psi(\rho^*(y))$$
$$\Leftrightarrow \quad \theta^*(x) \preceq_\theta \theta^*(y)$$
$$\Leftrightarrow \quad (x,y) \in \theta$$
$$\Leftrightarrow \quad (\rho^*(x), \rho^*(y)) \in \theta/\rho.$$

Hence, $k = \theta/\rho$ and by Remark 10.3, we have $k^* = (\theta/\rho)^* = ker\psi$.

$\square$

Definition 10.8. Let $(R, +, \cdot, \leq_R)$ and $(T, \uplus, \diamond, \leq_T)$ be two ordered Krasner hyperrings, ρ_1, ρ_2 be two pseudo-orders on R, T, respectively, and the map $f : R \to T$ be a homomorphism. Then, f is called a (ρ_1, ρ_2)-*homomorphism* if

$$(a,b) \in \rho_1 \Rightarrow (f(a), f(b)) \in \rho_2.$$

Lemma 10.1. *Let* $(R, +, \cdot, \leq_R)$ *and* $(T, \uplus, \diamond, \leq_T)$ *be two ordered Krasner hyperrings,* ρ_1, ρ_2 *be two pseudo-orders on* R, T, *respectively, and the map* $f : R \to T$ *be a* (ρ_1, ρ_2)-*homomorphism. Then, the map* $\overline{f} : R/\rho_1^* \to T/\rho_2^*$ *defined by*

$$\overline{f}(\rho_1^*(x)) = \rho_2^*(f(x)), \quad \text{for all } x \in R, \tag{10.1}$$

is a homomorphism of rings.

Proof. Suppose that $\rho_1^*(x) = \rho_1^*(y)$. Then, we have $(x, y) \in \rho_1$. Since f is a (ρ_1, ρ_2)-homomorphism, it follows that $(f(x), f(y)) \in \rho_2$. This implies that $\rho_2^*(f(x)) = \rho_2^*(f(y))$ or $\overline{f}(\rho_1^*(x)) = \overline{f}(\rho_1^*(y))$. Therefore, $\overline{f}$ is well defined. Now, we show that $\overline{f}$ is a homomorphism. Suppose that $\rho_1^*(x), \rho_1^*(y)$ are two arbitrary elements of R/ρ_1^*. Then,

$$\overline{f}(\rho_1^*(x) \oplus \rho_1^*(y)) = \overline{f}(\rho_1^*(z)) = \rho_2^*(f(z)), \quad \text{for all } z \in x + y.$$

Since $z \in x + y$, it follows that $f(z) \in f(x) \uplus f(y)$. Since ρ_2^* is a strongly regular relation, we obtain $\rho_2^*(f(z)) = \rho_2^*(f(x)) \boxplus \rho_2^*(f(y))$. Thus, we have

$$\overline{f}(\rho_1^*(x) \oplus \rho_1^*(y)) = \rho_2^*(f(x)) \boxplus \rho_2^*(f(y)) = \overline{f}(\rho_1^*(x)) \boxplus \overline{f}(\rho_1^*(y)).$$

Now, suppose that $\rho_1^*(x), \rho_1^*(y)$ are two arbitrary elements of R/ρ_1^*. Then,

$$\overline{f}(\rho_1^*(x) \odot \rho_1^*(y)) = \overline{f}(\rho_1^*(z)) = \rho_2^*(f(z)), \quad \text{for } z = x \cdot y.$$

Since $z = x \cdot y$, it follows that $f(z) = f(x) \diamond f(y)$. Since ρ_2^* is a strongly regular relation, we obtain $\rho_2^*(f(z)) = \rho_2^*(f(x)) \otimes \rho_2^*(f(y))$. Thus, we have

$$\overline{f}(\rho_1^*(x) \odot \rho_1^*(y)) = \rho_2^*(f(x)) \otimes \rho_2^*(f(y)) = \overline{f}(\rho_1^*(x)) \otimes \overline{f}(\rho_1^*(y)). \qquad \square$$

Theorem 10.6. *Let* $(R, +, \cdot, \leq_R)$ *and* $(T, \uplus, \diamond, \leq_T)$ *be two ordered Krasner hyperrings,* ρ_1, ρ_2 *be two pseudo-orders on* R, T, *respectively, and the map* $f : R \to T$ *be a* (ρ_1, ρ_2)-*homomorphism. Then, the relation* ρ_f *defined by*

$$\rho_f := \{(\rho_1^*(x), \rho_1^*(y)) \mid \rho_2^*(f(x)) \preceq_T \rho_2^*(f(y))\}$$

is a pseudo-order on R/ρ_1^*.

Proof. Suppose that $(\rho_1^*(x), \rho_1^*(y)) \in \preceq_R$. By Lemma 10.1, $\overline{f}$ is a homomorphism. Since $\rho_1^*(x) \preceq_R \rho_1^*(y)$, it follows that $\overline{f}(\rho_1^*(x)) \preceq_T \overline{f}(\rho_1^*(y))$. Thus, we have $\rho_2^*(f(x)) \preceq_T \rho_2^*(f(y))$. This means that $(\rho_1^*(x), \rho_1^*(y)) \in \rho_f$.

Let $(\rho_1^*(x), \rho_1^*(y)) \in \rho_f$ and $(\rho_1^*(y), \rho_1^*(z)) \in \rho_f$. Then, we have $\rho_2^*(f(x)) \preceq_T \rho_2^*(f(y))$ and $\rho_2^*(f(y)) \preceq_T \rho_2^*(f(z))$. So, $\rho_2^*(f(x)) \preceq_T \rho_2^*(f(z))$. This implies that $(\rho_1^*(x), \rho_1^*(z)) \in \rho_f$.

Now, let $(\rho_1^*(x), \rho_1^*(y)) \in \rho_f$ and $\rho_1^*(z) \in R/\rho_1^*$. We show that $(\rho_1^*(x) \oplus \rho_1^*(z), \rho_1^*(y) \oplus \rho_1^*(z)) \in \rho_f$. Since $(\rho_1^*(x), \rho_1^*(y)) \in \rho_f$, it follows that $\overline{f}(\rho_1^*(x)) \preceq_T \overline{f}(\rho_1^*(y))$. Thus, by definition, we get $\overline{f}(\rho_1^*(x)) \boxplus \overline{f}(\rho_1^*(z)) \preceq_T \overline{f}(\rho_1^*(y)) \boxplus \overline{f}(\rho_1^*(z))$. So, $\overline{f}(\rho_1^*(x) \oplus \rho_1^*(z)) \preceq_T \overline{f}(\rho_1^*(y) \oplus \rho_1^*(z))$. Hence, for all $u \in x + z$ and for all $v \in y + z$, we have $\overline{f}(\rho_1^*(u)) \preceq_T \overline{f}(\rho_1^*(v))$. This implies that $(\rho_1^*(u), \rho_1^*(v)) \in \rho_f$. Therefore, we have $(\rho_1^*(x) \oplus \rho_1^*(z), \rho_1^*(y) \oplus \rho_1^*(z)) \in \rho_f$. Similarly, we obtain $(\rho_1^*(z) \oplus \rho_1^*(x), \rho_1^*(z) \oplus \rho_1^*(y)) \in \rho_f$. Now, we show that $(\rho_1^*(x) \odot \rho_1^*(z), \rho_1^*(y) \odot \rho_1^*(z)) \in \rho_f$. Since $(\rho_1^*(x), \rho_1^*(y)) \in \rho_f$, we have $\overline{f}(\rho_1^*(x)) \preceq_T \overline{f}(\rho_1^*(y))$. Thus, by definition we get $\overline{f}(\rho_1^*(x)) \otimes \overline{f}(\rho_1^*(z)) \preceq_T \overline{f}(\rho_1^*(y)) \otimes \overline{f}(\rho_1^*(z))$. So, $\overline{f}(\rho_1^*(x) \odot \rho_1^*(z)) \preceq_T \overline{f}(\rho_1^*(y) \odot \rho_1^*(z))$. Hence, for $u' = x \cdot z$ and for $v' \in y \cdot z$, we have $\overline{f}(\rho_1^*(u')) \preceq_T \overline{f}(\rho_1^*(v'))$. This implies that $(\rho_1^*(u'), \rho_1^*(v')) \in \rho_f$. Therefore, we have $(\rho_1^*(x) \odot \rho_1^*(z), \rho_1^*(y) \odot \rho_1^*(z)) \in \rho_f$. Similarly, we obtain $(\rho_1^*(z) \odot \rho_1^*(x), \rho_1^*(z) \odot \rho_1^*(y)) \in \rho_f$. $\square$

Corollary 10.2. $\ker \overline{f} = \rho_f^*$.

Proof. It is straightforward. $\square$

Corollary 10.3. *Let $(R, +, \cdot, \leq_R)$ and $(T, \boxplus, \diamond, \leq_T)$ be two ordered Krasner hyperrings, ρ_1, ρ_2 be two pseudo-orders on R, T, respectively, and the map $f : R \to T$ be a (ρ_1, ρ_2)-homomorphism. Then, the following is commutative:*

$$
\begin{array}{ccc}
R & \xrightarrow{\ f\ } & T \\
\varphi_R \downarrow & & \downarrow \varphi_T \\
R/\rho_1^* & \xrightarrow{\ \phi\ } & T/\rho_2^*
\end{array}
$$

Proof. It is straightforward. $\square$

Now, we recall the definition of a regular ring. An element a in a ring R is said to be *regular* if $a \in aRa$. A ring R is called *regular* if every element of R is regular.

Definition 10.9. Let $(R, +, \cdot, \leq)$ be an ordered Krasner hyperring. An element $a \in R$ is said to be *right regular* if $a \in (a^2 \cdot R]$.

Example 10.11. The ordered Krasner hyperring $(R, \oplus, \odot)$ defined as in Example 10.3 is regular.

Example 10.12. Let $(\{0,1\}, \boxplus, \cdot)$ be an ordered Krasner hyperring defined as in Example 10.4. It can be easily seen that $(\{0,1\}, \boxplus, \cdot)$ is regular.

Theorem 10.7. *If I is a hyperideal of a regular ordered Krasner hyperring $(R,+,\cdot,\leq)$, then I is regular.*

Proof. Let $a \in I$. Since R is a regular ordered Krasner hyperring, there exists $x \in R$ such that $a \leq axa$. Thus, we have $a \leq a(xa) \leq (axa)(xa) = a(xax)a$. Since I is a hyperideal of R, it follows that $xax \in I$. Thus, there exists $y \in I$ such that $a \leq aya$. Therefore, I is a regular hyperideal of R. $\qquad\square$

Theorem 10.8. *If I and J are regular hyperideals of an ordered Krasner hyperring $(R,+,\cdot,\leq)$, then $I \cap J$ is also a regular hyperideal of R.*

Proof. Let I and J be regular hyperideals of R. It can be easily seen that $I \cap J$ is a hyperideal of R. Let $a \in I \cap J$. Since I, J are regular hyperideals of R, there exist $x \in I$ and $y \in J$ such that $a \leq axa$ and $a \leq aya$. Thus, we have $a \leq axa \leq (axa)x(aya) = a(xaxay)a$. Since I and J are hyperideals of R, we obtain $xaxay \in I \cap J$. Thus, there exists $z \in I \cap J$ such that $a \leq aza$. Therefore, $I \cap J$ is a regular hyperideal of R. $\qquad\square$

Theorem 10.9. *Let $(R,+,\cdot,\leq)$ be a regular ordered Krasner hyperring. If I is a hyperideal of J and J is a hyperideal of R, then I is a hyperideal of R.*

Proof. Since I is a subhyperring of R, we prove that for every $a \in I$ and $r \in R$, $r \cdot a \in I$ and $a \cdot r \in I$. Let $x = r \cdot a$. Since R is regular, there exists $y \in R$ such that $x \leq x \cdot y \cdot x = (r \cdot a) \cdot y \cdot (r \cdot a)$. Since $x = r \cdot a$, it follows that $x \cdot y = (r \cdot a) \cdot y$. Thus, we have $x \cdot y \cdot x = (x \cdot y) \cdot x = [(r \cdot a) \cdot y] \cdot (r \cdot a)$. Hence,

$$x \leq (r \cdot a) \cdot y \cdot (r \cdot a). \tag{10.2}$$

Since J is a hyperideal of R and $a \in J$, it follows that $r \cdot a \in J$. So, $[(r \cdot a) \cdot y] \cdot r \in J$. Thus, we have $(r \cdot a) \cdot y \cdot r \in J$. Since I is a hyperideal of J and $a \in I$, we obtain

$$(r \cdot a) \cdot y \cdot (r \cdot a) = [(r \cdot a) \cdot y \cdot r] \cdot a \in I. \tag{10.3}$$

So, from (1) and (2), we obtain $x \in I$. Thus, we have $r \cdot a \in I$. Similarly, we can show that $a \cdot r \in I$. Now, we show that for every

$a \in I$ and $b \in R$ such that $b \leq a$, we have $b \in I$. Since J is a hyperideal of R and $a \in J$, we obtain $b \in J$. Since I is a hyperideal of J and $b \in J$ such that $b \leq a$, we have $b \in I$. Therefore, I is a hyperideal of R. □

10.2 Hyperideals of Ordered Krasner Hyperrings

Theorem 10.10. *In any ordered Krasner hyperring* $(R, +, \cdot, \leq)$, *for each* $a, b \in R$, *we have*

$$a \leq b \Leftrightarrow -b \leq -a.$$

Proof. For each $a, b \in R$, we have

$$a \leq b \Leftrightarrow (-a + b) \cap R^+ \neq \emptyset$$
$$\Leftrightarrow (b - a) \cap R^+ \neq \emptyset$$
$$\Leftrightarrow (a - b) \cap R^- \neq \emptyset$$
$$\Leftrightarrow (-b + a) \cap R^- \neq \emptyset$$
$$\Leftrightarrow -b \leq -a.$$

□

Example 10.13. Let $R = \{a, b, c\}$ be a set with the hyperoperation $\oplus$ and the binary operation $\odot$ defined as follows:

$\oplus$	a	b	c
a	a	b	c
b	b	b	R
c	c	R	c

$\odot$	a	b	c
a	a	a	a
b	a	b	c
c	a	c	b

Then, $(R, \oplus, \odot)$ is a Krasner hyperring. We have that $(R, \oplus, \odot, \leq)$ is an ordered Krasner hyperring, where the order relation $\leq$ is defined by

$$\leq := \{(a, a), (b, b), (c, c), (a, b), (a, c)\}.$$

The covering relation and the figure of R are given by

$$\prec = \{(a, b), (a, c)\}.$$

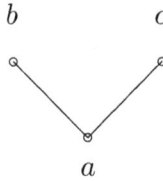

Example 10.14. If $(H, \leq, +)$ is a totally ordered group, then

$$x \oplus x = \{t \in H : t \leq x\} \quad \text{for all } x \in H,$$

$$x \oplus y = \{\max\{x, y\}\} \quad \text{for all } x, y \in H, x \neq y,$$

defines a structure of canonical hypergroup on H. If $(H, +, \cdot)$ is a totally ordered ring (for example, $\mathbb{R}$), then $(H, \oplus, \cdot)$ is a Krasner hyperring. Consider $(H, \oplus, \cdot)$ as a poset with natural ordering. Then, $(H, \oplus, \cdot)$ is an ordered Krasner hyperring.

Example 10.15. Let $(R, +, \cdot)$ be a Krasner hyperring and $M(R) = \{(a, b) : a, b \in R\}$. The hyperoperation $\oplus$ and the multiplication $\odot$ are defined on $M(R)$ by

$$(a, b) \oplus (c, d) = \{(x, y) : x \in a + c, y \in b + d\},$$

$$(a, b) \odot (c, d) = (ac, bd),$$

for all $(a, b), (c, d) \in M(R)$. Clearly, this hyperoperation is well defined and $(M(R), \oplus)$ is a canonical hypergroup. The element $(0, 0)$ is the additive identity of $M(R)$. Also, for each (a, b) of $M(R)$, there exists a unique element $(-a, -b) \in M(R)$ such that $(0, 0) \in (a, b) \oplus (-a, -b)$. Also, the multiplication $\odot$ is well defined and associative. Therefore, $(M(R), \odot)$ is a semigroup. Now, let $(a, b), (c, d), (e, f) \in M(R)$. Then,

$$(a, b) \odot \big((c, d) \oplus (e, f)\big) = (a, b) \odot \{(r, s) : r \in c + e, s \in d + f\}$$

$$= \{(ar, bs) : r \in c + e, s \in d + f\}.$$

Also,

$$\big((a, b) \odot (c, d)\big) \oplus \big((a, b) \odot (e, f)\big) = (ac, bd) \oplus (ae, bf)$$

$$= \{(g, h) : g \in ac + ae, h \in bd + bf\}.$$

By the left distributive axiom of R,

$$(a,b) \odot \Big((c,d) \oplus (e,f)\Big) = \Big((a,b) \odot (c,d)\Big) \oplus \Big((a,b) \odot (e,f)\Big).$$

Similarly, we can show that the right distributive law is also satisfied on $M(R)$. Thus, $(M(R), \oplus, \odot)$ is a Krasner hyperring. Now, let $(R, +, \cdot, \leq)$ be an ordered Krasner hyperring. Define the order relation $\preceq$ on $M(R)$ by

$$(a,b) \preceq (a',b') \quad \Leftrightarrow \quad a \leq a', b \leq b'$$

Then, $(M(R), \oplus, \odot, \preceq)$ is an ordered Krasner hyperring.

Theorem 10.11. *Let φ be a homomorphism from an ordered Krasner hyperring R onto an ordered Krasner hyperring T. Define $\theta : R/\mathrm{ker}\varphi \to T$ by $\theta(x + \mathrm{ker}\varphi) = \varphi(x)$, for all $x \in R$. Then, the following statements hold:*

(1) θ is a homomorphism from $R/\mathrm{ker}\varphi$ onto T;
(2) if φ is a good (strong) homomorphism, then θ is an isomorphism and hence $R/\mathrm{ker}\varphi \cong T$.

Proof. (1) We check the conditions of definition. Let $x, y \in R$ be such that $x + \mathrm{ker}\varphi = y + \mathrm{ker}\varphi$. Then, $x \in y + \mathrm{ker}\varphi$, so $x \in y + z$, for some $z \in \mathrm{ker}\varphi$. Thus, $\varphi(x) \in \varphi(y + z) \subseteq \varphi(y) + \varphi(z) = \varphi(y) + 0 = \{\varphi(y)\}$. So, $\varphi(x) = \varphi(y)$. Thus, the map θ is well defined. If $x, y \in R$, then we have

$$\theta((x + \mathrm{ker}\varphi) + (y + \mathrm{ker}\varphi)) = \theta(\{z + \mathrm{ker}\varphi : z \in x + y\})$$
$$= \{\theta(z + \mathrm{ker}\varphi) : z \in x + y\}$$
$$= \{\varphi(z) : z \in x + y\}.$$

Also,

$$\theta(x + \mathrm{ker}\varphi) + \theta(y + \mathrm{ker}\varphi) = \varphi(x) + \varphi(y)$$
$$\supseteq \varphi(x + y) = \{\varphi(z) : z \in x + y\}.$$

Thus, $\theta((x + \mathrm{ker}\varphi) + (y + \mathrm{ker}\varphi)) \subseteq \theta(x + \mathrm{ker}\varphi) + \theta(y + \mathrm{ker}\varphi)$. So, the first condition of the definition is verified. We have

$$\theta(x + \mathrm{ker}\varphi)(y + \mathrm{ker}\varphi) = \theta(xy + \mathrm{ker}\varphi) = \varphi(xy)$$
$$= \varphi(x)\varphi(y) = \theta(x + \mathrm{ker}\varphi) + \theta(y + \mathrm{ker}\varphi).$$

So, the second condition of the definition is verified. Now, let $x \leq_R y$. Since φ is a homomorphism, we have $\varphi(x) \leq_T \varphi(y)$. Thus,

$\theta(x + ker\varphi) \leq_T \theta(y + ker\varphi)$. So, the third condition of the definition is verified. Therefore, θ is a homomorphism.

(2) Assume that φ is a good (strong) homomorphism. It can be seen from the proof of (1) that θ is a good (strong) homomorphism. We know that $0 + ker\varphi \in ker\theta$. Let $x \in R$ be such that $\theta(x + ker\varphi) = 0$. Then, $\varphi(x) = 0$, so $x \in ker\varphi$. Hence, $x + ker\varphi = 0 + ker\varphi$. Thus, we have $ker\theta = \{0 + ker\varphi\}$. Hence, θ is one to one. Clearly, θ is onto. Thus, θ is a good (strong) isomorphism. That is, $R/ker\varphi$ is strongly isomorphic to T. $\qquad\square$

Theorem 10.12. *Let $(R, +, \cdot, \leq)$ be an ordered Krasner hyperring with positive cone P and $\varphi : R \to R$ be any good (strong) homomorphism of the canonical hypergroup $(R, +)$ such that $\varphi(P) \subseteq P$. Assume that for any $r \in R$, there exists an integer $n \geq 1$ such that $\varphi^n(r) = r$. Then, φ is the identity map.*

Proof. If $a < b$, then $b - a \subseteq P$. So, by hypothesis, $\varphi(b - a) \subseteq P$. Since φ is a good (strong) homomorphism of $(R, +)$, it follows that $\varphi(b - a) = \varphi(b) - \varphi(a)$. Therefore, $\varphi(a) < \varphi(b)$. Now, let $\varphi \neq id$. Then, $\varphi(r) \neq r$, for some $r \in R$. We have either $r < \varphi(r)$ or $\varphi(r) < r$. Say $r < \varphi(r)$. Fix an integer $n \geq 1$ such that $\varphi^n(r) = r$. Then, we have

$$r < \varphi(r) < \varphi^2(r) < \cdots < \varphi^n(r) = r,$$

which is a contradiction. If $\varphi(r) < r$, a similar contradiction results. Therefore, φ is the identity map. $\qquad\square$

In the following, we specialize our study to some of the basic facts concerning ordered Krasner hyperrings.

Definition 10.10. A non-empty subset P of an ordered Krasner hyperring $(R, +, \cdot, \leq)$ is called a *prime hyperideal* of R if the following conditions hold:

(1) $A \cdot B \subseteq P$ implies that $A \subseteq P$ or $B \subseteq P$ for any two hyperideal A and B of R;
(2) if $x \in P$ and $y \leq x$, then $y \in P$ for every $y \in R$.

Example 10.16. Define the hyperoperation $\oplus$ and the operation $\odot$ on the set $R = \{0, 1\}$ by

$\oplus$	0	1
0	0	1
1	1	$\{0,1\}$

$\odot$	0	1
0	0	0
1	0	1

Then, $(R, \oplus, \odot)$ is a commutative Krasner hyperring with the zero element 0. Consider $(R, \oplus, \odot)$ as a poset with natural ordering. Thus, $(R, \oplus, \odot)$ is an ordered Krasner hyperring. Now, it is easy to see that $\{0\}$ and $\{0, 1\}$ are hyperideals of R. It is obvious that $\{0\}$ is a prime hyperideal of R.

Example 10.17. Consider the hyperring $R = \{0, a, b\}$ with the hyperaddition $\oplus$ and the multiplication $\odot$ defined as follows:

$\oplus$	0	a	b
0	0	a	b
a	a	$\{a,b\}$	R
b	b	R	$\{a,b\}$

$\odot$	0	a	b
0	0	0	0
a	0	b	a
b	0	a	b

Then, $(R, \oplus, \odot)$ is a Krasner hyperring. We have that $(R, \oplus, \odot, \leq)$ is an ordered Krasner hyperring, where the order relation $\leq$ is defined by

$$\leq := \{(0,0), (a, a), (b, b), (0, a), (0, b)\}.$$

The covering relation and the figure of R are given by

$$\prec = \{(0, a), (0, b)\}.$$

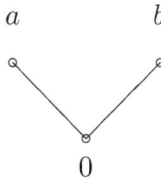

Now, it is easy to see that $\{0\}$ and $\{0, a, b\}$ are hyperideals of R. It is obvious that $\{0\}$ is a prime hyperideal of R.

Example 10.18. Let $R = \{0, a, b, c\}$ be a set with the hyperoperation $\oplus$ and the multiplication $\odot$ defined as follows:

$\oplus$	0	a	b	c
0	0	a	b	c
a	a	$\{0, b\}$	$\{a, c\}$	b
b	b	$\{a, c\}$	$\{0, b\}$	a
c	c	b	a	0

$\odot$	0	a	b	c
0	0	0	0	0
a	0	a	b	c
b	0	b	b	0
c	0	c	0	c

Then, $(R, \oplus, \odot)$ is a Krasner hyperring. We have that $(R, \oplus, \odot, \leq)$ is an ordered Krasner hyperring, where the order relation $\leq$ is defined by

$$\leq := \{(0, 0), (a, a), (b, b), (c, c), (0, b), (c, a)\}.$$

The covering relation and the figure of R are given by

$$\prec = \{(0, b), (c, a)\}.$$

$$
\begin{array}{cc}
b & a \\
\circ & \circ \\
| & | \\
\circ & \circ \\
0 & c
\end{array}
$$

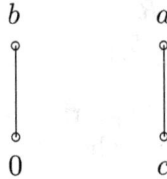

Now, it is easy to see that $I_1 = \{0\}$, $I_2 = \{0, b\}$, $I_3 = \{0, c\}$, $I_4 = \{0, b, c\}$ and $I_5 = \{0, a, b, c\}$ are hyperideals of R. Also, I_2, I_3 and I_4 are prime hyperideals of R. The hyperideal $I_1 = \{0\}$ is not a prime hyperideal of R. Indeed, $\{0, b\} \odot \{0, c\} = \{0\}$, but $\{0, b\} \nsubseteq \{0\}$ and $\{0, c\} \nsubseteq \{0\}$.

Example 10.19. Let $R = \{a, b, c, d, e, f\}$ be a set with the hyperoperation $\oplus$ and the multiplication $\odot$ defined as follows:

$\oplus$	a	b	c	d	e	f
a	a	b	c	d	e	f
b	b	$\{a, b\}$	d	$\{c, d\}$	f	$\{e, f\}$
c	c	d	c	d	$\{a, c, e\}$	$\{b, d, f\}$
d	d	$\{c, d\}$	d	$\{c, d\}$	$\{b, d, f\}$	R
e	e	f	$\{a, c, e\}$	$\{b, d, f\}$	e	f
f	f	$\{e, f\}$	$\{b, d, f\}$	R	f	$\{e, f\}$

and

$\odot$	a	b	c	d	e	f
a	a	a	a	a	a	a
b	a	b	a	b	a	b
c	a	a	c	c	e	e
d	a	b	c	d	e	f
e	a	a	e	e	c	c
f	a	b	e	f	c	d

Then, $(R, \oplus, \odot)$ is a Krasner hyperring. We have that $(R, \oplus, \odot, \leq)$ is an ordered Krasner hyperring, where the order relation $\leq$ is defined by

$$\leq := \{(a,a), (b,b), (c,c), (d,d), (e,e), (f,f), (a,b), (a,c),$$
$$(a,d), (a,e), (a,f), (b,d), (b,f), (c,d), (e,f)\}.$$

The covering relation and the figure of R are given by

$$\prec = \{(a,b), (a,c), (a,e), (b,d), (b,f), (c,d), (e,f)\}.$$

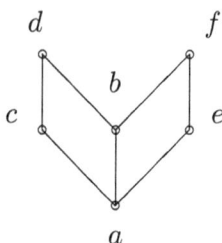

It is easy to see that $\{a\}$, $\{a,b\}$, $\{a,c,e\}$ and $\{a,b,c,d,e,f\}$ are hyperideals of R. It is obvious that $\{a,b\}$ and $\{a,c,e\}$ are prime hyperideals of R. The hyperideal $\{a\}$ is not a prime hyperideal of R. Indeed, $\{a,b\} \odot \{a,c,e\} = \{a\}$, but $\{a,b\} \not\subseteq \{a\}$ and $\{a,c,e\} \not\subseteq \{a\}$.

Definition 10.11. A non-empty subset I of an ordered Krasner hyperring $(R, +, \cdot, \leq)$ is called a *semiprime hyperideal* of R if the following conditions hold:

(1) $A \cdot A \subseteq I$ implies that $A \subseteq I$ for any hyperideal A of R;
(2) if $x \in I$ and $y \leq x$, then $y \in I$ for every $y \in R$.

Remark 10.4. Every prime hyperideal of R is a semiprime hyperideal of R.

Example 10.20. In Example 10.18, $I_1 = \{0\}$ is a semiprime hyperideal but not a prime hyperideal.

Definition 10.12. An ordered Krasner hyperring $(R, +, \cdot, \leq)$ is said to be a *prime hyperring* if $a \cdot R \cdot b = 0$, for $a, b \in R$, implies either $a = 0$ or $b = 0$. Equivalently, an ordered Krasner hyperring R is called *prime* if $a \cdot r \cdot b = 0$, for all $r \in R$, implies either $a = 0$ or $b = 0$.

Example 10.21. In Examples 10.16 and 10.17, R is prime; whereas, in Examples 10.18 and 10.19, R is not prime.

Definition 10.13. An ordered Krasner hyperring $(R, +, \cdot, \leq)$ is said to be a *semiprime hyperring* if $a \cdot R \cdot a = 0$, for $a \in R$, implies $a = 0$. Equivalently, an ordered Krasner hyperring R is called semiprime if $a \cdot r \cdot a = 0$, for all $r \in R$, implies $a = 0$.

Remark 10.5. Every prime ordered Krasner hyperring is a semiprime ordered Krasner hyperring.

Example 10.22. In Example 10.18, R is a semiprime ordered Krasner hyperring but not a prime ordered Krasner hyperring.

Definition 10.14. Let $(R, +, \cdot, \leq)$ be an ordered Krasner hyperring with the positive cone P. A subset $A \subseteq R$ is *convex* if $0 \leq p \leq q$, $q \in A$, implies that $p \in A$. Equivalently, A is convex if $p_1 + p_2 \subseteq A$, $p_i \in P$, implies that $p_i \in A$, $i = 1, 2$. A hyperideal A of an ordered Krasner hyperring $(R, +, \cdot, \leq)$ is said to be *convex* if it is convex as a subset.

Example 10.23.

(1) In Example 10.16, the hyperideals $\{0\}$ and $\{0, 1\}$ are convex.
(2) In Example 10.17, the hyperideals $\{0\}$ and $\{0, a, b\}$ are convex.
(3) In Example 10.18, the hyperideals I_1, I_2, I_3, I_4, and I_5 are convex.
(4) In Example 10.19, the hyperideals $\{a\}$, $\{a, b\}$, $\{a, c, e\}$, and $\{a, b, c, d, e, f\}$ are convex.

Theorem 10.13. *Let $(R, +, \cdot, \leq)$ be an ordered Krasner hyperring. A hyperideal $I \subseteq R$ is the kernel of a homomorphism in an ordered Krasner hyperring if and only if I is a convex hyperideal of R.*

Proof. Let $\varphi : R \to R$ be a homomorphism and $I = ker\varphi$. Let $0 \leq p \leq q$ in R with $\varphi(q) = 0$. Since φ is a homomorphism, it follows that $0 \leq \varphi(p) \leq \varphi(q) = 0$ in R. Thus, we have $\varphi(p) = 0$. Therefore, $I = ker\varphi$ is a convex hyperideal of R.

Conversely, suppose that I is a convex hyperideal of R. Consider the projection map $\pi : R \to R/I$. We can impose an order on R/I so that π is order preserving if, whenever $\sum_{i=1}^{n} p_i a_i^2 \subseteq I$, $p_i \in P$, $a_i \in R$, then $p_j a_j^2 \in I$, $1 \leq j \leq n$. This is the second characterization of convexity in Definition 10.14. The weakest order on R/I such that π is order preserving, namely $\pi_*(P) = \{p + I : p \in P\}$, is called the induced order. $\square$

Definition 10.15. A convex hyperideal $Q \subset R$ is a *maximal convex hyperideal* if $Q \neq R$ and whenever $Q \subseteq Q'$, where Q' a convex hyperideal, either $Q' = Q$ or $Q' = R$.

Now, we establish the existence of maximal convex hyperideals.

Theorem 10.14. *Let $(R, +, \cdot, \leq)$ be an ordered Krasner hyperring. Let $I \subsetneq R$ be a convex hyperideal. Then, I is contained in at least one maximal convex hyperideal.*

Proof. The family of all convex hyperideals containing I but not containing 1 is non-empty, partially ordered by inclusion, and satisfies the chain condition. Thus, by Zorn's lemma, the proof is complete. $\square$

Remark 10.6. Since $I = (0)$ is always a convex hyperideal of R, we conclude that any non-zero ordered Krasner hyperring $(R, +, \cdot, \leq)$ has a maximal convex hyperideal.

Remark 10.7. Maximal convex hyperideals are prime.

Definition 10.16. If $(R, +, \cdot, \leq)$ is an ordered Krasner hyperring and $A \subseteq R$, then $(A]$ is the subset of R defined as follows:

$$(A] = \{t \in R : t \leq a, \text{ for some } a \in A\}.$$

Lemma 10.2. *Let* $(R, +, \cdot, \leq)$ *be an ordered Krasner hyperring. If* A *and* B *are non-empty subsets of* R, *then we have:*

(1) $A \subseteq (A]$;

(2) *if* $A \subseteq B$, *then* $(A] \subseteq (B]$;

(3) $((A]] = (A]$;

(4) $(A \cup B] = (A] \cup (B]$;

(5) $(A] + (B] \subseteq (A + B]$;

(6) $(A] \cdot (B] \subseteq (A \cdot B]$;

(7) $((A] \cdot (B]] = (A \cdot B]$;

(8) *if* $A, B, C \subseteq R$ *such that* $A \subseteq B$, *then* $A + C \subseteq B + C$ *and* $C + A \subseteq C + B$;

(9) *if* $A, B, C \subseteq R$ *such that* $A \subseteq B$, *then* $A \cdot C \subseteq B \cdot C$ *and* $C \cdot A \subseteq C \cdot B$.

Proof. The proof is straightforward. □

Definition 10.17. Let $(R, +, \cdot, \leq)$ be an ordered Krasner hyperring. A non-empty subset S of R is called an *M-system* of R if for each $a, b \in S$, there exist $r \in R$ and $c \in S$ such that $c \leq a \cdot (r \cdot b)$, or, equivalently, $c \in (a \cdot (R \cdot b)]$.

Example 10.24.

(1) The set $\{0, 1\}$ is an M-system of an ordered Krasner hyperring defined in Example 10.16.

(2) The sets $\{0, a\}$, $\{0, b\}$, $\{a, b\}$, and $\{0, a, b\}$ are an M-system of an ordered Krasner hyperring defined in Example 10.17.

(3) The sets $\{0, a\}$, $\{0, b\}$, $\{0, c\}$, $\{a, b\}$, and $\{a, c\}$ are an M-system of an ordered Krasner hyperring defined in Example 10.18, whereas $\{b, c\}$ is not an M-system of an ordered Krasner hyperring defined in Example 10.18.

(4) The sets $\{a\}$, $\{a, b\}$, and $\{a, c, e\}$ are an M-system of an ordered Krasner hyperring defined in Example 10.19, whereas $\{b, c\}$ is not an M-system of an ordered Krasner hyperring defined in Example 10.19.

Definition 10.18. Let $(R, +, \cdot, \leq)$ be an ordered Krasner hyperring. A non-empty subset S of R is called an *N-system* of R if for each $a \in S$, there exist $r \in R$ and $c \in S$ such that $c \leq a \cdot (r \cdot a)$, or, equivalently, $c \in (a \cdot (R \cdot a)]$.

Remark 10.8. Every M-system of R is an N-system of R.

Example 10.25. The set $\{b, c\}$ is an N-system of an ordered Krasner hyperring defined in Example 10.18 but is not an M-system of an ordered Krasner hyperring defined in Example 10.18.

Definition 10.19. A non-empty subset I of an ordered Krasner hyperring $(R, +, \cdot, \leq)$ is called a *quasi-prime hyperideal* of R if for all left hyperideals A, B of R, $A \cdot B \subseteq I$ implies that $A \subseteq I$ or $B \subseteq I$.

Definition 10.20. A non-empty subset I of an ordered Krasner hyperring $(R, +, \cdot, \leq)$ is called a *quasi-semiprime hyperideal* of R if for any left hyperideal A of R, $A \cdot A \subseteq I$ implies that $A \subseteq I$.

Every quasi-prime hyperideal of R is a quasi-semiprime hyperideal of R.

Example 10.26. In Example 10.18, $\{0\}$ is a quasi-semiprime hyperideal of R but is not a quasi-prime hyperideal of R.

Definition 10.21. A non-empty subset I of an ordered Krasner hyperring $(R, +, \cdot, \leq)$ is called a *quasi-irreducible hyperideal* of R if for all left hyperideals A, B of R, $A \cap B \subseteq I$ implies that $A \subseteq I$ or $B \subseteq I$.

Example 10.27. In Example 10.18, $\{0, b\}$, $\{0, c\}$, and $\{0, b, c\}$ are quasi-irreducible hyperideals of R, whereas $\{0\}$ is not a quasi-irreducible hyperideal of R.

Lemma 10.3. *Let I be a left hyperideal of an ordered Krasner hyperring $(R, +, \cdot, \leq)$. Then, I is quasi-prime hyperideal if and only if for all $a, b \in R$, $a \cdot (R \cdot b) \subseteq I$ implies that $a \in I$ or $b \in I$.*

Proof. It is straightforward. $\square$

Theorem 10.15. *Let I be a left hyperideal of an ordered Krasner hyperring $(R, +, \cdot, \leq)$. Then, I is quasi-prime hyperideal if and only if $R \setminus I$ is an M-system.*

Proof. Let I be a quasi-prime hyperideal and $a, b \in R \setminus I$. Assume that $c \notin (a \cdot (R \cdot b)]$, for all $c \in R \setminus I$. Then, $(a \cdot (R \cdot b)] \subseteq I$. This implies that $a \cdot (R \cdot b) \subseteq I$. So, $a \in I$ or $b \in I$, which contradicts the assumption that $a, b \in R \setminus I$. Hence, $R \setminus I$ is an M-system.

Conversely, let $R \setminus I$ be an M-system and $a \cdot (R \cdot b) \subseteq I$, for some $a, b \in R \setminus I$. Then, there exist $c \in R \setminus I$ and $x \in R$ such that $c \leq a \cdot (x \cdot b)$, which implies that $c \in I$, it contradicts the assumption $c \in R \setminus I$. Hence, I is a quasi-prime hyperideal of R.   □

Lemma 10.4. *Let I be a left hyperideal of an ordered Krasner hyper-ring $(R, +, \cdot, \leq)$. Then, I is quasi-semiprime hyperideal if and only if for all $a \in R$, $a \cdot (R \cdot a) \subseteq I$ implies that $a \in I$.*

Proof. It is straightforward.   □

Theorem 10.16. *Let I be a left hyperideal of an ordered Krasner hyperring $(R, +, \cdot, \leq)$. Then, I is quasi-semiprime hyperideal if and only if $R \setminus I$ is an N-system.*

Proof. Let I be a quasi-semiprime hyperideal and $a \in R \setminus I$. Assume that $c \notin (a \cdot (R \cdot a)]$, for all $c \in R \setminus I$. Then, $(a \cdot (R \cdot a)] \subseteq I$. This implies that $a \cdot (R \cdot a) \subseteq I$. So, $a \in I$, which contradicts the assumption that $a \in R \setminus I$. Hence, $R \setminus I$ is an N-system.

Conversely, let $R \setminus I$ be an N-system and $a \cdot (R \cdot a) \subseteq I$ with $a \notin I$. Then, there exist $c \in R \setminus I$ and $r \in R$ such that $c \leq a \cdot (r \cdot a)$, which implies that $c \in I$, it contradicts the assumption $c \in R \setminus I$. Hence, $a \in I$. Therefore, I is a quasi-semiprime hyperideal of R.   □

Theorem 10.17. *If N is an N-system of an ordered Krasner hyper-ring $(R, +, \cdot, \leq)$ and $a \in N$, then there exists an M-system M of R such that $a \in M \subseteq N$.*

Proof. Let N be an N-system of an ordered Krasner hyperring R and $a \in N$. Then, by the definition of an N-system, there exist some $c_1 \in N$ such that $c_1 \in (a \cdot (R \cdot a)]$, so $(a \cdot (R \cdot a)] \cap N \neq \emptyset$. Take $a_1 \in (a \cdot (R \cdot a)] \cap N$, and again using the definition of N-system, there exist $c_2 \in N$ such that $c_2 \in (a_1 \cdot (R \cdot a_1)]$, so $(a_1 \cdot (R \cdot a_1)] \cap N \neq \emptyset$. Continuing in this way, we take $a_i \in (a_{i-1} \cdot (R \cdot a_{i-1})] \cap N \neq \emptyset$. Take $a_0 = a$ and define $M = \{a_0, a_1, \ldots\}$. Then, M is an M-system and $a \in M \subseteq N$.   □

Theorem 10.18. *Let $(R, +, \cdot, \leq)$ be an ordered Krasner hyperring. Then, R is a regular ordered Krasner hyperring if and only if $(A \cdot B] = (A \cap B]$ for the right hyperideal A and left hyperideal B of R.*

Proof. Let R be regular. It is clear that $(A \cdot B] \subseteq (A \cap B]$. If $c \in (A \cap B]$, then $c \leq z$, for some $z \in A \cap B$. Since R is regular, there exists an element $x \in R$ such that $c \leq (c \cdot x) \cdot c$. We have $c \leq (c \cdot x) \cdot c \subseteq (c \cdot x) \cdot z \subseteq ((A \cdot R) \cdot B]$. Thus, $c \in ((A \cdot R) \cdot B] \subseteq (A \cdot B]$. Hence, $(A \cap B] \subseteq (A \cdot B]$. Therefore, we have $(A \cdot B] = (A \cap B]$.

Conversely, let $a \in R$. Then, we have $a \in (a \cdot R] \cap (R \cdot a] = ((a \cdot R) \cdot (R \cdot a)] = (a \cdot R \cdot a]$. So, there exists an element $x \in R$ such that $a \leq (a \cdot x) \cdot a$. Therefore, R is a regular ordered Krasner hyperring. $\square$

Theorem 10.19. *Every hyperideal of a regular ordered Krasner hyperring R is a prime hyperideal if and only if it is an irreducible hyperideal of R.*

Proof. Suppose that P is a prime hyperideal of R and $(A \cap B] \subseteq P$. By Theorem 10.18, $(A \cdot B] = (A \cap B]$, so $(A \cdot B] \subseteq P$ which implies that $(A] \subseteq P$ or $(B] \subseteq P$. Therefore, P is an irreducible hyperideal of R.

Conversely, suppose that P is an irreducible hyperideal of R. Then, $(A \cap B] \subseteq P$ implies that $(A] \subseteq P$ or $(B] \subseteq P$. By Theorem 10.18, $(A \cdot B] = (A \cap B]$, and so P is a prime hyperideal of R. $\square$

Definition 10.22. An ordered Krasner hyperring $(R, +, \cdot, \leq)$ is called *intra-regular* if for every $a \in R$, there exist $x, y \in R$ such that $a \leq x \cdot a^2 \cdot y$, or, equivalently, $a \in (R \cdot a^2 \cdot R]$.

Example 10.28. The ordered Krasner hyperring $(R, \oplus, \odot)$ defined, as in Example 10.16, to be intra-regular.

Definition 10.23. Let $(R, +, \cdot, \leq_R)$ and $(T, \oplus, \otimes, \leq_T)$ be two ordered Krasner hyperrings. Under coordinatewise multiplication, i.e.,

$$(r_1, t_1) \boxplus (r_2, t_2) = (r_1 + r_2, t_1 \oplus t_2),$$

$$(r_1, t_1) \star (r_2, t_2) = (r_1 \cdot r_2, t_1 \otimes t_2),$$

where $(r_1, t_1), (r_2, t_2) \in R \times T$, the Cartesian product $R \times T$ of R and T forms a Krasner hyperring. Define a partial order $\leq$ on $R \times T$ by $(r_1, t_1) \leq (r_2, t_2)$ if and only if $r_1 \leq_R r_2$ and $t_1 \leq_T t_2$, where $(r_1, t_1), (r_2, t_2) \in R \times T$. Then, $(R \times T, \boxplus, \star, \leq)$ is an ordered Krasner hyperring.

Definition 10.24. Let $(R, +, \cdot, \leq_R)$ and $(T, \oplus, \otimes, \leq_T)$ be two ordered Krasner hyperrings and ρ_1, ρ_2 be two pseudo-orders on R and T, respectively. On $R \times T$, we define

$$(r_1, t_1)\rho(r_2, t_2) \Leftrightarrow r_1 \rho_1 r_2 \quad \text{and} \quad t_1 \rho_2 t_2.$$

Lemma 10.5. *In Definition* 10.24, ρ *is pseudo-order on* $R \times T$.

Proof. It is straightforward. $\qquad\qquad\qquad\qquad\qquad\qquad\square$

Theorem 10.20. *Let* $(R, +, \cdot, \leq_R)$ *and* $(T, \oplus, \otimes, \leq_T)$ *be two ordered Krasner hyperrings and* ρ_1, ρ_2 *be two pseudo-orders on* R, T, *respectively. Then,*

$$(R \times T)/\rho^* \cong R/\rho_1^* \times T/\rho_2^*.$$

Proof. We consider the map $\psi : (R \times T)/\rho^* \to R/\rho_1^* \times T/\rho_2^*$ by $\psi(\rho^*(r, t)) = (\rho_1^*(r), \rho_2^*(t))$. Suppose that $\rho^*(r_1, t_1) = \rho^*(r_2, t_2)$. Then, $(r_1, t_1)\rho^*(r_2, t_2)$, which implies that $(r_1, t_1)\rho(r_2, t_2)$ and $(r_2, t_2)\rho(r_1, t_1)$. Hence, $r_1 \rho_1 r_2$, $t_1 \rho_2 t_2$, $r_2 \rho_1 r_1$ and $t_2 \rho_2 t_1$, which imply that $r_1 \rho_1^* r_2$ and $t_1 \rho_2^* t_2$. So, $(\rho_1^*(r_1), \rho_2^*(t_1)) = (\rho_1^*(r_2), \rho_2^*(t_2))$. This means that $\psi(\rho^*(r_1, t_1)) = \psi(\rho^*(r_2, t_2))$. Therefore, ψ is well defined. Now, we show that ψ is a homomorphism. Suppose that $\rho^*(r_1, t_1)$ and $\rho^*(r_2, t_2)$ are two arbitrary elements of $(R \times T)/\rho^*$. Then,

$$\psi(\rho^*(r_1, t_1) \uplus \rho^*(r_2, t_2)) = \psi(\rho^*(r, t)), \quad \text{for all } (r, t) \in (r_1, t_1) \boxplus (r_2, t_2)$$

$$= (\rho_1^*(r), \rho_2^*(t)), \quad \text{for all } r \in r_1 + r_2, \ t \in t_1 \oplus t_2$$

$$= (\rho_1^*(r_1) + \rho_1^*(r_2), \rho_2^*(t_1) \oplus \rho_2^*(t_2))$$

$$= (\rho_1^*(r_1), \rho_2^*(t_1)) \boxplus (\rho_1^*(r_2), \rho_2^*(t_2))$$

$$= \psi(\rho^*(r_1, t_1)) \boxplus \psi(\rho^*(r_2, t_2)).$$

So, the first condition of the definition of homomorphism is verified. Suppose that $\rho^*(r_1, t_1)$ and $\rho^*(r_2, t_2)$ are two arbitrary elements of $(R \times T)/\rho^*$. Then,

$$\psi(\rho^*(r_1, t_1) \triangledown \rho^*(r_2, t_2)) = \psi(\rho^*(r, t)), \quad \text{for } (r, t) = (r_1, t_1) \star (r_2, t_2)$$

$$= (\rho_1^*(r), \rho_2^*(t)), \quad \text{for } r = r_1 \cdot r_2, \ t = t_1 \otimes t_2$$

$$= (\rho_1^*(r_1) \odot \rho_1^*(r_2), \rho_2^*(t_1) \diamond \rho_2^*(t_2))$$
$$= (\rho_1^*(r_1), \rho_2^*(t_1)) \times (\rho_1^*(r_2), \rho_2^*(t_2))$$
$$= \psi(\rho^*(r_1, t_1)) \times \psi(\rho^*(r_2, t_2)).$$

So, the second condition of the definition of homomorphism is verified. Now, suppose that $\rho^*(r_1, t_1) \preceq \rho^*(r_2, t_2)$. Then, $(r_1, t_1)\rho(r_2, t_2)$, which implies that $r_1\rho_1 r_2$ and $t_1\rho_2 t_2$. Thus, $\rho_1^*(r_1) \preceq_R \rho_1^*(r_2)$ and $\rho_2^*(t_1) \preceq_T \rho_2^*(t_2)$. Hence, $(\rho_1^*(r_1), \rho_2^*(t_1)) \preceq_{R\times T} (\rho_1^*(r_2), \rho_2^*(t_2))$. This means that $\psi(\rho^*(r_1, t_1)) \preceq_{R\times T} \psi(\rho^*(r_2, t_2))$, and so the third condition of the definition of homomorphism is verified. Therefore, ψ is a homomorphism. Clearly, ψ is onto. So, we show that it is one to one. Suppose that $\psi(\rho^*(r_1, t_1)) = \psi(\rho^*(r_2, t_2))$. Then, $(\rho_1^*(r_1), \rho_2^*(t_1)) = (\rho_1^*(r_2), \rho_2^*(t_2))$, and so $\rho_1^*(r_1) = \rho_1^*(r_2)$ and $\rho_2^*(t_1) = \rho_2^*(t_2)$. Hence, $(r_1, r_2) \in \rho_1^*$ and $(t_1, t_2) \in \rho_2^*$. This implies that $r_1\rho_1 r_2$, $r_2\rho_1 r_1$, $t_1\rho_2 t_2$ and $t_2\rho_2 t_1$. Thus, $(r_1, t_1)\rho(r_2, t_2)$ and $(r_2, t_2)\rho(r_1, t_1)$. Therefore, $(r_1, t_1)\rho^*(r_2, t_2)$ or $\rho^*(r_1, t_1) = \rho^*(r_2, t_2)$. Therefore, ψ is an isomorphism and so the proof is complete. $\square$

10.3 Bi-hyperideals and Prime Bi-hyperideals in Ordered Krasner Hyperrings

Lemma 10.6. *Let* $(R_1, +_1, \cdot_1, \leq_1)$ *and* $(R_2, +_2, \cdot_2, \leq_2)$ *be two ordered Krasner hyperrings. Then:*

(1) *if* $\varphi : R_1 \to R_2$ *is an order homomorphism, then we have* $\varphi(R_1^+) \subseteq R_2^+$;
(2) *if* $\varphi : R_1 \to R_2$ *is a good (strong) hyperring homomorphism and* $\varphi(R_1^+) \subseteq R_2^+$, *then* φ *is isotone.*

Proof. (1) If $x \in R_1^+$, i.e., $0 \leq_1 x$, then $0_{R_2} = \varphi(0_{R_1}) \leq_2 \varphi(x)$. This means that $\varphi(x) \in R_2^+$. Hence, $\varphi(R_1^+) \subseteq R_2^+$.

(2) Assume that $x \leq_1 y$. Then, $-x + y \subseteq R_1^+$, and so $-\varphi(x) + \varphi(y) = \varphi(-x + y) \subseteq \varphi(R_1^+) \subseteq R_2^+$. Hence, $\varphi(x) \leq_2 \varphi(y)$. This implies that φ is isotone. $\square$

Definition 10.25. Let $(R_1, +_1, \cdot_1, \leq_1)$ and $(R_2, +_2, \cdot_2, \leq_2)$ be two ordered Krasner hyperrings. A function $\varphi : R_1 \to R_2$ is said to be an *exact* if $\varphi(R_1^+) = R_2^+$. Also, R_1 is *strongly isomorphic* to R_2 if there is a good (strong) order isomorphism $\varphi : R_1 \to R_2$. If R_1 is strongly isomorphic to R_2, then it is denoted by $R_1 \cong R_2$.

Theorem 10.21. *Let* $(R_1, +_1, \cdot_1, \leq_1)$ *and* $(R_2, +_2, \cdot_2, \leq_2)$ *be two ordered Krasner hyperrings. Then, the following assertions are equivalent:*

(1) $R_1 \cong R_2$.
(2) *There is an exact hyperring isomorphism* $\varphi : R_1 \to R_2$.

Proof. $(1 \Rightarrow 2)$: Assume that (1) holds. Then, there is a good (strong) order isomorphism $\varphi : R_1 \to R_2$. By Lemma 10.6(1), $\varphi(R_1^+) \subseteq R_2^+$. Let $\psi = \varphi^{-1}$. Then, ψ satisfies the condition (1) of Lemma 10.6, and so $\psi(R_2^+) \subseteq R_1^+$. Hence, $R_2^+ = \varphi(\psi(R_2^+)) \subseteq \varphi(R_1^+)$. Thus, $\varphi(R_1^+) = R_2^+$.
$(2 \Rightarrow 1)$ It is clear. $\qquad\square$

Our aim in the following is to introduce and study the concept of a quasi-hyperideal of ordered Krasner hyperrings.

Definition 10.26. A non-empty subset Q of an ordered Krasner hyperring $(R, +, \cdot, \leq)$ is called a *quasi-hyperideal* of R if the following conditions hold:

(1) $(Q, +)$ is a canonical subhypergroup of $(R, +)$;
(2) $(Q \cdot R) \cap (R \cdot Q) \subseteq Q$;
(3) when $x \in Q$ and $y \in R$ such that $y \leq x$, imply that $y \in Q$.

Example 10.29. Let $R = \{a, b, c, d, e, f, g, h\}$ be a set with the hyperaddition $\oplus$ and the multiplication $\odot$ defined as follows:

$\oplus$	a	b	c	d	e	f	g	h
a	a	b	c	d	e	f	g	h
b	b	$\{a,b\}$	d	$\{c,d\}$	f	$\{e,f\}$	h	$\{g,h\}$
c	c	d	$\{a,e\}$	$\{b,f\}$	$\{c,g\}$	$\{d,h\}$	e	f
d	d	$\{c,d\}$	$\{b,f\}$	$\{a,b,e,f\}$	$\{d,h\}$	$\{c,d,g,h\}$	f	$\{e,f\}$
e	e	f	$\{c,g\}$	$\{d,h\}$	$\{a,e\}$	$\{b,f\}$	c	d
f	f	$\{e,f\}$	$\{d,h\}$	$\{c,d,g,h\}$	$\{b,f\}$	$\{a,b,e,f\}$	d	$\{c,d\}$
g	g	h	e	f	c	d	a	b
h	h	$\{g,h\}$	f	$\{e,f\}$	d	$\{c,d\}$	b	$\{a,b\}$

and

$\odot$	a	b	c	d	e	f	g	h
a	a	a	a	a	a	a	a	a
b	a	b	a	b	a	b	a	b
c	a	a	c	c	e	e	g	g
d	a	b	c	d	e	f	g	h
e	a	a	e	e	e	e	a	a
f	a	b	e	f	e	f	a	b
g	a	a	g	g	a	a	g	g
h	a	b	g	h	a	b	g	h

Then, $(R, \oplus, \odot)$ is a Krasner hyperring. We have that $(R, \oplus, \odot, \leq)$ is an ordered Krasner hyperring where the order relation $\leq$ is defined by

$$\leq := \{(a,a), (b,b), (c,c), (d,d), (e,e), (f,f), (g,g),$$
$$(h,h), (a,b), (a,e), (a,f), (b,f), (c,d), (e,f),$$
$$(g,c), (g,d), (g,h), (h,d)\}.$$

The covering relation and the figure of R are given by

$$\prec = \{(a,b), (a,e), (b,f), (c,d), (e,f), (g,c), (g,h), (h,d)\}.$$

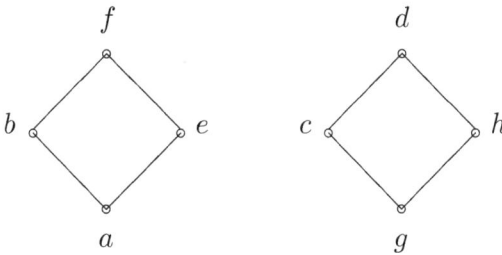

It is easy to see that $\{a\}$, $\{a,b\}$, $\{a,e\}$, $\{a,g\}$, $\{a,b,e,f\}$, $\{a,b,g,h\}$, $\{a,c,e,g\}$ and $\{a,b,c,d,e,f,g,h\}$ are quasi-hyperideals of R.

Every left, right, and two-sided hyperideal of an ordered Krasner hyperring R is a quasi-hyperideal of R. The converse is not true, in general, i.e., a quasi-hyperideal may not be a left, right, or two-sided hyperideal of R.

Example 10.30. Let $R = \{a, b, c, d\}$ be a set with a hyperoperation $\oplus$ and a binary operation $\odot$ as follows:

$\oplus$	a	b	c	d
a	a	b	c	d
b	b	$\{a, b\}$	d	c
c	c	d	$\{a, c\}$	b
d	d	c	b	$\{a, d\}$

$\odot$	a	b	c	d
a	a	a	a	a
b	a	b	b	b
c	a	c	c	c
d	a	d	d	d

Then, $(R, \oplus, \odot)$ is a Krasner hyperring. We have that $(R, \oplus, \odot, \leq)$ is an ordered Krasner hyperring where the order relation $\leq$ is defined by

$$\leq := \{(a, a), (b, b), (c, c), (d, d)\}.$$

Now, it is easy to see that $Q_1 = \{a, b\}$, $Q_2 = \{a, c\}$, and $Q_3 = \{a, d\}$ are quasi-hyperideals of R, but they are not left hyperideals of R.

Theorem 10.22. *Let* $(R, +, \cdot, \leq)$ *be an ordered Krasner hyperring. Then, the following statements hold:*

(1) *Every quasi-hyperideal of R is a subhyperring of R.*
(2) *The intersection of quasi-hyperideals set of R is a quasi-hyperideal of R.*
(3) *If I is a left hyperideal and J a right hyperideal of R, then $Q = I \cap J$ is a quasi-hyperideal of R.*
(4) *If Q is a quasi-hyperideal of R and T is a subhyperring of R, then $Q \cap T$ is a quasi-hyperideal of T.*

Proof. (1) Let Q be a quasi-hyperideal of an ordered Krasner hyperring $(R, +, \cdot, \leq)$. Then, $Q \cdot Q \subseteq R \cdot Q$ and $Q \cdot Q \subseteq Q \cdot R$. Hence, $Q \cdot Q \subseteq R \cdot Q \cap Q \cdot R \subseteq Q$. Therefore, Q is a subhyperring of R.

(2) Let $\{Q_k : k \in \Lambda\}$ be a family of quasi-hyperideals of R and $Q = \bigcap_{k \in \Lambda} Q_k$. Since $0 \in \bigcap_{k \in \Lambda} Q_k$, it follows that $\bigcap_{k \in \Lambda} Q_k \neq \emptyset$. We show that Q is a quasi-hyperideal of R. Let $x, y \in Q$. Then, $x, y \in Q_k$ for every $k \in \Lambda$. By assumption, we have $x + y \subseteq Q_k$ and $-x \in Q_k$, for each $k \in \Lambda$. So, we have $x + y \subseteq Q$ and $-x \in Q$. Thus, $(Q, +)$ is a canonical subhypergroup of $(R, +)$. Also, for all Q_k, $k \in \Lambda$, we have $(R \cdot Q) \cap (Q \cdot R) \subseteq (R \cdot Q_k) \cap (Q_k \cdot R) \subseteq Q_k$. Now, let $x \in Q$ and $y \in R$ such that $y \leq x$. Then, for every $k \in \Lambda$, $y \in Q_k$. Hence, $y \in Q$. Therefore, $Q = \bigcap_{k \in \Lambda} Q_k$ is a quasi-hyperideal of R.

(3) Let $x, y \in Q = I \cap J$. Then, $x, y \in I$ and $x, y \in J$. So, we have $x + y \subseteq I \cap J = Q$ and $-x \in I \cap J = Q$. Thus, $(Q, +)$ is a canonical subhypergroup of $(R, +)$. Since I is a left hyperideal and J a right hyperideal of R, we have $IJ \subseteq I$ and $IJ \subseteq J$. So, $IJ \subseteq I \cap J$. Thus, $I \cap J = Q \neq \emptyset$. Also, we have

$$(R \cdot Q) \cap (Q \cdot R) = (R \cdot (I \cap J)) \cap ((I \cap J) \cdot R)$$
$$\subseteq (R \cdot I) \cap (J \cdot R)$$
$$\subseteq I \cap J = Q.$$

Now, let $x \in Q$ and $y \in R$ such that $y \leq x$. Then, we have $x \in I$ and $x \in J$. So, $y \in I$ and $y \in J$. Hence, $y \in I \cap J = Q$. Therefore, Q is a quasi-hyperideal of R.

(4): Let $Q_1 = Q \cap T$. We show that Q_1 is a quasi-hyperideal of T. Clearly, Q_1 is a canonical subhypergroup of T. Since $Q_1 \subseteq Q$, it follows that $(Q_1 \cdot T) \cap (T \cdot Q_1) \subseteq (Q \cdot R) \cap (R \cdot Q) \subseteq Q$. Since $Q_1 \subseteq T$ and T is a subhyperring of R, we have $(Q_1 \cdot T) \cap (T \cdot Q_1) \subseteq T \cdot T \subseteq T$. So, we have checked that $(Q_1 \cdot T) \cap (T \cdot Q_1) \subseteq Q_1$. If $x \in Q_1$ and $y \in T$ such that $y \leq x$, then since $x \in Q$, it follows that $y \in Q$. Hence, $y \in Q_1$. Therefore, Q_1 is a quasi-hyperideal of T. □

Definition 10.27. Let $(R, +, \cdot, \leq)$ be an ordered Krasner hyperring. A non-empty subset A of R is called a *bi-hyperideal* of R if the following conditions hold:

(1) $(A, +)$ is a canonical subhypergroup of $(R, +)$ and $A \cdot A \subseteq A$;
(2) $A \cdot R \cdot A \subseteq A$;
(3) when $x \in A$ and $y \in R$ such that $y \leq x$, imply that $y \in A$.

Example 10.31. Let $R = \{a, b, c\}$ be a set with the hyperaddition $\oplus$ and the multiplication $\odot$ defined as follows:

$\oplus$	a	b	c
a	a	b	c
b	b	R	b
c	c	b	$\{a, c\}$

$\odot$	a	b	c
a	a	a	a
b	a	b	c
c	a	c	a

Then, $(R, \oplus, \odot)$ is a Krasner hyperring. We have that $(R, \oplus, \odot, \leq)$ is an ordered Krasner hyperring where the order relation $\leq$ is defined

by

$$\leq := \{(a,a),(b,b),(c,c),(a,b),(a,c),(c,b)\}.$$

The covering relation and the figure of R are given by

$$\prec = \{(a,c),(c,b)\}.$$

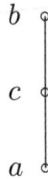

It is easy to see that $\{a\}$, $\{a,c\}$ and $\{a,b,c\}$ are bi-hyperideals of R.

Example 10.32. Let $R = \{a,b,c,d,e,f\}$ be a set with the hyper-addition $\oplus$ and the multiplication $\odot$ defined as follows:

$\oplus$	a	b	c	d	e	f
a	a	b	c	d	e	f
b	b	$\{a,b\}$	d	$\{c,d\}$	f	$\{e,f\}$
c	c	d	$\{a,c,e\}$	$\{b,d,f\}$	c	d
d	d	$\{c,d\}$	$\{b,d,f\}$	R	d	$\{c,d\}$
e	e	f	c	d	$\{a,e\}$	$\{b,f\}$
f	f	$\{e,f\}$	d	$\{c,d\}$	$\{b,f\}$	$\{a,b,e,f\}$

and

$\odot$	a	b	c	d	e	f
a	a	a	a	a	a	a
b	a	b	a	b	a	b
c	a	a	c	c	e	e
d	a	b	c	d	e	f
e	a	a	e	e	a	a
f	a	b	e	f	a	b

Then, $(R,\oplus,\odot)$ is a Krasner hyperring. We have that $(R,\oplus,\odot,\leq)$ is an ordered Krasner hyperring where the order relation $\leq$ is defined

by

$$\leq := \{(a,a),(b,b),(c,c),(d,d),(e,e),(f,f),(a,b),$$
$$(a,c),(a,d),(a,e),(a,f),(b,d),(b,f),(c,d),$$
$$(e,c),(e,d),(e,f),(f,d)\}.$$

The covering relation and the figure of R are given by

$$\prec = \{(a,b),(a,e),(b,f),(c,d),(e,c),(e,f),(f,d)\}.$$

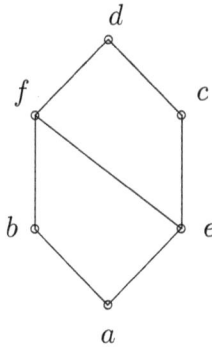

It is easy to see that $\{a\}$, $\{a,b\}$, $\{a,e\}$, $\{a,c,e\}$, $\{a,b,e,f\}$, and $\{a,b,c,d,e,f\}$ are bi-hyperideals of R.

The concept of bi-hyperideals of an ordered Krasner hyperring is a generalization of the concept of hyperideals (left and right hyperideals) of an ordered Krasner hyperring. Obviously, every left (resp., right) hyperideal of an ordered Krasner hyperring R is a bi-hyperideal of R; however, the converse need not be true. Indeed, If A is a left (right) hyperideal of R, then $(A,+)$ is a canonical subhypergroup of $(R,+)$. Since $AA \subseteq RA \subseteq A$, it follows that A is a subhyperring of R.

Example 10.33. Let $(R,+,\cdot)$ be a Krasner hyperring and $M(R) = \left\{ \begin{pmatrix} a & b \\ 0 & 0 \end{pmatrix} : a,b \in R \right\}$ be a collection of 2×2 matrices over R. The hyperaddition $\oplus$ and the multiplication $\odot$ are defined on $M(R)$ by

$$\begin{pmatrix} a & b \\ 0 & 0 \end{pmatrix} \oplus \begin{pmatrix} c & d \\ 0 & 0 \end{pmatrix} = \left\{ \begin{pmatrix} x & y \\ 0 & 0 \end{pmatrix} : x \in a + c, y \in b + d \right\},$$

$$\begin{pmatrix} a & b \\ 0 & 0 \end{pmatrix} \odot \begin{pmatrix} c & d \\ 0 & 0 \end{pmatrix} = \begin{pmatrix} ac & ad \\ 0 & 0 \end{pmatrix},$$

for all $\begin{pmatrix} a & b \\ 0 & 0 \end{pmatrix}, \begin{pmatrix} c & d \\ 0 & 0 \end{pmatrix} \in M(R)$. Then, $(M(R), \oplus, \odot)$ is a Krasner hyperring. Moreover, $(M(R), \oplus, \odot, \preceq)$ is an ordered Krasner hyperring, where $A = (a_{ij}) \preceq B = (b_{ij}) \Leftrightarrow a_{ij} = b_{ij}$, for all $1 \leqslant i, j \leqslant 2$. Here, $A = \left\{ \begin{pmatrix} a & 0 \\ 0 & 0 \end{pmatrix} : a \in R \right\}$ is a bi-hyperideal of $M(R)$, but it is not a right hyperideal of $M(R)$.

Lemma 10.7. *Let* $(R, +, \cdot, \leq)$ *be an ordered Krasner hyperring. Then:*

(1) *If* A_k *is a bi-hyperideal of* R *for all* $k \in \Lambda$, *then* $\bigcap_{k \in \Lambda} A_k$ *is a bi-hyperideal of* R.
(2) *If* $A_0 \subseteq A_1 \subseteq A_2 \subseteq \cdots \subseteq A_n \subseteq \cdots$ *is a ascending chain of bi-hyperideals of* R *and* A *is the union of these bi-hyperideals, then* A *is a bi-hyperideal of* R.

Proof. (1) Let $\{A_k : k \in \Lambda\}$ be a family of bi-hyperideals of R and $A = \bigcap_{k \in \Lambda} A_k$. Since $0 \in \bigcap_{k \in \Lambda} A_k$, it follows that $\bigcap_{k \in \Lambda} A_k \neq \emptyset$. It is easy to check that $(A, +)$ is a canonical subhypergroup of $(R, +)$ and $A \cdot A \subseteq A$. Now, let $x \in A \cdot R \cdot A$. Then, $x = a_1 \cdot r \cdot a_2$, for some $a_1, a_2 \in A$ and $r \in R$. Since each A_k is a bi-hyperideal of R, we have $x \in A_k \cdot R \cdot A_k \subseteq A_k$, for all $k \in \Lambda$. Thus, $x \in A_k$, for all $k \in \Lambda$. Hence, $x \in \bigcap_{k \in \Lambda} A_k = A$. Since x was chosen arbitrarily, we have $A \cdot R \cdot A \subseteq A$. If $x \in A$ and $y \in R$ such that $y \leq x$, then $x \in A_k$ for all $k \in \Lambda$. Since each A_k is a bi-hyperideal of R, we have $y \in A_k$, for all $k \in \Lambda$. Thus, $y \in \bigcap_{k \in \Lambda} A_k = A$. Therefore, A is a bi-hyperideal of R.

(2) It is easy to show that $(A, +)$ is a canonical subhypergroup of $(R, +)$ and $A \cdot A \subseteq A$. Now, let $x \in A \cdot R \cdot A$. Then, $x \in A_n \cdot R \cdot A_n$, for some bi-hyperideal A_n of R. Hence, $x \in A_n \subseteq A$. Thus, we have $A \cdot R \cdot A \subseteq A$. If $x \in A$ and $y \in R$ such that $y \leq x$, then $x \in A_n$ for

some bi-hyperideal A_n of R. Since each A_n is a bi-hyperideal of R, it follows that $y \in A_n$ for some bi-hyperideal A_n of R. Thus, $y \in A$. Therefore, A is a bi-hyperideal of R. □

Lemma 10.8. *Let $(R, +, \cdot, \leq)$ be an ordered Krasner hyperring. Then:*

(1) *Every quasi-hyperideal Q of a two-sided hyperideal I of R is a bi-hyperideal of R. In particular, every quasi-hyperideal Q of R is a bi-hyperideal of R.*
(2) *If A is a bi-hyperideal of R and T is a subhyperring of R, then $A \cap T$ is a bi-hyperideal of T.*
(3) *If B is a hyperideal of R and Q is a quasi-hyperideal of R, then $B \cap Q$ is a bi-hyperideal and a quasi-hyperideal of B.*

Proof. (1) It is easy to see that $(Q, +)$ is a canonical subhypergroup of $(R, +)$ and $Q \cdot Q \subseteq Q$. Since $Q \subseteq I$, we have

$$Q \cdot R \cdot Q \subseteq Q \cdot R \cdot I \cap I \cdot R \cdot Q \subseteq Q \cdot I \cap I \cdot Q \subseteq Q.$$

Now, let $x \in Q$ and $y \in R$ such that $y \leq x$. Since $Q \subseteq I$, it follows that $x \in I$. By assumption, I is a hyperideal of R. Thus, we have $y \in I$. Since Q is a quasi-hyperideal of I, it follows that $y \in Q$. Hence, Q is a bi-hyperideal of R.

(2) Let $A_1 = A \cap T$. We show that A_1 is a bi-hyperideal of T. Clearly, A_1 is a canonical subhypergroup of $(T, +)$ and $A_1 \cdot A_1 \subseteq A_1$. Since $A_1 \subseteq A$, it follows that $A_1 \cdot T \cdot A_1 \subseteq A \cdot R \cdot A \subseteq A$. Since $A_1 \subseteq T$ and T is a subhyperring of R, we have $A_1 \cdot T \cdot A_1 \subseteq T$. So, we have checked that $A_1 \cdot T \cdot A_1 \subseteq A_1$. Now, if $x \in A_1$ and $y \in T$ such that $y \leq x$, then since $x \in A$, it follows that $y \in A$. Hence, $y \in A_1$. Therefore, A_1 is a bi-hyperideal of T.

(3) It is straightforward. □

First, we give certain definitions needed for our purpose. The set of bi-hyperideals of R is totally ordered under inclusion if for all bi-hyperideals A, J, either $A \subseteq J$ or $J \subseteq A$. In Example 10.31, the set of bi-hyperideals of R is totally ordered under inclusion, whereas in Example 10.32, the set of bi-hyperideals of R is not totally ordered under inclusion. In Example 10.29, $\{a\}$, $\{a, b\}$, $\{a, e\}$, $\{a, g\}$, $\{a, b, e, f\}$, $\{a, b, g, h\}$, $\{a, c, e, g\}$, and $\{a, b, c, d, e, f, g, h\}$ are bi-hyperideals of R. So, the set of bi-hyperideals of R is not totally ordered under inclusion. A non-empty subset P of an ordered Krasner hyperring $(R, +, \cdot, \leq)$ is called a *prime hyperideal* of R if the following conditions hold: (1) $A \cdot B \subseteq P$ implies that $A \subseteq P$ or $B \subseteq P$, for any two hyperideal A, B of R, and (2) if $x \in P$ and $y \leq x$, then $y \in P$, for every $y \in R$. In Example 10.31, $\{a, c\}$ is a prime hyperideal of R. A non-empty subset I of an ordered Krasner hyperring $(R, +, \cdot, \leq)$ is called a *semiprime hyperideal* of R if the following conditions hold: (1) $A \cdot A \subseteq I$ implies that $A \subseteq I$, for any hyperideal A of R, and (2) if $x \in I$ and $y \leq x$, then $y \in I$, for every $y \in R$. In Example 10.31, $\{a, c\}$ is a semiprime hyperideal of R, but $\{a\}$ is not a semiprime hyperideal of R. Indeed, $\{a, c\} \odot \{a, c\} = \{a\}$, but $\{a, c\} \nsubseteq \{a\}$.

Definition 10.28. A bi-hyperideal A of an ordered Krasner hyperring $(R, +, \cdot, \leq)$ is called a *prime bi-hyperideal* of R if $A_1 \cdot A_2 \subseteq A$ implies either $A_1 \subseteq A$ or $A_2 \subseteq A$, for any bi-hyperideals A_1 and A_2 of R.

Example 10.34.

(1) In Example 10.31, $\{a, c\}$ is a prime bi-hyperideal of R, but $\{a\}$ is not a prime bi-hyperideal of R.
(2) In Example 10.32, $\{a, b\}$, $\{a, c, e\}$, and $\{a, b, e, f\}$ are prime bi-hyperideals of R. The bi-hyperideal $\{a\}$ is not prime. Indeed, $\{a, b\} \odot \{a, e\} = \{a\}$, but $\{a, b\} \nsubseteq \{a\}$ and $\{a, e\} \nsubseteq \{a\}$. Also, $\{a, e\}$ is not a prime bi-hyperideal of R. Indeed, $\{a, b, e, f\} \odot \{a, c, e\} = \{a, e\}$, but $\{a, b, e, f\} \nsubseteq \{a, e\}$ and $\{a, c, e\} \nsubseteq \{a, e\}$.
(3) In Example 10.29, $\{a, b, e, f\}$, $\{a, b, g, h\}$, and $\{a, c, e, g\}$ are prime bi-hyperideals of R, but $\{a\}$, $\{a, b\}$, $\{a, e\}$, and $\{a, g\}$ are not prime bi-hyperideals of R.

Theorem 10.23. *Let A be a prime bi-hyperideal of an ordered Krasner hyperring R. Then, A is a prime one-sided hyperideal of R.*

Proof.   Let A be a prime bi-hyperideal of R. Let I be a right hyperideal of R and J a left hyperideal of R such that $IJ \subseteq A$. Suppose that $I \nsubseteq A$. Assume that $x \in J$ and $s \in I \setminus A$. Then, $sIx \subseteq IJ \subseteq A$. Since A is a prime bi-hyperideal of R and $s \notin A$, we have $x \in A$. Hence, $J \subseteq A$. So, for any right hyperideal I and left hyperideal J of R, $IJ \subseteq A$ implies $I \subseteq A$ or $J \subseteq A$. Now, we show that A is a one-sided hyperideal of R. Since A is a bi-hyperideal of R, it follows that $(AR)(RA) \subseteq ARA \subseteq A$. Since AR is a right hyperideal and RA is a left hyperideal of R, we have $AR \subseteq A$ or $RA \subseteq A$. Therefore, A is a right hyperideal or a left hyperideal of R. □

Definition 10.29. A bi-hyperideal A of an ordered Krasner hyperring $(R, +, \cdot, \leq)$ is called a *semiprime bi-hyperideal* of R if $A_1 \cdot A_1 \subseteq A$ implies $A_1 \subseteq A$, for any bi-hyperideal A_1 of R.

Note that every prime bi-hyperideal is a semiprime bi-hyperideal. A semiprime bi-hyperideal is not necessarily prime. In Example 10.29, $\{a\}$ and $\{a, b\}$ are semiprime bi-hyperideals of R, but $\{a\}$ and $\{a, b\}$ are not prime bi-hyperideals of R.

Theorem 10.24. *The intersection of any family of prime bi-hyperideals of an ordered Krasner hyperring $(R, +, \cdot, \leq)$ is a semiprime bi-hyperideal of R.*

Proof. Let $\{A_k : k \in \Lambda\}$ be a family of prime bi-hyperideals of R and $A = \bigcap_{k \in \Lambda} A_k$. By (1) of Lemma 10.7, A is a bi-hyperideal of R. Let B be any bi-hyperideal of R such that $B^2 \subseteq A$. Then, $B^2 \subseteq A_k$, for all $k \in \Lambda$. Since each A_k is a prime bi-hyperideal of R, it follows that $B \subseteq A_k$, for all $k \in \Lambda$. Hence, $B \subseteq A$. Therefore, A is a semiprime bi-hyperideal of R. □

Theorem 10.25. *Let A be a semiprime bi-hyperideal and B a left (right) hyperideal of an ordered Krasner hyperring $(R, +, \cdot, \leq)$ such that $B^2 \subseteq A$. Then, we have $B \subseteq A$.*

Proof. Suppose that $B \nsubseteq A$. Then, there exists $a \in B$ such that $a \notin A$. Since B is a left (right) hyperideal of R, we have $aRa \subseteq BRB \subseteq BB \subseteq A$. Since A is a semiprime bi-hyperideal of R, it follows that $a \in A$; that is a contradiction. Hence, $B \subseteq A$, and so the proof is complete. □

Definition 10.30. If $(R, +, \cdot, \leq)$ is an ordered Krasner hyperring and $A \subseteq R$, then $(A]$ is a subset of R defined as follows:

$$(A] = \{t \in R : t \leq a, \text{ for some } a \in A\}.$$

Note that condition (3) in Definition 10.4 is equivalent to $A = (A]$. If A and B are non-empty subsets of R, then we have:

(1) $A \subseteq (A]$. Hence, $R = (R]$.
(2) $(A] \cdot (B] \subseteq (A \cdot B]$.
(3) $((A] \cdot (B]] = (A \cdot B]$.
(4) $((A]] = (A]$.
(5) $A \subseteq B$ implies $(A] \subseteq (B]$.
(6) If A and B are left (resp., right, two-sided) hyperideals of R, then $(AB]$ is a left (resp., right, two-sided) hyperideal of R.

Remark 10.9. Let $(R, +, \cdot, \leq)$ be an ordered Krasner hyperring. We recall that an element $a \in R$ is said to be regular if there exists an element $x \in R$ such that $a \leq (a \cdot x) \cdot a$. An ordered Krasner hyperring R is called regular if all elements of R are regular.

Equivalent definitions:

(1) $a \in (aRa]$, for all $a \in R$.
(2) $A \subseteq (ARA]$, for all $A \subseteq R$.

Example 10.35. In Example 10.29, $(R, \oplus, \odot, \leq)$ is a regular ordered Krasner hyperring.

Definition 10.31. An ordered Krasner hyperring $(R, +, \cdot, \leq)$ is called *left* (resp., *right*) *regular* if for every $a \in R$ there exists an element $x \in R$ such that $a \leq x \cdot a^2$ (resp., $a \leq a^2 \cdot x$). An ordered Krasner hyperring R is called left (resp., right) regular if all elements of R are left (resp., right) regular.

Equivalent definitions:

(1) $a \in (Ra^2]$, (resp., $a \in (a^2R]$), for all $a \in R$.
(2) $A \subseteq (RA^2]$, (resp., $A \subseteq (A^2R]$), for all $A \subseteq R$.

Example 10.36.

(1) In Example 10.29, $(R, \oplus, \odot, \leq)$ is a left regular ordered Krasner hyperring.
(2) In Examples 10.31 and 10.32, $(R, \oplus, \odot, \leq)$ is not a left regular ordered Krasner hyperring.

Definition 10.32. An ordered Krasner hyperring is called *completely regular* if it is regular, left regular, and right regular.

Example 10.37. In Example 10.29, $(R, \oplus, \odot, \leq)$ is a completely regular ordered Krasner hyperring, whereas in Examples 10.31 and 10.32, $(R, \oplus, \odot, \leq)$ is not a completely regular ordered Krasner hyperring.

Theorem 10.26. *Let $(R, +, \cdot, \leq)$ be an ordered Krasner hyperring. Then, R is regular if and only if every bi-hyperideal A of R is a semiprime bi-hyperideal.*

Proof. Assume that R is a regular ordered Krasner hyperring. Let A be a bi-hyperideal of R. Let $x \in R$ be such that $xRx \subseteq A$. Since R is regular, there exists $y \in R$ such that $x \leq xyx$. Thus, we have $x \leq xyx \in xRx \subseteq A$. Hence, $x \in (A] = A$. Therefore, A is a semiprime bi-hyperideal of R. We remark that since A is a bi-hyperideal of R, we have $(xRx] \subseteq A$ if and only if $xRx \subseteq A$.

Conversely, suppose that every bi-hyperideal of R is a semiprime bi-hyperideal. Let $a \in R$. It is easy to check that aRa is a bi-hyperideal of R. By assumption, aRa is a semiprime bi-hyperideal of R. Since $aRa \subseteq (aRa]$, it follows that $a \in (aRa]$. Thus, there exists $x \in R$ such that $a \leq axa$. Hence, R is a regular ordered Krasner hyperring. □

Theorem 10.27. *Let $(R, +, \cdot, \leq)$ be an ordered Krasner hyperring. Then, R is a regular ordered Krasner hyperring if and only if $(A \cdot B] = (A \cap B]$, for every right hyperideal A and left hyperideal B of R.*

Proof. It is straightforward. □

In the following, some properties and relationships between bi-hyperideals and quasi-hyperideals are investigated.

Theorem 10.28. *Let $(R, +, \cdot, \leq)$ be an ordered Krasner hyperring. Then, the following conditions are equivalent:*

(1) *R is regular.*
(2) *$A = A \cdot R \cdot A$, for every bi-hyperideal A of R.*
(3) *$Q = Q \cdot R \cdot Q$, for every quasi-hyperideal Q of R.*

Proof. (1⇒2): Assume that (1) holds. Let A be any bi-hyperideal of R and a any element of A. Then, there exists $x \in R$ such that $a \leq (a \cdot x) \cdot a$. It is easy to see that $(a \cdot x) \cdot a \in A \cdot R \cdot A$. Hence, $A \subseteq A \cdot R \cdot A$. Since A is a bi-hyperideal of R, it follows that $A \cdot R \cdot A \subseteq A$. Therefore, we have $A = A \cdot R \cdot A$.

(2⇒3): Evidently, every quasi-hyperideal of R is a bi-hyperideal of R. Then, by the assumption, we have $Q = Q \cdot R \cdot Q$, for every quasi-hyperideal Q of R.

(3⇒1): Assume that (3) holds. Let I and J be any right hyperideal and any left hyperideal of R, respectively. Then, we have $(I \cap J) \cdot R \cap R \cdot (I \cap J) \subseteq I \cdot R \cap R \cdot J \subseteq I \cap J$, and so it is easy to see that $I \cap J$ is a quasi-hyperideal of R. By the assumption and Theorem 10.27, we have $I \cap J = (I \cap J) \cdot R \cdot (I \cap J) \subseteq I \cdot R \cdot J \subseteq I \cdot J \subseteq I \cap J$. Hence, $I \cdot J = I \cap J$, and so R is regular.   □

Corollary 10.4. *Let $(R, +, \cdot, \leq)$ be an ordered Krasner hyperring. Then, the following statements are true:*

(1) *If $Q \cap A = Q \cdot A \cdot Q$, for every quasi-hyperideal Q and every hyperideal A of R, then R is a regular ordered Krasner hyperring.*
(2) *If $B \cap A = B \cdot A \cdot B$, for every bi-hyperideal B and every hyperideal A of R, then R is a regular ordered Krasner hyperring.*
(3) *If the set of all quasi-hyperideals of R is regular, then R is a regular ordered Krasner hyperring.*

Proof.   (1) Let Q be any quasi-hyperideal of R. Since R itself is a hyperideal of R, it follows that

$$Q = Q \cap R = Q \cdot R \cdot Q.$$

Then, it follows from Theorem 10.28 that R is regular.

(2) It is obvious.

(3) By assumption, for every quasi-hyperideal Q of R, there exists a quasi-hyperideal B of R such that $Q = Q \cdot B \cdot Q \subseteq Q \cdot R \cdot Q \subseteq (Q \cdot R) \cap (R \cdot Q) \subseteq Q$. Hence, for every quasi-hyperideal Q of R, we have $Q = Q \cdot R \cdot Q$. Then, it follows from Theorem 10.28 that R is regular.   □

The following theorem shows that the notions of quasi-hyperideal and bi-hyperideal in a regular ordered Krasner hyperring coincide.

Theorem 10.29. *Let $(R, +, \cdot, \leq)$ be a regular ordered Krasner hyperring. Then, the following statements are hold:*

(1) *Every bi-hyperideal A of R is a quasi-hyperideal.*
(2) *For every bi-hyperideal A of a two-sided hyperideal I of R, A is a quasi-hyperideal of R.*

Proof. (1) Let A be a bi-hyperideal of R. It is easy to see that $R \cdot A$ is a left hyperideal and $A \cdot R$ is a right hyperideal of R. By Theorem 10.27, we have $(A \cdot R) \cap (R \cdot A) = A \cdot R \cdot R \cdot A \subseteq A \cdot R \cdot A \subseteq A$. Hence, A is a quasi-hyperideal of R.

(2) First, we show that I is a regular subhyperring of R. Let $a \in I$. Then, there exists $x \in R$ such that $a \leq a \cdot x \cdot a \leq a \cdot x \cdot (a \cdot x \cdot a) = a \cdot (x \cdot a \cdot x) \cdot a$. Since $x \cdot a \cdot x \in I$, it follows that a is a regular element of I. Hence, I is a regular subhyperring of R. By (1), the bi-hyperideal A of I is a quasi-hyperideal of I. By (1) of Lemma 10.8, A is a bi-hyperideal of R. So by (1), A is a quasi-hyperideal of R. □

A subset A of an ordered Krasner hyperring $(R, +, \cdot, \leq)$ is called idempotent if $A = (A^2]$. In Example 10.29, $\{a, b\}$ is a idempotent subset of R.

Theorem 10.30. *Let $(R, +, \cdot, \leq)$ be an ordered Krasner hyperring. Then, the following are equivalent:*

(1) *R is regular.*
(2) *Every hyperideal of R is idempotent.*
(3) *Every hyperideal of R is semiprime.*

Proof. (1$\Rightarrow$2): Assume that (1) holds. Let A be any hyperideal of R. Since $A^2 = A \cdot A \subseteq A \cdot R \subseteq A$, implies $A^2 \subseteq A$. Now, let $a \in A$. Since R is regular, there exists an element $x \in R$ such that $a \leq (a \cdot x) \cdot a$. Since $a \in A$, it follows that $a \cdot x \in A$, for all $x \in R$. It is easy to see that $(a \cdot x) \cdot a \in A \cdot A = A^2$. Hence, $A \subseteq A^2$. Thus, we have $A = A^2$. Therefore, every hyperideal of R is idempotent.

(2$\Rightarrow$3): Let A and J be any hyperideals of R such that $A^2 \subseteq J$. Since $A = A \cdot A \subseteq J$, it follows that $A \subseteq J$.

(3$\Rightarrow$1): Assume that (3) holds. Let A and J be any right hyperideal and any left hyperideal of R, respectively. It is obvious that $A \cdot J \subseteq A \cap J$. Also, $A \cap J \subseteq A$ and $A \cap J \subseteq J$. Thus, we have $(A \cap J)^2 \subseteq A \cdot J$. By the assumption and $(A \cap J)^2 \subseteq A \cdot J$, we have $A \cap J \subseteq A \cdot J$. Hence $A \cap J = A \cdot J$. By Theorem 10.27, R is regular. □

Definition 10.33. A bi-hyperideal A of an ordered Krasner hyperring $(R, +, \cdot, \leq)$ is called an *irreducible bi-hyperideal* if for any bi-hyperideals I and J of R, $I \cap J = A$ implies that either $I = A$ or

$J = A$. The bi-hyperideal A is *strongly irreducible* if for bi-hyperideals I and J of R, $I \cap J \subseteq A$ implies that either $I \subseteq A$ or $J \subseteq A$.

Theorem 10.31. *Let* $(R, +, \cdot, \leq)$ *be an ordered Krasner hyperring. Then, the following conditions are equivalent:*

(1) *Every bi-hyperideal of R is strongly irreducible.*
(2) *Every bi-hyperideal of R is irreducible.*
(3) *The bi-hyperideals of R form a chain under inclusion.*

Proof. (1⇒2): Assume that (1) holds. Let I and J be any two bi-hyperideals of R. Let A be a bi-hyperideal of R such that $I \cap J = A$. Then, we have $A \subseteq I$ and $A \subseteq J$. Since A is strongly irreducible, it follows that $I \subseteq A$ or $J \subseteq A$. So, we have $I = A$ or $J = A$. Hence, A is an irreducible bi-hyperideal of R.

(2⇒3): Let I and J be any two bi-hyperideals of R. By Lemma 10.7, $I \cap J$ is a bi-hyperideal of R. Since $I \cap J = I \cap J$, by assumption we have $I = I \cap J$ or $J = I \cap J$. This implies that $I \subseteq J$ or $J \subseteq I$. Therefore, the set of bi-hyperideals of R form a chain under inclusion.

(3⇒1): Assume that (3) holds. Let A be a bi-hyperideal of R. Let I and J be any two bi-hyperideals of R such that $I \cap J \subseteq A$. By assumption, we have $I \subseteq J$ or $J \subseteq I$. Thus, either $I \cap J = I$ or $I \cap J = J$. This implies that $I \subseteq A$ or $J \subseteq A$. Hence, A is a strongly irreducible bi-hyperideal of R. □

Definition 10.34. A bi-hyperideal A of an ordered Krasner hyperring $(R, +, \cdot, \leq)$ is called a *strongly prime bi-hyperideal* of R if $A_1 \cdot A_2 \cap A_2 \cdot A_1 \subseteq A$ implies either $A_1 \subseteq A$ or $A_2 \subseteq A$, for any bi-hyperideals A_1 and A_2 of R.

Theorem 10.32. *Every strongly irreducible, semiprime bi-hyperideal of an ordered Krasner hyperring* $(R, +, \cdot, \leq)$ *is a strongly prime bi-hyperideal of R.*

Proof. Let A be a strongly irreducible, semiprime bi-hyperideal of an ordered Krasner hyperring R. Let A_1 and A_2 be any two bi-hyperideals of R such that $A_1 \cdot A_2 \cap A_2 \cdot A_1 \subseteq A$. Since $(A_1 \cap A_2)^2 \subseteq A_1 \cdot A_2$ and $(A_1 \cap A_2)^2 \subseteq A_2 \cdot A_1$, we have $(A_1 \cap A_2)^2 \subseteq A_1 \cdot A_2 \cap A_2 \cdot A_1 \subseteq A$. Since A is a semiprime bi-hyperideal, it follows that $A_1 \cap A_2 \subseteq A$. Since A is a strongly irreducible bi-hyperideal of R, either $A_1 \subseteq A$ or $A_2 \subseteq A$. Therefore, A is a strongly prime bi-hyperideal of R. □

Theorem 10.33. *Let A be a bi-hyperideal of an ordered Krasner hyperring $(R, +, \cdot, \leq)$ and $a \in R$ such that $a \notin A$. Then, there exists a strongly irreducible bi-hyperideal I of R such that $A \subseteq I$ and $a \notin I$.*

Proof. Let $\mathcal{C} = \{I : I$ be a bi-hyperideal of R, $A \subseteq I$, $a \notin I\}$. Since $A \in \mathcal{C}$, it follows that $\mathcal{C} \neq \emptyset$. Also, $\mathcal{C}$ is a partially ordered set under the usual inclusion. Let $\{I_k : k \in \Lambda\}$ be a chain in $\mathcal{C}$. Consider $B = \bigcup_{k \in \Lambda} I_k$. We show that B is a bi-hyperideal of R and $A \subseteq B$. If $x, y \in B$, then $x \in I_i$ and $y \in I_j$, for some $i, j \in \Lambda$. Since $\{I_k : k \in \Lambda\}$ is a totally ordered set, it follows that $I_i \subseteq I_j$ or $I_j \subseteq I_i$. If, say, $I_i \subseteq I_j$, then both x, y are inside I_j, so we have $x + y \subseteq I_j \subseteq B$. This implies that $x + y \subseteq B$. If, say, $I_j \subseteq I_i$, then both x, y are inside I_i, so we have $x + y \subseteq I_i \subseteq B$. This implies that $x + y \subseteq B$. If $x \in B$, then $x \in I_i$, for some $i \in \Lambda$. Thus, we have $-x \in I_i \subseteq B$. This implies that $-x \in B$. Now, let $x, y \in B$, then $x \in I_i$ and $y \in I_j$, for some $i, j \in \Lambda$. Since $I_i \subseteq I_j$ or $I_j \subseteq I_i$, it follows that $x, y \in I_i$ or $x, y \in I_j$. Thus, $x \cdot y \in I_i \subseteq B$ or $x \cdot y \in I_j \subseteq B$. This implies that $B \cdot B \subseteq B$. Therefore, $(B, +)$ is a canonical subhypergroup of $(R, +)$ and $B \cdot B \subseteq B$. Let $x, y \in B$ and $z \in R$. Then, $x \in I_r$ and $y \in I_s$, for some $r, s \in \Lambda$. Without any loss of generality, we assume that $x, y \in I_s$. Hence, $x \cdot z \cdot y \in I_s \subseteq B$. This implies that $B \cdot R \cdot B \subseteq B$. If $x \in B$ and $y \in R$ such that $y \leq x$, then $x \in I_k$ for some $k \in \Lambda$. So, we have $y \in I_k \subseteq B$. This implies that $y \in B$. Therefore, $B = \bigcup_{k \in \Lambda} I_k$ is a bi-hyperideal of R. Since each $I_k \in \mathcal{C}$ contains A and $a \notin I_k$, we have $A \subseteq \bigcup_{k \in \Lambda} I_k = B$ and $a \notin B$. Hence, $B \in \mathcal{C}$ is an upper bound for the chain $\{I_k : k \in \Lambda\}$. By Zorn's lemma, $\mathcal{C}$ has a maximal element, say N. We show that N is a strongly irreducible bi-hyperideal of R. Let A_1 and A_2 be two bi-hyperideals of R such that $A_1 \not\subseteq N$ and $A_2 \not\subseteq N$. By the maximality of N, we have $a \in A_1$ and $a \in A_2$. Thus, $a \in A_1 \cap A_2$, which implies $A_1 \cap A_2 \not\subseteq N$. This shows that N is strongly irreducible. Hence, the proof is complete. $\square$

Corollary 10.5. *A bi-hyperideal A of an ordered Krasner hyperring $(R, +, \cdot, \leq)$ is the intersection of all strongly irreducible bi-hyperideals of R containing A.*

Proof. Let $\mathcal{I} = \{I_k : k \in \Lambda\}$ be the set of all strongly irreducible bi-hyperideals of R containing A. Clearly, $A \subseteq \bigcap_{k \in \Lambda} I_k$. Suppose $B = \bigcap_{k \in \Lambda} I_k$ and $A \in \mathcal{I}$. Let $0 \neq a \in R$ such that $a \notin A$. By Theorem 10.33, there exists a strongly irreducible bi-hyperideal J of R such that $A \subseteq J$ and $a \notin J$. Hence, $J \in \mathcal{I}$, and so $a \notin B$. Thus, we have $B = \bigcap_{k \in \Lambda} I_k \subseteq A$. Hence, $A = \bigcap_{k \in \Lambda} I_k$, and so the proof is complete. $\qquad \square$

Let $\mathcal{P}_R$ denote the set of strongly irreducible proper bi-hyperideals of an ordered Krasner hyperring $(R, +, \cdot, \leq)$. For a bi-hyperideal A of R, define the set $\mathcal{E}_A = \{J \in \mathcal{P}_R : A \not\subseteq J\}$ and $\xi(\mathcal{P}_R) = \{\mathcal{E}_A : A$ is a bi-hyperideal of $R\}$.

Theorem 10.34. *Let $(R, +, \cdot, \leq)$ be an ordered Krasner hyperring. If the set of bi-hyperideals of R form a chain under inclusion, then the set $\xi(\mathcal{P}_R)$ forms a topology on the set $\mathcal{P}_R$.*

Proof. Since $\{0\}$ is a bi-hyperideal of R and $\mathcal{E}_{\{0\}} = \{J \in \mathcal{P}_R : \{0\} \not\subseteq J\} = \emptyset$, we have $\emptyset \in \xi(\mathcal{P}_R)$. Thus, $\mathcal{E}_{\{0\}}$ is an empty subset of $\xi(\mathcal{P}_R)$. Since R is a bi-hyperideal of itself and strongly irreducible bi-hyperideals are proper, we have $\mathcal{E}_R = \{J \in \mathcal{P}_R : R \not\subseteq J\} = \mathcal{P}_R$. So, we obtain $\mathcal{P}_R \in \xi(\mathcal{P}_R)$. Hence, the first axiom for the topology holds. Now, let $\mathcal{E}_{A_1}, \mathcal{E}_{A_2} \in \xi(\mathcal{P}_R)$. We shall show that $\mathcal{E}_{A_1} \cap \mathcal{E}_{A_2} = \mathcal{E}_{A_1 \cap A_2}$. If $J \in \mathcal{E}_{A_1} \cap \mathcal{E}_{A_2}$, then $J \in \mathcal{P}_R$ and $A_1 \not\subseteq J$, $A_2 \not\subseteq J$. Suppose $A_1 \cap A_2 \subseteq J$. Since J is a strongly irreducible bi-hyperideal of R, we have $A_1 \subseteq J$ or $A_2 \subseteq J$, which is a contradiction. Hence, $A_1 \cap A_2 \not\subseteq J$, which implies that $J \in \mathcal{E}_{A_1 \cap A_2}$. Thus, $\mathcal{E}_{A_1} \cap \mathcal{E}_{A_2} \subseteq \mathcal{E}_{A_1 \cap A_2}$. Now, if $J \in \mathcal{E}_{A_1 \cap A_2}$, then $J \in \mathcal{P}_R$ and $A_1 \cap A_2 \not\subseteq J$. Thus, $A_1 \not\subseteq J$ and $A_2 \not\subseteq J$. Hence, $J \in \mathcal{E}_{A_1}$ and $J \in \mathcal{E}_{A_2}$, which implies that $J \in \mathcal{E}_{A_1} \cap \mathcal{E}_{A_2}$. Therefore, we have $\mathcal{E}_{A_1 \cap A_2} \subseteq \mathcal{E}_{A_1} \cap \mathcal{E}_{A_2}$. Now, consider an arbitrary family $\{A_\alpha : \alpha \in \Lambda\}$ of bi-hyperideals of R. Let $\{\mathcal{E}_{A_\alpha} : \alpha \in \Lambda\} \subseteq \xi(\mathcal{P}_R)$. Then, we have $\bigcup_{\alpha \in \Lambda} \mathcal{E}_{A_\alpha} = \bigcup_{\alpha \in \Lambda}\{J \in \mathcal{P}_R : A_\alpha \not\subseteq J\} = \{J \in \mathcal{P}_R : A_\alpha \not\subseteq J$ for some $\alpha \in \Lambda\} = \{J \in \mathcal{P}_R : \langle \bigcup A_\alpha \rangle \not\subseteq J\} = \mathcal{E}_{\bigcup A_\alpha}$, where $\langle \bigcup A_\alpha \rangle$ is a bi-hyperideal of R generated by $\bigcup A_\alpha$, and by the proof of Theorem 10.33, $\bigcup A_\alpha$ is a bi-hyperideal of R. Thus, we have $\bigcup_{\alpha \in \Lambda} \mathcal{E}_{A_\alpha} \in \xi(\mathcal{P}_R)$. Hence, $\xi(\mathcal{P}_R)$ forms a topology on the set $\mathcal{P}_R$, and so the proof is complete. $\qquad \square$

Definition 10.35. Let $(R, +, \cdot, \leq)$ be an ordered Krasner hyperring. A non-zero bi-hyperideal A of R is called a *minimal bi-hyperideal* of R if there is no non-zero bi-hyperideal B of R such that $B \subset A$. A bi-hyperideal A of R is said to be *maximal* if for any proper bi-hyperideal B of R, $A \subseteq B$ implies that $A = B$.

Theorem 10.35. *Let A be a proper bi-hyperideal of an ordered Krasner hyperring $(R, +, \cdot, \leq)$. Then, A is contained in a maximal bi-hyperideal of R.*

Proof. Let $\mathcal{C} = \{J : J$ be a proper bi-hyperideal of R and $A \subseteq J\}$. Since $A \in \mathcal{C}$, it follows that $\mathcal{C} \neq \emptyset$. Also, $\mathcal{C}$ is an ordered set by inclusion. Let $\{J_k : k \in \Lambda\}$ be a totally ordered subset in $\mathcal{C}$. Consider $B = \bigcup_{k\in\Lambda} J_k$. It is easy to see that B is in $\mathcal{C}$. Hence, by Zorn's lemma, $\mathcal{C}$ has a maximal element, say M. Now, let K be a proper bi-hyperideal of R containing M. Then, K contains A, and so it belongs to $\mathcal{C}$. Since M is maximal in $\mathcal{C}$, it follows that $K = M$. Therefore, M is a maximal bi-hyperideal of R. $\square$

Theorem 10.36. *Let $(R, +, \cdot, \leq)$ be an ordered Krasner hyperring. If the set of bi-hyperideals of R is totally ordered under inclusion, then any maximal bi-hyperideal of R is strongly irreducible.*

Proof. Let A be a maximal bi-hyperideal of R. Let I and J be bi-hyperideals of R such that $I \cap J \subseteq A$ and $I \nsubseteq A$. Consider $B = I \cup A$. We show that B is a bi-hyperideal of R. Let $b_1, b_2 \in B$. By the assumption, we have $A \subseteq I$. So, both b_1, b_2 are inside I. Thus, we have $b_1 + b_2 \subseteq I \subseteq B$, $-b_1 \in I \subseteq B$ and $b_1 \cdot b_2 \in I \subseteq B$. This implies that $b_1 + b_2 \subseteq B$, $-b_1 \in B$ and $b_1 \cdot b_2 \in B$. Therefore, $(B, +)$ is a canonical subhypergroup of $(R, +)$ and $B \cdot B \subseteq B$. Now, let $b_1, b_2 \in B$ and $r \in R$. Then, $b_1, b_2 \in I$. Hence, $b_1 \cdot r \cdot b_2 \in I \subseteq B$. This implies that $B \cdot R \cdot B \subseteq B$. If $x \in B$ and $y \in R$ such that $y \leq x$, then $x \in I$. So, we have $y \in I \subseteq B$. This implies that $y \in B$. Therefore, $B = I \cup A$ is a bi-hyperideal of R such that $A \subset I \cup A$, so $I \cup A = R$. Thus, we have $J = J \cap R = J \cap (I \cup A) = (J \cap I) \cup (J \cap A) \subseteq A$. Hence, A is a strongly irreducible bi-hyperideal of R, and so the proof is complete. $\square$

10.4 Operations on Hyperideals

Example 10.38. Let $R = \{0, a, b, c\}$ be a set with the hyperaddition $\oplus$ and the multiplication $\odot$ defined as follows:

$\oplus$	0	a	b	c
0	0	a	b	c
a	a	$\{0,b\}$	$\{a,c\}$	b
b	b	$\{a,c\}$	$\{0,b\}$	a
c	c	b	a	0

$\odot$	0	a	b	c
0	0	0	0	0
a	0	a	b	c
b	0	b	b	0
c	0	c	0	c

Then, $(R, \oplus, \odot)$ is a Krasner hyperring. We have that $(R, \oplus, \odot, \leq)$ is an ordered Krasner hyperring, where the order relation $\leq$ is defined

by

$$\leq := \{(0,0),(a,a),(b,b),(c,c),(0,b),(c,a)\}.$$

The covering relation and the figure of R are given by

$$\prec = \{(0,b),(c,a)\}.$$

b \qquad\qquad a

0 \qquad\qquad c

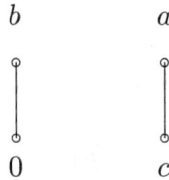

It is easy to see that $I_1 = \{0\}$, $I_2 = \{0,b\}$, $I_3 = \{0,c\}$, $I_4 = \{0,b,c\}$, and $I_5 = \{0,a,b,c\}$ are hyperideals of R.

Lemma 10.9. *Let $(R,+,\cdot,\leq)$ be an ordered Krasner hyperring. Then:*

(1) *if $\{A_k : k \in \Lambda\}$ is a family of hyperideals of R, then $\bigcup_{k\in\Lambda} A_k$ is a hyperideal of R.*
(2) *if $\{A_k : k \in \Lambda\}$ is a family of hyperideals of R, then $\bigcap_{k\in\Lambda} A_k$ is a hyperideal of R.*

Proof. (1) Since $0 \in \bigcup_{k\in\Lambda} A_k$, it follows that $\bigcup_{k\in\Lambda} A_k \neq \emptyset$. Let $a,b \in \bigcup_{k\in\Lambda} A_k$. Then, $a,b \in A_k$, for some $k \in \Lambda$. Since A_k is a hyperideal of R, we obtain $a-b \subseteq A_k$, for some $k \in \Lambda$. Thus, $a-b \subseteq \bigcup_{k\in\Lambda} A_k$. Also, we have $(\bigcup_{k\in\Lambda} A_k)\cdot R = \bigcup_{k\in\Lambda} A_k \cdot R \subseteq \bigcup_{k\in\Lambda} A_k$ and $R\cdot(\bigcup_{k\in\Lambda} A_k) = \bigcup_{k\in\Lambda} R\cdot A_k \subseteq \bigcup_{k\in\Lambda} A_k$. So, for each $a \in \bigcup_{k\in\Lambda} A_k$ and $r \in R$, $a\cdot r \in \bigcup_{k\in\Lambda} A_k$. Similarly, $r\cdot a \in \bigcup_{k\in\Lambda} A_k$. Now, let $x \in \bigcup_{k\in\Lambda} A_k$, $y \in R$ and $y \leq x$. Then, $x \in A_k$, for some $k \in \Lambda$. Since A_k is a hyperideal of R, it follows that $y \in A_k \subseteq \bigcup_{k\in\Lambda} A_k$. Therefore, $\bigcup_{k\in\Lambda} A_k$ is a hyperideal of R, as desired.

(2) Since $0 \in \bigcap_{k\in\Lambda} A_k$, it follows that $\bigcap_{k\in\Lambda} A_k \neq \emptyset$. Let $a,b \in \bigcap_{k\in\Lambda} A_k$ and $r \in R$. Then, $a,b \in A_k$, for each $k \in \Lambda$. By assumption, we obtain $a-b \subseteq A_k$, for each $k \in \Lambda$. Thus, $a-b \subseteq \bigcap_{k\in\Lambda} A_k$. Similarly, $r\cdot a, a\cdot r \in \bigcap_{k\in\Lambda} A_k$. Now, let $x \in \bigcap_{k\in\Lambda} A_k$ and $y \in R$ such that $y \leq x$. Then, for every $k \in \Lambda$, $y \in A_k$. Hence, $y \in \bigcap_{k\in\Lambda} A_k$. Therefore, $\bigcap_{k\in\Lambda} A_k$ is a hyperideal of R. $\square$

We define a preorder relation as a relation which satisfies the conditions of reflexivity and transitivity. We continue this section with the following theorem.

Theorem 10.37. *Let $(R, +, \cdot, \leq)$ be a preordered Krasner hyperring and ρ be a strongly regular relation on R. Then, $(R/\rho, \oplus, \odot, \preceq)$ is a preordered ring with respect to the following hyperoperations on the quotient set R/ρ:*

$$\overline{a} \oplus \overline{b} = \{\overline{c} \mid c \in a + b\},$$

$$\overline{a} \odot \overline{b} = \overline{a \cdot b},$$

where for all $\overline{a}, \overline{b} \in R/\rho$, a preorder relation $\preceq$ is defined by

$$\overline{a} \preceq \overline{b} \Leftrightarrow \forall a_1 \in \overline{a} \exists b_1 \in \overline{b} \quad \text{such that } a_1 \leq b_1.$$

Proof. Since ρ is a strongly regular relation on R, it follows that $(R/\rho, \oplus, \odot)$ is a ring. First, we show that the binary relation $\preceq$ is a preorder relation on R/ρ. Since $\leq$ is reflexive, it follows that $(a, a) \in \leq$. So, $\overline{a} \preceq \overline{a}$, for every $\overline{a} \in R/\rho$. Thus, $\preceq$ is reflexive. Now, let $\overline{a} \preceq \overline{b}$ and $\overline{b} \preceq \overline{c}$. Then, for every $a_1 \in \overline{a}$, there exists $b_1 \in \overline{b}$ such that $a_1 \leq b_1$. Since $b_1 \in \overline{b} \preceq \overline{c}$, there exists $c_1 \in \overline{c}$ such that $b_1 \leq c_1$. So, $\overline{a} \preceq \overline{c}$. Hence, $\preceq$ is transitive. Therefore, the binary relation $\preceq$ is a preorder relation on R/ρ. Now, let $\overline{a}, \overline{b}, \overline{x} \in R/\rho$ such that $\overline{a} \preceq \overline{b}$. If $\overline{u} = \overline{x} \oplus \overline{a}$, then for every $u_1 \in \overline{u}$, there exist $x_1 \in \overline{x}$ and $a_1 \in \overline{a}$ such that $u_1 \in x_1 + a_1$. Since $a_1 \in \overline{a} \preceq \overline{b}$, there exists $b_1 \in \overline{b}$ such that $a_1 \leq b_1$. Hence, $x_1 + a_1 \leq x_1 + b_1$. Thus, there exists $v_1 \in x_1 + b_1$ such that $u_1 \leq v_1$. Hence, $\overline{u} = \overline{u_1} \preceq \overline{v_1} = \overline{x} \oplus \overline{b}$. So, we have $\overline{x} \oplus \overline{a} \preceq \overline{x} \oplus \overline{b}$. If $\overline{s} = \overline{x} \odot \overline{a}$, then for every $s_1 \in \overline{s}$, there exist $x_1 \in \overline{x}$ and $a_1 \in \overline{a}$ such that $s_1 = x_1 \cdot a_1$. Since $a_1 \in \overline{a} \preceq \overline{b}$, there exists $b_1 \in \overline{b}$ such that $a_1 \leq b_1$. Hence, $x_1 \cdot a_1 \leq x_1 \cdot b_1$. Thus, for $t_1 = x_1 \cdot b_1$, we have $s_1 \leq t_1$. Hence, $\overline{s} = \overline{s_1} \preceq \overline{t_1} = \overline{x} \odot \overline{b}$. Therefore, $(R/\rho, \oplus, \odot, \preceq)$ is a preordered ring. $\square$

Theorem 10.38. *Let φ be a homomorphism from an ordered Krasner hyperring $(R, +, \cdot, \leq)$ onto an ordered Krasner hyperring $(T, \oplus, \odot, \preceq)$. If I is a hyperideal of T, then $\varphi^{-1}(I) = \{a \in R : \varphi(a) \in I\}$ is a hyperideal of R containing $\ker\varphi$.*

Proof. Since $0 \in \varphi^{-1}(I)$, it follows that $\varphi^{-1}(I) \neq \emptyset$. Let $x \in R$. Since φ is a homomorphism and $0 \in x - x$, we have $0 = \varphi(0) \in \varphi(x - x) \subseteq \varphi(x) \oplus \varphi(-x)$. So, $0 \in \varphi(x) \oplus \varphi(-x)$. Thus, $\varphi(-x)$ is the inverse of $\varphi(x)$ in the canonical hypergroup $(T, \oplus)$. Since $0 \in \varphi(x) \oplus \varphi(-x)$, it follows that $\varphi(-x) = -\varphi(x)$. Now, let $a_1, a_2 \in \varphi^{-1}(I)$. Then, $\varphi(a_1), \varphi(a_2) \in I$. Since I is a hyperideal of T, we have $\varphi(a_1 - a_2) \subseteq \varphi(a_1) \ominus \varphi(a_2) \subseteq I$. Hence, $a_1 - a_2 \subseteq \varphi^{-1}(I)$. Let $x \in R$ and $a \in \varphi^{-1}(I)$. Then, $\varphi(a) \in I$. Since φ is a homomorphism, it follows that $\varphi(x \cdot a) = \varphi(x) \odot \varphi(a) \in I$. Thu, $x \cdot a \in \varphi^{-1}(I)$. Similarly, $a \cdot x \in \varphi^{-1}(I)$. Now, let $a \in \varphi^{-1}(I)$ and $b \in R$ such that $b \leq a$. Then, $\varphi(a) \in I$. Since $b \leq a$ and φ is a homomorphism, we have $\varphi(b) \preceq \varphi(a)$. Since I is a hyperideal of T, it follows that $\varphi(b) \in I$. So, $b \in \varphi^{-1}(I)$. This proves that $\varphi^{-1}(I)$ is a hyperideal of R, as desired. Moreover, if $x \in ker\varphi$, then $\varphi(x) = 0 \in I$. Hence, $x \in \varphi^{-1}(I)$. Therefore, $ker\varphi \subseteq \varphi^{-1}(I)$. $\qquad\square$

We continue this section with the following definition.

Definition 10.36. Let I and J be two hyperideals of an ordered Krasner hyperring $(R, +, \cdot, \leq)$. The hyperideal quotient is $(I : J) = \{x \in R : x \cdot J \subseteq I\}$. The hyperideal quotient $(0 : J)$ is called the *annihilator* of J and denoted by $Ann(J)$. The set $Ann_l(x) = \{a \in R : a \cdot x = 0\}$ is called the *left annihilator* of x in R. Similarly, the set $Ann_r(x) = \{a \in R : x \cdot a = 0\}$ is called the *right annihilator* of x in R. In a commutative ordered Krasner hyperring R, we have $Ann_l(x) = Ann_r(x)$. In this case, we denote it by $Ann(x)$.

Lemma 10.10. *In Definition 10.36, $Ann(x)$ is a hyperideal of R.*

Proof. Since $0 \in Ann(x)$, it follows that $Ann(x) \neq \emptyset$. Let $a, b \in Ann(x)$. Then, $a \cdot x = 0$ and $b \cdot x = 0$. So, we have $(a + b) \cdot x = a \cdot x + b \cdot x = 0 + 0 = 0$. Thus, $c \cdot x = 0$, for all $c \in a + b$. Hence, $a + b \subseteq Ann(x)$. Also, we have $(-a) \cdot x = -(a \cdot x) = -0 = 0$. So, $-a \in Ann(x)$. Now, let $a \in Ann(x)$ and $r \in R$. Since $a \cdot x = 0$, it follows that $(r \cdot a) \cdot x = r \cdot (a \cdot x) = r \cdot 0 = 0$. So, we have $r \cdot a \in Ann(x)$. Let $a \in Ann(x)$, $b \in R$ and $b \leq a$. Then, we have $b \cdot x \leq a \cdot x$. Since $a \cdot x = 0$ and $\{0\}$ is a hyperideal of R, we obtain

$b \cdot x = 0$. So, $b \in Ann(x)$. Therefore, $Ann(x)$ is a hyperideal of R, as desired. □

Example 10.39. In Example 10.38, $Ann(0) = \{0, a, b, c\}$, $Ann(a) = \{0\}$, $Ann(b) = \{0, c\}$, and $Ann(c) = \{0, b\}$, which are hyperideals of R.

Theorem 10.39. *In Definition* 10.36, $(I : J)$ *is a hyperideal of* R.

Proof. This proof is straightforward. □

Definition 10.37. Let $(R, +, \cdot, \leq)$ be an ordered Krasner hyperring (resp., Krasner hyperring). R is said to be a *reduced ordered Krasner hyperring* (resp., *reduced Krasner hyperring*) if it has no nilpotent elements, i.e., if $a^n = 0$, for $a \in R$ and a natural number n, then $a = 0$. In a reduced ordered Krasner hyperring R, if $a \cdot b = 0$ for all $a, b \in R$, then $b \cdot a = 0$. So, we have $Ann_l(x) = Ann_r(x)$. In this case, we denote it by $Ann(x)$.

Remark 10.10. In Definition 10.36, we can replace the commutative ordered Krasner hyperring with the reduced ordered Krasner hyperring.

10.5 Properties of Interior Hyperideals in Ordered Krasner Hyperrings

In this section, we introduce the notion of interior hyperideals in ordered Krasner hyperrings and investigate some related results. We provide the conditions for an interior hyperideal to be a hyperideal. In particular, we prove that the concepts of interior hyperideals and hyperideals coincide in the case of regular (resp., intra-regular) ordered Krasner hyperrings.

Definition 10.38. A non-empty subset A of an ordered Krasner hyperring $(R, +, \cdot, \leq)$ is called an *interior hyperideal* of R if the following conditions hold:

(1) $(A, +)$ is a canonical subhypergroup of $(R, +)$ and $A \cdot A \subseteq A$;
(2) $R \cdot A \cdot R \subseteq A$;
(3) when $x \in A$ and $y \in R$ such that $y \leq x$, it implies that $y \in A$.

Example 10.40. Let $R = \{a,b,c,d,e,f,g,h\}$ be a set with the hyperaddition $\oplus$ and the multiplication $\odot$ defined as follows:

$\oplus$	a	b	c	d	e	f	g	h
a	a	b	c	d	e	f	g	h
b	b	b	$\{a,b,c,d\}$	b	f	f	$\{e,f,g,h\}$	f
c	c	$\{a,b,c,d\}$	c	c	g	$\{e,f,g,h\}$	g	g
d	d	b	c	a	h	f	g	e
e	e	f	g	h	$\{a,e\}$	$\{b,f\}$	$\{c,g\}$	$\{d,h\}$
f	f	f	$\{e,f,g,h\}$	f	$\{b,f\}$	$\{b,f\}$	R	$\{b,f\}$
g	g	$\{e,f,g,h\}$	g	g	$\{c,g\}$	R	$\{c,g\}$	$\{c,g\}$
h	h	f	g	e	$\{d,h\}$	$\{b,f\}$	$\{c,g\}$	$\{a,e\}$

and

$\odot$	a	b	c	d	e	f	g	h
a	a	a	a	a	a	a	a	a
b	a	b	c	d	a	b	c	d
c	a	c	b	d	a	c	b	d
d	a	a	a	a	a	a	a	a
e	a	a	a	a	e	e	e	e
f	a	b	c	d	e	f	g	h
g	a	c	b	d	e	g	f	h
h	a	a	a	a	e	e	e	e

Then, $(R, \oplus, \odot)$ is a Krasner hyperring. We have $(R, \oplus, \odot, \leq)$ is an ordered Krasner hyperring where the order relation $\leq$ is defined by:

$$\leq \; := \{(a,a),(b,b),(c,c),(d,d),(e,e),(f,f),(g,g),$$

$$(h,h),(a,e),(b,f),(c,g),(d,h)\}.$$

The covering relation and the figure of R are given by

$$\prec = \{(a,e),(b,f),(c,g),(d,h)\}.$$

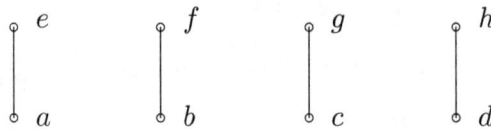

It is easy to see that $\{a\}$, $\{a,d\}$, $\{a,e\}$ $\{a,b,c,d\}$, $\{a,d,e,h\}$, and R are interior hyperideals of R.

Obviously, every hyperideal of an ordered Krasner hyperring R is an interior hyperideal; however, the converse is not true in general, i.e., an interior hyperideal may not be a hyperideal of R.

Example 10.41. Let R be the set $\left\{ \begin{pmatrix} a & 0 \\ b & c \end{pmatrix} : a,b,c \in \mathbb{Z} \right\}$. We define the binary hyperoperation $\oplus$ as $A \oplus B = \{A + B\}$. Consider the operation $\odot$ as usual matrix multiplication. Then, $(R, \oplus, \odot)$ is a Krasner hyperring. Moreover, $(R, \oplus, \odot, \leq)$ is an ordered Krasner hyperring, where $A = (a_{ij}) \leq B = (b_{ij}) \Leftrightarrow a_{ij} = b_{ij}$, for all $1 \leqslant i,j \leqslant 2$. Let $A = \left\{ \begin{pmatrix} a & 0 \\ b & 0 \end{pmatrix} : a,b \in \mathbb{Z} \right\}$. It is easy to check that A is an interior hyperideal of R. Since $\begin{pmatrix} 1 & 0 \\ 0 & 0 \end{pmatrix} \odot \begin{pmatrix} 0 & 1 \\ 0 & 0 \end{pmatrix} = \begin{pmatrix} 0 & 1 \\ 0 & 0 \end{pmatrix} \notin A$, it follows that A is not a right hyperideal of R. Thus, A is not a hyperideal of R.

Lemma 10.11. *Let* $(R, +, \cdot, \leq)$ *be an ordered Krasner hyperring. If* A_k *is an interior hyperideal of R, for all $k \in \Lambda$, then $\bigcap_{k \in \Lambda} A_k$ is an interior hyperideal of R.*

Proof. Let $\{A_k : k \in \Lambda\}$ be a family of interior hyperideals of R and $A = \bigcap_{k \in \Lambda} A_k$. Since $0 \in \bigcap_{k \in \Lambda} A_k$, it follows that $\bigcap_{k \in \Lambda} A_k \neq \emptyset$. It is easy to check that $(A, +)$ is a canonical subhypergroup of $(R, +)$ and $A \cdot A \subseteq A$. Now, let $x \in R \cdot A \cdot R$. Then, $x = r_1 \cdot a \cdot r_2$, for some $r_1, r_2 \in R$ and $a \in A$. Since each A_k is an interior hyperideal of R, it follows that $x \in R \cdot A_k \cdot R \subseteq A_k$, for all $k \in \Lambda$. Thus, $x \in A_k$, for all $k \in \Lambda$. Hence, $x \in \bigcap_{k \in \Lambda} A_k = A$. Since x was chosen arbitrarily, it follows that $R \cdot A \cdot R \subseteq A$. If $x \in A$ and $y \in R$ such that $y \leq x$, then $x \in A_k$, for all $k \in \Lambda$. Since each A_k is an interior hyperideal of R, it follows that $y \in A_k$, for all $k \in \Lambda$. Thus, we have $y \in \bigcap_{k \in \Lambda} A_k = A$. Therefore, A is an interior hyperideal of R. $\square$

Theorem 10.40. *Let* $(R, +, \cdot, \leq)$ *and* $(T, \oplus, \odot, \preceq)$ *be two ordered Krasner hyperrings. If $\varphi : R \to T$ is a homomorphism and A is an interior hyperideal of T, then $\varphi^{-1}(A) = \{r \in R : \varphi(r) \in A\}$ is an interior hyperideal of R.*

Proof. Since $0 \in \varphi^{-1}(A)$, it follows that $\varphi^{-1}(A) \neq \emptyset$. Now, let $r_1, r_2 \in \varphi^{-1}(A)$. Then, $\varphi(r_1), \varphi(r_2) \in A$. Since A is an interior hyperideal of T, we have $\varphi(r_1 + r_2) \subseteq \varphi(r_1) \oplus \varphi(r_2) \subseteq A$ and $\varphi(r_1 \cdot r_2) = \varphi(r_1) \odot \varphi(r_2) \in A$. Thus, $r_1 + r_2 \subseteq \varphi^{-1}(A)$ and $r_1 \cdot r_2 \in \varphi^{-1}(A)$. Obviously, other properties of a hyperring hold for $\varphi^{-1}(A)$ since $\varphi^{-1}(A)$ is a subset of R. Let $r_1, r_2 \in R$ and $a \in \varphi^{-1}(A)$. Then, $\varphi(r_1), \varphi(r_2) \in T$ and $\varphi(a) \in A$. Since A is an interior hyperideal of T, it follows that $\varphi(r_1 \cdot a \cdot r_2) = \varphi(r_1) \odot \varphi(a) \odot \varphi(r_2) \subset A$. So, we have $r_1 \cdot a \cdot r_2 \in \varphi^{-1}(A)$. Hence, $R \cdot \varphi^{-1}(A) \cdot R \subseteq \varphi^{-1}(A)$. Now, let $a \in \varphi^{-1}(A)$ and $r \in R$ such that $r \leq a$. Since φ is a homomorphism, it follows that $\varphi(r) \preceq \varphi(a)$. Since $\varphi(a) \in A$ and A is an interior hyperideal of T, we obtain $\varphi(r) \in A$. So, we have $r \in \varphi^{-1}(A)$. Therefore, $\varphi^{-1}(A)$ is an interior hyperideal of R. $\square$

In the following, we provide the conditions for an interior hyperideal to be a hyperideal.

Theorem 10.41. *Let $(R, +, \cdot, \leq)$ be an ordered Krasner hyperring. Then:*

(1) *If R is a regular ordered Krasner hyperring, then every interior hyperideal of R is a hyperideal of R.*
(2) *If R is an intra-regular ordered Krasner hyperring, then every interior hyperideal of R is a hyperideal of R.*

Proof. (1) Let A be an interior hyperideal of R and $a \in A$. Since R is regular, there exists $x \in R$ such that $a \leq a \cdot x \cdot a$. Now, let $r \in R$. Since A is an interior hyperideal of R, it follows that $a \cdot r \leq (a \cdot x \cdot a) \cdot r = (a \cdot x) \cdot a \cdot r \in A \cdot A \subseteq A$. Since r was chosen arbitrarily, we have $A \cdot R \subseteq A$. Hence A is a right hyperideal of R. Similarly, we can prove that A is a left hyperideal of R. Therefore, A is a hyperideal of R.

(2): Let A be an interior hyperideal of R. Let $r \in R$ and $a \in A$. Since R is intra-regular, there exist $x, y \in R$ such that $a \leq x \cdot a^2 \cdot y$. Since A is an interior hyperideal of R, it follows that $a \cdot r \leq (x \cdot a^2 \cdot y) \cdot r = (x \cdot a) \cdot a \cdot (y \cdot r) \in A$. Since r was chosen arbitrarily, we have $A \cdot R \subseteq A$. Hence, A is a right hyperideal of R. Similarly, we can prove that A is a left hyperideal of R. Therefore, A is a hyperideal of R. $\square$

As a consequence, we obtain the following corollary.

Corollary 10.6. *Let $(R, +, \cdot, \leq)$ be a regular (resp., intra-regular) ordered Krasner hyperring. Then, the following assertions are equivalent:*

(1) *A is a hyperideal of R.*
(2) *A is an interior hyperideal of R.*

The concepts of regular (resp., intra-regular) ordered Krasner hyperrings generalize the corresponding concepts of regular (resp., intra-regular) Krasner hyperrings as each regular (resp., intra-regular) Krasner hyperring endowed with the order $\leq := \{(a, b) : a = b\}$ is a regular (resp., intra-regular) ordered Krasner hyperring. Let $(R, +, \cdot, \leq)$ be an ordered Krasner hyperring and $\emptyset \neq A \subseteq R$. Then, $(A]$ is the subset of R defined as follows: $(A] = \{x \in R : x \leq a,$ for some $a \in A\}$. Let $A, B \subseteq R$. Then, $A \subseteq (A]$, $(A](B] \subseteq (AB]$ and $((A]] = (A]$. An element a of an ordered Krasner hyperring $(R, +, \cdot, \leq)$ is said to be *regular* if there exists an element $x \in R$ such that $a \leq (a \cdot x) \cdot a$, i.e., $a \in (aRa]$, for all $a \in R$, or $A \subseteq (ARA]$, for all $A \subseteq R$. An ordered Krasner hyperring $(R, +, \cdot, \leq)$ is said to be regular if every element of R is regular. An element a of an ordered Krasner hyperring $(R, +, \cdot, \leq)$ is said to be *intra-regular* if there exist $x, y \in R$ such that $a \leq x \cdot a^2 \cdot y$, i.e., $a \in (Ra^2R]$, for all $a \in R$, or $A \subseteq (RA^2R]$, for all $A \subseteq R$. An ordered Krasner hyperring $(R, +, \cdot, \leq)$ is said to be intra-regular if every element of R is intra-regular. An ordered Krasner hyperring $(R, +, \cdot, \leq)$ is called *left* (resp., *right*) *regular* if for every $a \in R$, there exists an element $x \in R$ such that $a \leq x \cdot a^2$ (resp., $a \leq a^2 \cdot x$). An ordered Krasner hyperring R is called left (resp., right) regular if all elements of R are left (resp., right) regular, i.e., $a \in (Ra^2]$ (resp., $a \in (a^2R]$), for all $a \in R$, or $A \subseteq (RA^2]$ (resp., $A \subseteq (A^2R]$), for all $A \subseteq R$. An ordered Krasner hyperring is called *completely regular* if it is regular, left regular, and right regular.

Asokkumar [7] studied the idempotent elements of Krasner hyperrings.

Definition 10.39. Let $(R, +, \cdot, \leq)$ be an ordered Krasner hyperring (resp., Krasner hyperring). An element x of R is said to be *idempotent* if $\{x\} = x \cdot x = x^2$. An ordered Krasner hyperring (resp., Krasner hyperring) is called *idempotent* if every element x of R is an idempotent. An ordered Krasner hyperring (resp., Krasner hyperring) is called a *Boolean ordered Krasner hyperring* (resp., *Boolean Krasner hyperring*) if every element x of R is an idempotent.

The following theorem was inspired by Corollary 10.6.

Theorem 10.42. *Let $(R, +, \cdot, \leq)$ be an ordered Krasner hyperring. Then, the following assertions hold:*

(1) *If R is a Boolean ordered Krasner hyperring, then every interior hyperideal of R is a hyperideal of R.*
(2) *If R is left (resp., right) regular, then every interior hyperideal of R is a hyperideal of R.*

Proof. (1) Let R be a Boolean ordered Krasner hyperring and $x \in R$. Then, $\{x\} = x^2 = x \cdot x = x^2 \cdot x^2 = x \cdot x^2 \cdot x$. Since $\leq$ is reflexive, it follows that $(x, x) \in \leq$. So, $(x, x \cdot x^2 \cdot x) \in \leq$. Thus, $x \leq x \cdot x^2 \cdot x$ for every $x \in R$. Hence, R is intra-regular. Hence, by Corollary 10.6, every interior hyperideal of R is a hyperideal of R.

(2) Let R be a left regular ordered Krasner hyperring and $x \in R$. Then, there exists an element $y \in R$ such that $x \leq y \cdot x^2 \leq y \cdot (y \cdot x^2) \cdot x \in Rx^2R$. So, R is intra-regular. Hence, by Corollary 10.6, every interior hyperideal of R is a hyperideal of R. Similarly, we can prove that in a right regular ordered Krasner hyperring R, every interior hyperideal of R is a hyperideal of R. $\square$

The concepts of hyperideals, prime hyperideals, and semiprime hyperideals of ordered Krasner hyperrings generalize the corresponding concepts of Krasner hyperrings. Let $(R, +, \cdot, \leq)$ be an ordered Krasner hyperring. A hyperideal A of R is called a *prime hyperideal* of R if for any hyperideals A_1, A_2 of R such that $A_1 \cdot A_2 \subseteq A$, we have $A_1 \subseteq A$ or $A_2 \subseteq A$. Note that if a hyperideal A of R is prime, then $A \neq R$. A hyperideal A of R is called a *semiprime hyperideal* of R if for any hyperideal B of R such that $B^2 \subseteq A$, we have $B \subseteq A$. Note that every prime hyperideal of R is a semiprime hyperideal of R; however, the converse is not true in general, i.e., a semiprime hyperideal may not be a prime hyperideal of R.

Example 10.42.

(1) In Example 10.40, $\{a, b, c, d\}$ and $\{a, d, e, h\}$ are prime hyperideals of R, whereas $\{a\}$, $\{a, d\}$, and $\{a, e\}$ are not prime hyperideals of R.

(2) In Example 10.38, $\{0\}$ is not a prime hyperideal of R. Indeed, $\{0, b\} \odot \{0, c\} = \{0\}$, whereas $\{0, b\} \not\subseteq \{0\}$, and $\{0, c\} \not\subseteq \{0\}$.
(3) In Example 10.38, $\{0\}$ is a semiprime hyperideal of R but not a prime hyperideal of R.
(4) In Example 10.40, $\{a, d\}$ is a semiprime hyperideal of R but not a prime hyperideal of R. Indeed, $\{a, b, c, d\} \odot \{a, d, e, h\} = \{a, d\}$, but $\{a, b, c, d\} \not\subseteq \{a, d\}$ and $\{a, d, e, h\} \not\subseteq \{a, d\}$.

Theorem 10.43. *Let* $(R, +, \cdot, \leq)$ *be an ordered Krasner hyperring. Then, the following statements are equivalent:*

(1) *R is intra-regular.*
(2) *Every interior hyperideal of R is semiprime.*
(3) *Every hyperideal of R is semiprime.*

Proof. $(1 \Rightarrow 2)$: Assume that (1) holds. Let A be an interior hyperideal of R and $B \subseteq R$ such that $B^2 \subseteq A$. Since R is intra-regular, we obtain $B \subseteq (RB^2R] \subseteq (RAR] \subseteq (A] = A$. So, A is semiprime.
$(2 \Rightarrow 3)$: It is straightforward.
$(3 \Rightarrow 1)$: Assume that (3) holds. Let $A \subseteq R$. It is easy to see that $(RA^2R]$ is a hyperideal of R. By assumption, $(RA^2R]$ is semiprime. Since $A^4 \subseteq (RA^2R]$, it follows that $A \subseteq (RA^2R]$. So, R is intra-regular. $\square$

Let $(R, +, \cdot, \leq)$ be an ordered Krasner hyperring. A subhyperring A of R is a bi-hyperideal of R if $A \cdot R \cdot A \subseteq A$ and $(A] \subseteq A$. For every left hyperideal, right hyperideal, hyperideal, and bi-hyperideal A of R, we have $(A] = A$.

Theorem 10.44. *Let* $(R, +, \cdot, \leq)$ *and* $(T, \oplus, \odot, \preceq)$ *be two ordered Krasner hyperrings. If $\varphi : R \to T$ is a homomorphism and A is a bi-hyperideal of T, then $\varphi^{-1}(A) = \{r \in R : \varphi(r) \in A\}$ is a bi-hyperideal of R.*

Proof. Since $0 \in \varphi^{-1}(A)$, it follows that $\varphi^{-1}(A) \neq \emptyset$. Now, let $r_1, r_2 \in \varphi^{-1}(A)$. Then, $\varphi(r_1), \varphi(r_2) \in A$. Since A is a bi-hyperideal, of T, we have $\varphi(r_1 + r_2) \subseteq \varphi(r_1) \oplus \varphi(r_2) \subseteq A$ and $\varphi(r_1 \cdot r_2) = \varphi(r_1) \odot \varphi(r_2) \in A$. Thus, $r_1 + r_2 \subseteq \varphi^{-1}(A)$ and $r_1 \cdot r_2 \in \varphi^{-1}(A)$. Obviously, other properties of a hyperring hold for $\varphi^{-1}(A)$ since $\varphi^{-1}(A)$ is a subset of R. Let $a_1, a_2 \in \varphi^{-1}(A)$ and $x \in R$. Then, $\varphi(a_1), \varphi(a_2) \in A$ and $\varphi(x) \in T$. Since A is a bi-hyperideal of T, it follows that $\varphi(a_1 \cdot x \cdot a_2) = \varphi(a_1) \odot \varphi(x) \odot \varphi(a_2) \in A$. So, we have

$a_1 \cdot x \cdot a_2 \in \varphi^{-1}(A)$. Hence, $\varphi^{-1}(A) \cdot R \cdot \varphi^{-1}(A) \subseteq \varphi^{-1}(A)$. Now, let $a \in \varphi^{-1}(A)$ and $x \in R$ such that $x \leq a$. Since φ is a homomorphism, it follows that $\varphi(x) \preceq \varphi(a)$. Since $\varphi(a) \in A$ and A is a bi-hyperideal of T, we obtain $\varphi(x) \in A$. Thus, we have $x \in \varphi^{-1}(A)$. Therefore, $\varphi^{-1}(A)$ is a bi-hyperideal of R. $\qquad\square$

Theorem 10.45. *Let $(R, +, \cdot, \leq)$ be an ordered Krasner hyperring with a non-zero proper bi-hyperideal. Then, every non-zero proper bi-hyperideal of R is minimal if and only if the intersection of any two distinct non-zero proper bi-hyperideals is $\{0\}$.*

Proof. Assume that every non-zero proper bi-hyperideal of R is minimal. Let A_1 and A_2 be two distinct non-zero proper bi-hyperideals of R and $A_1 \cap A_2 \neq \{0\}$. It is easy to see that $A_1 \cap A_2$ is a bi-hyperideal of R. By hypothesis, A_1 and A_2 are minimal. Since $\{0\} \neq A_1 \cap A_2 \subseteq A_1$ and $\{0\} \neq A_1 \cap A_2 \subseteq A_1$, we obtain $A_1 = A_2$, which is a contradiction. So, $A_1 \cap A_2 = \{0\}$, as desired. $\qquad\square$

Theorem 10.46. *Let $(R, +, \cdot, \leq)$ be an ordered Krasner hyperring. Then:*

(1) *If R is left regular, then the left hyperideals of R are semiprime.*
(2) *If R is completely regular, then every bi-hyperideal of R is semiprime.*

Proof. (1) Let A be a left hyperideal of R and $T \subseteq R$ such that $T^2 \subseteq A$. Since R is left regular, we have $T \subseteq (RT^2] \subseteq (RA] \subseteq (A] = A$. So, A is semiprime.

(2) Let A be a bi-hyperideal of R. Let T be a bi-hyperideal of R such that $T^2 \subseteq A$. Since R is completely regular, we have

$$T \subseteq (TRT] \subseteq ((T^2 R]R(RT^2]] = ((T^2 R](R](RT^2]]$$
$$\subseteq ((T^2 R)R(RT^2)] \subseteq (T^2 RT^2].$$

So, $T \subseteq (T^2 RT^2] \subseteq (ARA] \subseteq (A] = A$. Hence, A is a semiprime bi-hyperideal of R. $\qquad\square$

Theorem 10.47. *Let $(R, +, \cdot, \leq)$ be a regular ordered Krasner hyperring. Then, the following statements hold:*

(1) $A \cap K_1 \cap K_2 \subseteq AK_1 K_2$, *for every left hyperideals K_1, K_2 and right hyperideal A of R.*
(2) $B \cap K_1 \cap K_2 \subseteq BK_1 K_2$, *for every left hyperideals K_1, K_2 and interior hyperideal B of R.*

(3) $C \cap K_1 \cap K_2 \subseteq CK_1K_2$, *for every left hyperideals K_1, K_2 and bi-hyperideal C of R.*

(4) $D \cap E \cap K \subseteq DEK$, *for every bi-hyperideal D, interior hyperideal E, and left hyperideal K of R.*

Proof. (1) Let $a \in A \cap K_1 \cap K_2$. Since R is regular, there exists $x \in R$ such that $a \leq axa$. Thus, we have $a \leq axa \leq ax(axa) \in (AR)K_1RK_2 \subseteq AK_1K_2$. Therefore, we have $A \cap K_1 \cap K_2 \subseteq AK_1K_2$.

(2) Let B be an interior hyperideal of R. By Corollary 10.6, B is a hyperideal of R. The rest of this proof is similar to the proof of (1).

(3) Let $a \in C \cap K_1 \cap K_2$. Since R is regular, there exists $x \in R$ such that $a \leq axa$. Since C is a bi-hyperideal of R, it follows that $a \leq axa \leq ax(axa) \leq axax(axa) \in (CRC)RK_1RK_2 \subseteq CK_1K_2$. Therefore, we have $C \cap K_1 \cap K_2 \subseteq CK_1K_2$.

(4) By Corollary 10.6, E is a hyperideal of R. The rest of this proof is similar to the proof of (3). □

At the end of the chapter, we prove the following theorem.

Theorem 10.48. *Let P be a prime left hyperideal of an ordered Krasner hyperring $(R, +, \cdot, \leq)$. Then, the set $(P : a) = \{x \in R : x \cdot a \in P\}$ is a prime left hyperideal of R, for any $a \in R \setminus P$.*

Proof. Since $0 \in (P : a)$, it follows that $(P : a) \neq \emptyset$. Let $x, y \in (P : a)$. Then, $x \cdot a \in P$, $y \cdot a \in P$. Thus, we have $(x + y) \cdot a = x \cdot a + y \cdot a \subseteq P$ and $(-x) \cdot a = -(x \cdot a) \in P$. So, $x + y \subseteq (P : a)$ and $-x \in (P : a)$. Now, let $x \in (P : a)$ and $r \in R$. Since $x \cdot a \in P$, it follows that $(r \cdot x) \cdot a = r \cdot (x \cdot a) \in R \cdot P \subseteq P$. So, $r \cdot x \subseteq (P : a)$. Let $x \in (P : a)$, $y \in R$ and $y \leq x$. Then, $x \cdot a \in P$ and $y \cdot a \leq x \cdot a$. Since P is a left hyperideal of R, it follows that $y \cdot a \in P$. So, $y \in (P : a)$. Hence, $(P : a)$ is a left hyperideal of R.

Finally, let A and B be any left hyperideals of R such that $AB \subseteq (P : a)$. Then, $(AB)a \subseteq P$. Since P is a left hyperideal of R, it follows that $(P] = P$. It is easy to see that $(Aa]$ and $(Ba]$ are left hyperideals of R. Thus, we have $(Aa] \cdot (Ba] \subseteq (A(aB)a] \subseteq ((AB)a] \subseteq (P] = P$. Since P is a prime left hyperideal of R, it follows that $(Aa] \subseteq P$ or $(Ba] \subseteq P$. Thus, $Aa \subseteq P$ or $Ba \subseteq P$. So, we have $A \subseteq (P : a)$ or $B \subseteq (P : a)$. Therefore, $(P : a)$ is a prime left hyperideal of R. □

10.6   Clean Ordered Krasner Hyperrings

Let $(R, +, \cdot, \leq)$ be a commutative ordered hyperring with identity in the sense of Krasner. Denote the set of all invertible elements in R by $U(R)$ and the set of all idempotent elements in R by $Id(R)$. We start with the following definition.

Definition 10.40. Let $(R, +, \cdot, \leq)$ be an ordered Krasner hyperring. Then, an element $a \in R$ is said to be *clean* if $a \leq u + e$, where $u \in U(R)$ and $e \in Id(R)$. Also, we say that R is a *clean ordered Krasner hyperring* if all the elements in R are clean elements.

In the following, we present several examples of clean ordered Krasner hyperrings with different covering relations.

Example 10.43. Every clean Krasner hyperring induces a clean ordered Krasner hyperring. Indeed, let $(R, +, \cdot)$ be a clean Krasner hyperring. Define the order on R by $\leq := \{(a, b) \mid a = b\}$. Then, $(R, +, \cdot, \leq)$ is a clean ordered Krasner hyperring.

Example 10.44. Consider the hyperring $R = \{0, 1, -1\}$ with the hyperaddition $+$ and the multiplication $\cdot$ defined as follows:

+	0	1	−1
0	0	1	−1
1	1	1	R
−1	−1	R	−1

·	0	1	−1
0	0	0	0
1	0	1	−1
−1	0	−1	1

We have that $(R, +, \cdot, \leq)$ is an ordered Krasner hyperring, where the order relation $\leq$ is defined by

$$\leq := \{(0, 0), (1, 1), (-1, -1), (0, 1), (0, -1)\}.$$

The covering relation and the figure of R are given by

$$\prec = \{(0,1),(0,-1)\}.$$

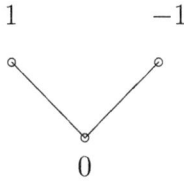

$$\begin{array}{ccc} 1 & & -1 \\ & 0 & \end{array}$$

Now, it is easy to see that R is a clean ordered Krasner hyperring.

Example 10.45. Let $R = \{0,1,a\}$. Consider the following tables:

+	0	1	a
0	0	1	a
1	1	R	1
a	a	1	$\{0,a\}$

$\cdot$	0	1	a
0	0	0	0
1	0	1	a
a	0	a	0

Then, $(R,+,\cdot)$ is a Krasner hyperring. We have that $(R,+,\cdot,\leq)$ is an ordered Krasner hyperring, where the order relation $\leq$ is defined by

$$\leq := \{(0,0),(1,1),(a,a),(0,1),(0,a),(a,1)\}.$$

The covering relation and the figure of R are given by

$$\prec = \{(0,a),(a,1)\}.$$

$$\begin{array}{c} 1 \\ a \\ 0 \end{array}$$

We can easily verify that R is a clean ordered Krasner hyperring.

Example 10.46. Let $R = \{0, a, b, c\}$ be a set with the hyperoperation $+$ and the multiplication $\cdot$ defined as follows:

+	0	a	b	c
0	0	a	b	c
a	a	$\{0,b\}$	$\{a,c\}$	b
b	b	$\{a,c\}$	$\{0,b\}$	a
c	c	b	a	0

$\cdot$	0	a	b	c
0	0	0	0	0
a	0	a	b	c
b	0	b	b	0
c	0	c	0	c

Then, $(R, +, \cdot)$ is a Krasner hyperring. We have that $(R, +, \cdot, \leq)$ is an ordered Krasner hyperring, where the order relation $\leq$ is defined by

$$\leq := \{(0,0), (a,a), (b,b), (c,c), (0,b), (c,a)\}.$$

The covering relation and the figure of R are given by

$$\prec = \{(0,b), (c,a)\}.$$

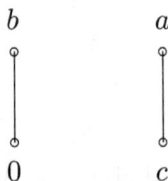

The following can easily be verified: $0 \leq a + c$, $a \leq b + c$, $b \leq a + c$ and $c \leq b + c$, where $a, b \in Id(R)$ and $c \in U(R)$. Hence, R is a clean ordered Krasner hyperring.

Example 10.47. Let $(R, +, \cdot, \leq)$ be a clean ordered Krasner hyperring. We consider

$$\mathbf{M} = \left\{ \begin{pmatrix} a & 0 \\ 0 & b \end{pmatrix} \,\middle|\, a, b \in R \right\}$$

and define the hyperoperation $\boxplus$ and operation $\boxdot$ on $\mathbf{M}$ as

$$\begin{pmatrix} a & 0 \\ 0 & b \end{pmatrix} \boxplus \begin{pmatrix} c & 0 \\ 0 & d \end{pmatrix} = \left\{ \begin{pmatrix} p & 0 \\ 0 & q \end{pmatrix} \,\middle|\, p \in a + c, q \in b + d \right\},$$

$$\begin{pmatrix} a & 0 \\ 0 & b \end{pmatrix} \boxdot \begin{pmatrix} c & 0 \\ 0 & d \end{pmatrix} = \begin{pmatrix} ac & 0 \\ 0 & bd \end{pmatrix},$$

where $A = \begin{pmatrix} a & 0 \\ 0 & b \end{pmatrix}$ and $B = \begin{pmatrix} c & 0 \\ 0 & d \end{pmatrix}$ are two arbitrary elements of $\mathbf{M}$. We define $A \preceq B$ if and only if $a \leq c$ and $b \leq d$. Then,

$(\mathbf{M}, \boxplus, \Box, \preceq)$ is an ordered Krasner hyperring. Let $A = \begin{pmatrix} a & 0 \\ 0 & b \end{pmatrix} \in$ $\mathbf{M}$. Since R is clean, it follows that $a \leq u + e$ and $b \leq v + f$, where $u, v \in U(R)$ and $e, f \in Id(R)$. Thus, there exist $x \in u + e$ and $y \in v + f$ such that $a \leq x$ and $b \leq y$. This means

$$\begin{pmatrix} a & 0 \\ 0 & b \end{pmatrix} \preceq \begin{pmatrix} x & 0 \\ 0 & y \end{pmatrix} \in \begin{pmatrix} u & 0 \\ 0 & v \end{pmatrix} \boxplus \begin{pmatrix} e & 0 \\ 0 & f \end{pmatrix}$$

Now, we have $A \preceq U \boxplus E$, where $U = \begin{pmatrix} u & 0 \\ 0 & v \end{pmatrix} \in U(\mathbf{M})$ and $E = \begin{pmatrix} e & 0 \\ 0 & f \end{pmatrix} \in Id(\mathbf{M})$. Hence, $(\mathbf{M}, \boxplus, \Box, \preceq)$ is a clean ordered Krasner hyperring.

Theorem 10.49. *Let $(R_i, +_i, \cdot_i, \leq_i)$ be a clean ordered Krasner hyperring, for all $i \in I$. Then, $\prod_{i \in I} R_i = \{(r_i)_{i \in I} \mid r_i \in R_i\}$ is a clean ordered Krasner hyperring.*

Proof. For all $(x_i)_{i \in I}, (y_i)_{i \in I} \in \prod_{i \in I} R_i$, we define:

(1) $(x_i)_{i \in I} + (y_i)_{i \in I} = \{(z_i)_{i \in I} \mid z_i \in x_i +_i y_i\}$,
(2) $(x_i)_{i \in I} \cdot (y_i)_{i \in I} = (x_i \cdot_i y_i)_{i \in I}$,
(3) $(x_i)_{i \in I} \leq (y_i)_{i \in I}$ if and only if $x_i \leq_i y_i$ for all $i \in I$.

First, we show that $(\prod_{i \in I} R_i, +, \cdot, \leq)$ is an ordered Krasner hyperring. Suppose that $(x_i)_{i \in I} \leq (y_i)_{i \in I}$, for $(x_i)_{i \in I}, (y_i)_{i \in I} \in \prod_{i \in I} R_i$. If $(t_i)_{i \in I} \in (a_i)_{i \in I} + (x_i)_{i \in I}$, where $(a_i)_{i \in I} \in \prod_{i \in I} R_i$, then $t_i \in a_i +_i x_i$. Since $(x_i)_{i \in I} \leq (y_i)_{i \in I}$, it follows that $x_i \leq_i y_i$, for all $i \in I$. By hypothesis, we have $t_i \in a_i +_i x_i \leq_i a_i +_i y_i$. So, there exists $s_i \in a_i +_i y_i$ such that $t_i \leq_i s_i$. Thus, we have $(t_i)_{i \in I} \leq (s_i)_{i \in I}$. This implies that $(a_i)_{i \in I} + (x_i)_{i \in I} \leq (a_i)_{i \in I} + (y_i)_{i \in I}$. Also, we have $a_i \cdot_i x_i \leq_i a_i \cdot_i y_i$, where $(0) \leq (a_i)_{i \in I}$. This means that $(a_i)_{i \in I} \cdot (x_i)_{i \in I} \leq (a_i)_{i \in I} \cdot (y_i)_{i \in I}$. Therefore, $\prod_{i \in I} R_i$ is an ordered Krasner hyperring.

Now, let $\{R_i\}_{i \in I}$ be clean for each $i \in I$ and $(a_i)_{i \in I} \in \prod_{i \in I} R_i$. We have $a_i \leq_i u_i +_i e_i$, where $u_i \in U(R_i)$ and $e_i \in Id(R_i)$. Thus, there exists $b_i \in u_i +_i e_i$ such that $a_i \leq_i b_i$. This implies that $(a_i)_{i \in I} \leq (b_i)_{i \in I}$, where $(b_i)_{i \in I} \in (u_i)_{i \in I} + (e_i)_{i \in I}$. Then, $(a_i)_{i \in I} \leq (u_i)_{i \in I} + (e_i)_{i \in I}$, where $(u_i)_{i \in I} \in U(\prod_{i \in I} R_i)$ and $(e_i)_{i \in I} \in Id(\prod_{i \in I} R_i)$. Hence, $\prod_{i \in I} R_i$ is a clean ordered Krasner hyperring. $\square$

Theorem 10.50. *Any homomorphic image of a clean ordered Krasner hyperring is a clean ordered Krasner hyperring.*

Proof. Suppose that φ is a surjective homomorphism from an ordered Krasner hyperring $(R, +, \cdot, \leq)$ onto an ordered Krasner hyperring $(T, \boxplus, \boxdot, \preceq)$. Take any $t \in T$; then, there exists $x \in R$ such that $\varphi(x) = t$. Since R is clean, we have $x \leq u + e$, where $u \in U(R)$ and $e \in Id(R)$. Thus, there exists $y \in u + e$ such that $x \leq y$. So, we have

$$\varphi(x) \preceq \varphi(y) \in \varphi(u + e) \subseteq \varphi(u) \boxplus \varphi(e),$$

where $\varphi(u) \in U(T)$ and $\varphi(e) \in Id(T)$. This completes the proof. $\square$

Theorem 10.51. *A clean ordered Krasner hyperring $(R, +, \cdot, \leq)$ is a clean ordered ring if and only if $1 + (-1) = \{0\}$.*

Proof. The necessity follows easily, so we concentrate on the sufficiency. To that aim, Suppose that $x, y \in R$. Let $u, v \in x + y$. Then, we have

$$
\begin{aligned}
u - v &\subseteq (a + b) - (a + b) \\
&= (a + b) - a - b \\
&= (a + (-a)) + (b + (-b)) \\
&= a \cdot (1 + (-1)) + b \cdot (1 + (-1)) \\
&= a \cdot \{0\} + b \cdot \{0\} \\
&= 0 + 0 \\
&= \{0\}.
\end{aligned}
$$

Thus, $u - v = \{0\}$ and hence $u = v$. It follows that $a + b = \{u\}$, and so $+$ is a binary operation. Therefore, $(R, +, \cdot, \leq)$ is an ordered ring. By hypothesis, every $a \in R$ can be written as $a \leq u + e$, where $u \in U(R)$, $e \in Id(R)$ and $u + e$ is a singleton set. Thus, R is a clean ordered ring. $\square$

Theorem 10.52. *Let $(R, +, \cdot, \leq)$ be a clean ordered Krasner hyperring and σ a pseudo-order on R. Then, there exists a strongly regular equivalence relation $\sigma^* = \{(a, b) \in R \times R \,|\, a\sigma b \text{ and } b\sigma a\}$ on R such that $(R/\sigma^*, \oplus, \odot, \preceq_{\sigma^*})$ is a clean ordered ring, where $\preceq_{\sigma^*} := \{(\sigma^*(x), \sigma^*(y)) \in R/\sigma^* \times R/\sigma^* \,|\, \exists a \in \sigma^*(x), \exists b \in \sigma^*(y) \text{ such that } (a, b) \in \sigma\}$.*

Proof. We divide the proof into three steps:

Step 1. We first construct an ordered ring from an ordered Krasner hyperring.

Suppose that σ^* is the relation on R defined as follows:

$$\sigma^* = \{(a,b) \in R \times R \,|\, a\sigma b \text{ and } b\sigma a\}.$$

Clearly, σ^* is a strongly regular relation on $(R,+)$ and $(R, \cdot)$. Hence, by Theorem 10.27, R/σ^* with the following operations is a ring:

$$\sigma^*(x) \oplus \sigma^*(y) = \sigma^*(z), \quad \text{for all } z \in x + y;$$
$$\sigma^*(x) \odot \sigma^*(y) = \sigma^*(x \cdot y).$$

Now, for each $\sigma^*(x), \sigma^*(y) \in R/\sigma^*$, define the order relation $\preceq_{\sigma^*}$ on R/σ^* by

$$\preceq_{\sigma^*} := \{(\sigma^*(x), \sigma^*(y)) \in R/\sigma^* \times R/\sigma^* \,|\, \exists a \in \sigma^*(x),$$
$$\exists b \in \sigma^*(y) \text{ such that } (a,b) \in \sigma\}.$$

We have

$$\sigma^*(x) \preceq_{\sigma^*} \sigma^*(y) \quad \Leftrightarrow \quad x\sigma y.$$

Now, we prove that $(R/\sigma^*, \oplus, \odot, \preceq_{\sigma^*})$ is an ordered ring. Let $a, b, c \in R$. Since $(a,a) \in \leq \subseteq \sigma$, we have $\sigma^*(a) \preceq_{\sigma^*} \sigma^*(a)$. If $\sigma^*(a) \preceq_{\sigma^*} \sigma^*(b)$ and $\sigma^*(b) \preceq_{\sigma^*} \sigma^*(a)$, then $(a,b) \in \sigma$ and $(b,a) \in \sigma$. This means that $(a,b) \in \sigma^*$, and so $\sigma^*(a) = \sigma^*(b)$. Let $\sigma^*(a) \preceq_{\sigma^*} \sigma^*(b)$ and $\sigma^*(b) \preceq_{\sigma^*} \sigma^*(c)$. Then, $(a,b) \in \sigma$ and $(b,c) \in \sigma$. This means that $(a,c) \in \sigma$, and so $\sigma^*(a) \preceq_{\sigma^*} \sigma^*(c)$. Therefore, $\preceq_{\sigma^*}$ is an order on R/σ^*. Now, let $\sigma^*(x) \preceq_{\sigma^*} \sigma^*(y)$ and $\sigma^*(z) \in R/\sigma^*$. Then, $x\sigma y$ and $z \in R$. Since σ is a pseudo-order on R, we have $x + z \overline{\overline{\sigma}} y + z$. So, for all $a \in x + z$ and $b \in y + z$, we have $a\sigma b$. This implies that $\sigma^*(a) \preceq_{\sigma^*} \sigma^*(b)$. Hence, $\sigma^*(x) \oplus \sigma^*(z) \preceq_{\sigma^*} \sigma^*(y) \oplus \sigma^*(z)$. Similarly, we have $\sigma^*(x) \odot \sigma^*(z) \preceq_{\sigma^*} \sigma^*(y) \odot \sigma^*(z)$.

Step 2. The following hold for an ordered Krasner hyperring R:

(1) If $e \in Id(R)$, then $\sigma^*(e) \in Id(R/\sigma^*)$.
(2) If $u \in U(R)$, then $\sigma^*(u) \in U(R/\sigma^*)$.

Step 3. We finally show that R/σ^* is clean.

Suppose that $(R, +, \cdot, \leq)$ is a clean ordered Krasner hyperring. Let $\sigma^*(a) \in R/\sigma^*$, where $a \in R$. Since R is clean, there exist $u \in U(R)$

and $e \in Id(R)$ such that $a \leq u + e$. Hence, there exists $x \in u + e$ such that $a \leq x$. So, $(a, x) \in \leq \subseteq \sigma$. Thus, $a\sigma x$. Since $a \in \sigma^*(a)$ and $x \in \sigma^*(x)$, we have $\sigma^*(a) \preceq_{\sigma^*} \sigma^*(x)$. Since $\sigma^*(x) \in \sigma^*(u) \oplus \sigma^*(e)$, it follows that $\sigma^*(a) \preceq_{\sigma^*} \sigma^*(u) \oplus \sigma^*(e)$. Now, by the previous step, $\sigma^*(a)$ is clean. Hence, R/σ^* is clean. $\qquad\square$

The following example illustrates this result.

Example 10.48. Let $(R, +, \cdot, \leq)$ be a clean ordered Krasner hyper-ring defined as in Example 10.46. Consider the pseudo-order

$$\sigma = \{(0,0), (a,a), (b,b), (c,c), (0,b), (b,0), (a,c), (c,a)\}.$$

Note that $\sigma^* = \sigma$ and that

$$R/\sigma^* = \{u_1, u_2\}, \text{ where } u_1 = \{0, b\} \text{ and } u_2 = \{a, c\}.$$

Now, $(R/\sigma^*, \oplus, \odot, \preceq_{\sigma^*})$ is a clean ordered ring, where $\oplus$ and $\odot$ are defined in the following tables:

$\oplus$	u_1	u_2
u_1	u_1	u_2
u_2	u_2	u_1

$\odot$	u_1	u_2
u_1	u_1	u_1
u_2	u_1	u_2

and $\preceq_{\sigma^*} = \{(u_1, u_1), (u_2, u_2)\}$.

Corollary 10.7. *Let us follow the notations and definitions used in Theorem* 10.52. *If R is regular, then R/σ^* is regular.*

Proof. Let R be regular and $\sigma^*(x) \in R/\sigma^*$, where $x \in R$. Then, there exists $a \in R$ such that $x \leq x \cdot a \cdot x$. Clearly, $\leq \subseteq \sigma$, so $x\sigma x \cdot a \cdot x$. Since $x \in \sigma^*(x)$ and $x \cdot a \cdot x \in \sigma^*(x \cdot a \cdot x)$, clearly, we obtain $\sigma^*(x) \preceq_{\sigma^*} \sigma^*(x \cdot a \cdot x)$. This shows that $\sigma^*(x) \preceq_{\sigma^*} \sigma^*(x) \odot \sigma^*(a) \odot \sigma^*(x)$, so R/σ^* is regular. $\qquad\square$

10.7   Associated Hyperringoid to a Krasner Hyperring

We create new hyperringoids by the modified "Ends Lemma." The results are obtained by Mirabdollahi *et al.* [38].

Lemma 10.12. *Let $(R, \leq)$ be an ordered set and $A \subseteq R$. Then, we have:*

(1) $A \leq [A)_{\leq}$,
(2) *the set $[A)_{\leq}$ is the maximum element of the family of subsets A' of R such that $A \leq A'$.*

Proof. (1) We have $A \subseteq [A)_{\leq} = \bigcup_{a \in A} [a)_{\leq}$. So, according to the reflexivity of the (partial) order relation $\leq$, for every $a \in A$, there exists $y(= a) \in [A)_{\leq}$ such that $a \leq y$. On the other hand, if $y \in [A)_{\leq}$ is an arbitrary element, then there exists $a \in A$ such that $y \in [a)_{\leq}$, or, equivalently, there exists $a \in A$ such that $a \leq y$.

(2) Suppose to the contrary there exists $B \subseteq R$ such that $[A)_{\leq} \subset B$ and $A \leq B$. Since $[A)_{\leq} \subset B$, there exists $x_0 \in B$ such that $x_0 \notin [A)_{\leq} = \bigcup_{a \in A} [a)_{\leq}$. As a result, for every $a \in A$, $x_0 < a$. On the other hand, from $A \leq B$ and $x_0 \in B$, it follows that there exists $a \in A$ such that $a \leq x_0$. It is a contradiction. $\square$

Lemma 10.13. *Let $(R, \leq)$ be an ordered set and $A \subseteq R$. Then, we have $[A)_{\leq} = [[A))_{\leq}$.*

Proof. According to the maximality of $[A)_{\leq}$ among all the subsets X of R satisfying the property $A \leq X$, it is sufficient to prove that $A \leq [[A))_{\leq}$. Since $A \subseteq [A)_{\leq} \subseteq [[A))_{\leq}$, from $a \in A$, it follows that $a \in [[A))_{\leq}$, and so by reflexivity, for every $a \in A$, there exists $a \in [[A))_{\leq}$ such that $a \leq a$. On the other hand, if $x \in [[A))_{\leq} = \bigcup_{y \in [A)_{\leq}} [y)_{\leq}$, then there exists $y \in [A)_{\leq}$ such that $y \leq x$. Also, from $y \in [A)_{\leq}$, it follows that there exists $a \in A$ such that $a \leq y$. As a result, according to the transitivity of the (partial) order relation, there exists $a \in A$ such that $a \leq x$. $\square$

Due to the maximality of $[A)_{\leq}$ among all the subsets X of $(R, \leq)$ satisfying the property $A \leq X$, the following corollary is easy to see.

Corollary 10.8. *For non-empty subsets A and B of an ordered canonical hypergroup $(H, +, \leq)$, we have:*

(1) $[[A)_{\leq} + B)_{\leq} = [A + B)_{\leq}$ and $[A + [B)_{\leq})_{\leq} = [A + B)_{\leq}$,
(2) $[[A + B)_{\leq})_{\leq} = [A + B)_{\leq}$,
(3) $[A)_{\leq} + [B)_{\leq} = [A + B)_{\leq}$.

Lemma 10.14. *Let $(R, +, \cdot, \leq)$ be an ordered Krasner hyperring, $a \in R^+$, and $B \subseteq R$. Then, we have $a \cdot [B)_{\leq} \subseteq [a \cdot B)_{\leq}$. If R is an ordered Krasner hyperfield, then $a \cdot [B)_{\leq} = [a \cdot B)_{\leq}$.*

Proof. If $y \in a \cdot [B)_{\leq}$, then there exists $x \in [B)_{\leq}$ such that $y = ax$. Since $x \in [B)_{\leq}$, so there exists $b \in B$ such that $b \leq x$. Therefore, $ab \leq ax$ and $ab \in a.B$. Thus, $y = ax \in [a.B)_{\leq}$. Now, if R is an ordered Krasner hyperfield and $y \in [a.B)_{\leq}$, then there exists $b \in B$ such that $ab \leq y$. Since a and therefore a^{-1} are positive, we have $b = a^{-1}ab \leq a^{-1}y$. Also, we have $ab \leq a(a^{-1}y) = y$. Now, by putting $x = a^{-1}y$, we have $b \leq x$, and as a result, $y \in a.[B)_{\leq}$. $\square$

In the following, we construct new hyperringoids from ordered Krasner hyperrings through what we have achieved so far.

Definition 10.41. Let $(R, +, \cdot, \leq)$ be an ordered Krasner hyperring. For $a, b \in R$, we define the new hyperoperation $\oplus : R \times R \longrightarrow \wp^*(R)$ as follows:

$$a \oplus b = [|a| + |b|)_{\leq} = \bigcup_{m \in |a| + |b|} [m)_{\leq}.$$

Remark 10.11.

(1) It is easy to see that Definition 10.41 is not useful for satisfying the reproduction principles. Indeed, there exists no positive ordered Krasner hyperring. On the other hand, this definition is meaningless for Krasner hyperrings which are not ordered. So, we improve Definition 10.41 as follows:

$$a \oplus_1 b = \{a, b\} \bigcup [|a| + |b|)_{\leq} = \{a, b\} \bigcup \bigcup_{m \in |a| + |b|} [m)_{\leq}.$$

(2) If . is the multiplicative operation in an ordered Krasner hyperring, then due to the definition, we define the new multiplicative operation as follows:

$$a.'b = |a| \cdot |b|.$$

Theorem 10.53. *Let $(H, +, \leq)$ be an ordered canonical hypergroup. Then, the hyperoperation $\oplus$ is associative.*

Proof. Suppose that $0 \le a, c$ and $b \le 0$. Then,

$$(a \oplus b) \oplus c = [|a| + |b|)_\le \oplus c = [[|a| + |b|)_\le + |c|)_\le = [[a - b)_\le + c)_\le$$
$$= [(a - b) + c)_\le = [a + (c - b))_\le = [a + [c - b)_\le)_\le$$
$$= a \oplus (b \oplus c).$$

Other states are proved in the same way. □

Theorem 10.54. *Let* $(H, +, \le)$ *be an ordered canonical hypergroup.
Then,* EL^2*-hypergroupoid* $(H, \oplus_1)$ *is a hypergroup.*

Proof. Suppose that $0 \le a, c$ and $b \le 0$. Then,

$$(a \oplus_1 b) \oplus_1 c = ([|a| + |b|)_\le \bigcup \{a, b\}) \oplus_1 c = ([a - b)_\le \bigcup \{a, b\}) \oplus_1 c$$
$$= ([a - b)_\le \oplus_1 c) \bigcup a \oplus_1 c \bigcup b \oplus_1 c$$
$$= ([a - b)_\le \bigcup \{c\} \bigcup [(a - b) + c)_\le \bigcup [a + c)_\le \bigcup \{a, c\}$$
$$\times \bigcup [-b + c)_\le \bigcup \{b, c\}$$
$$= a \oplus_1 (b \oplus_1 c).$$

Now, if $0 \le a$, $c \le 0$ and $\{(0, b), (b, 0)\} \not\subseteq \le$, then we have

$$(a \oplus_1 b) \oplus_1 c = ([|a| + |b|)_\le \bigcup \{a, b\}) \oplus_1 c = (\emptyset \bigcup \{a, b\}) \oplus_1 c$$
$$= (a \oplus_1 c) \bigcup (b \oplus_1 c).$$

On the other hand,

$$a \oplus_1 (b \oplus_1 c) = a \oplus_1 (\emptyset \bigcup \{b, c\}) = (a \oplus_1 c) \bigcup (b \oplus_1 c).$$

Other states are proved in the same way. Due to the definition $\oplus_1$,
it is easy to see that the reproduction principle is established. □

Theorem 10.55. *Let* $(R, +, ., \le)$ *be an ordered Krasner hyperring.
Then, the algebraic structure* $(R, .')$ *is a semigroup.*

Proof. Suppose that $0 \le a$ and $b, c \le 0$. Then,

$$(a.'b).'c = (|a|.|b|).'c = (-ab).'c = (-ab)(-c)$$
$$= abc = a(bc) = |a|(|b||c|) = a.'(b.'c).$$

□

Theorem 10.56. *Let $(R, +, \cdot, \leq)$ be an ordered Krasner hyperring. Then, the hyperstructure $(R, \oplus_1, \cdot')$ is an additive hyperring.*

Proof.   It is sufficient that we show the distributivity $\cdot'$ by the ratio of $\oplus_1$ on the left. So, suppose that $b \leq 0$ and $0 \leq a, c$:

$$a \cdot' (b \oplus_1 c) = a \cdot' ([-b + c)_\leq \bigcup \{b, c\}) = a \cdot' [-b + c)_\leq \bigcup \{a \cdot' b, a \cdot' c\}$$

$$\subseteq [a(-b + c)) \leq \bigcup \{-ab, ac\} = [ac - ab)_\leq \bigcup \{-ab, ac\}.$$

On the other hand,

$$(a \cdot' b) \oplus_1 (a \cdot' c) = (-ab) \oplus_1 (ac) = [-ab + ac)_\leq \bigcup \{-ab, ac\}.$$

The distributivity $\cdot'$ by the ratio of $\oplus_1$ on the right is proved similarly. $\square$

Theorem 10.57. *Let $(R, +, \cdot, \leq)$ be an ordered Krasner hyperfield. Then, the hyperstructure $(R, \oplus_1, \cdot')$ is a good additive hyperring.*

Proof.   It is straightforward. $\square$

Bibliography

[1] J. Alajbegovic and J. Mockor, Valuation on multirings, *Comm Math. Univ. St. Pauli* (Tokyo), 34, 201–225 (1985).

[2] J. Alajbegovic and J. Mockor, *Approximations Theorems in Commutative Algebra*. Kluwer Academic Publishers, MA, United States (1992).

[3] R. Ameri and M.M. Zahedi, Hyperalgebraic system, *Ital. J. Pure Appl. Math.*, 6, 21–32 (1999).

[4] F.W. Anderson, Lattice-ordered rings of quotients, *Can. J. Math.*, 17, 434–448 (1965).

[5] D.F. Anderson and A. Badawi, On n-absorbing ideals of commutative rings, *Commun. Algebra*, 39, 1646–1672 (2011).

[6] A. Asokkumar, Hyperlattice formed by the idempotents of a hyperring, *Tamkang J. Math.*, 38(3), 209–215 (2007).

[7] A. Asokkumar, Derivations in hyperrings and prime hyperrings, *Iran. J. Math. Sci. Inf.*, 8(1), 1–13 (2013).

[8] A. Asokkumar and M. Velrajan, Characterizations of regular hyperrings, *Ital. J. Pure Appl. Math.*, 22, 115–124 (2007).

[9] A. Badawi, On 2-absorbing ideals of commutative rings, *Bull. Aust. Math. Soc.*, 75(3), 417–429 (2007).

[10] L. Berardi, F. Eugeni and S. Innamorati, Remarks on hypergroupoids and cryptography, *J. Comb. Inf. Syst. Sci.*, 17, 217–231 (1992).

[11] R. Biswas and S. Nanda, Rough groups and rough subgroups, *Bull. Polish Acad. Sci. Math.*, 42, 251–254 (1994).

[12] J. Chvalina and L. Chvalinová, Multistructures determined by diferencial rings, *Arch. Math.* (Berbo), 36, 429–434 (2000).

[13] S. Comer, A remark on Cromatic Polygroups, *Congr. Numer.*, 38, 85–95 (1983).

[14] S. Comer, Extension of polygroups by polygroups and their representations using color schemes. Lecture Notes in Mathematics, Vol. 1004, pp. 91–103 (1984).

[15] S. Comer, Polygroups derived from cogroups, *J. Algebra*, 89(2), 397–405 (1984).

[16] A. Connes and C. Consani, The hyperring of adéle classes, *J. Number Theory*, 131, 159–194 (2011).

[17] P. Corsini, (i.p.s.) ipergruppi di ordine 7, *Atti Sem. Mat. Fis. Un. Modena* (Modena, Italy), 34, 199–216 (1986).

[18] P. Corsini, (i.p.s.) *Hypergroups of Order 8*. Aviani Editore, Udine, Italy (1989).

[19] P. Corsini, *Prolegomena of Hypergroup Theory*. Aviani Editore, Udine, Italy (1993).

[20] P. Corsini, Feebly canonical and 1-hypergroupes, *Acta Un. Carolinae-Math. Ph.*, 24(2), 49–56 (1983).

[21] P. Corsini, Sugli ipergruppi canonici finiti con identita parziali scalari, *Rend. Circolo Mat. Palermo* (Palermo, Italy), S.II, T. 36, 205–219 (1987).

[22] P. Corsini and V. Leoreanu, *Applications of Hyperstructure Theory*. Kluwer Academic Publishers, Dordrecht/Boston/London (2003).

[23] B. Davvaz, Isomorphism theorems of hyperrings, *Indian J. Pure Appl. Math.*, 35(3), 321–331 (2004).

[24] B. Davvaz, Roughness in rings, *Inf. Sci.*, 164, 147–163 (2004).

[25] B. Davvaz, Approximations in hyperrings, *J. Mult. Valued Log. Soft Comput.*, 15(5–6), 471–488 (2009).

[26] B. Davvaz, Rough algebraic structures: Corresponding to ring theory, in *Algebraic Methods in General Rough Sets*. Trends in Mathematics. Springer Nature Switzerland AG (2018).

[27] B. Davvaz and and V. Leoreanu-Fotea, *Hyperring Theory and Applications*. International Academic Press, USA (2007).

[28] B. Davvaz and A. Salasi, A realization of hyperrings, *Commun. Algebra*, 34(12), 4389–4400 (2006).

[29] B. Davvaz and T. Vougiouklis, Commutative rings obtained from hyperrings (H_v-rings) with α^*-relations, *Commun. Algebra*, 35(11), 3307–3320 (2007).

[30] B. Davvaz and T. Vougiouklis, *A Walk through Weak Hyperstructures: H_v-Structures*. World Scientific Publishing Co. Pte. Ltd., Hackensack, NJ (2019), xi+334 pp.

[31] M. De Salvo, Feebly canonical hypergroups, *J. Comb. Inf. Syst. Sci.*, 15, 133–150 (1990).

[32] H. Durna, Symmetric bi-derivation on hyperrings, *Cumhuriyet Univ. Fac. Sci. Sci. J. (CSJ)*, 37(4), 391–397 (2016).

[33] F. Farzalipour and P. Ghiasvand, On graded hyperrings and graded hypermodules, *J. Algebraic Struct. Their Appl.*, 7(2), 15–28 (2020).

[34] D. Freni, A note on the core of a hypergroup and the transitive closure β^* of β, *Riv. Mat. Pura Appl.*, 8, 153–156 (1991).

[35] D. Freni, A new characterization of the derived hypergroup via strongly regular equivalences, *Commun. Algebra*, 30(8), 3977–3989 (2002).

[36] S.H. Ghazavi, S.M. Anvarieh and S. Mirvakili, EL-hyperstructures derived from (partially) quasi ordered hyperstructures, *Iran J. Math. Sci. Inf.*, 10(2), 99–114 (2015).

[37] D. Hankerson, A. Menezes and S.A. Vanstone, *Guide to Elliptic Curve Cryptography.* Springer-Verlag, Berline (2004).

[38] H. Mirabdollahi, S.M. Anvariyeh and S. Mirvakili, The associated hyperringoid to a Krasner hyperring, *J. Taibah Univ. Sci.*, 12(3), 348–356 (2018).

[39] T.W. Hungerford, *Algebra* (Reprint of the 1974 original). Graduate Texts in Mathematics, Vol. 73. Springer-Verlag, New York-Berlin (1980).

[40] H.M. Jafarabadi, N.H. Sarmin and M.R. Molaei, Completely simple and regular semihypergroups, *Bull. Malays. Math. Sci. Soc.*, 35(2), 335–343 (2012).

[41] L. Kamali Ardekani and B. Davvaz, A generalization of prime hyperideals in Krasner hyperrings, *J. Algebraic Syst.*, 7(2), 205–216 (2020).

[42] L. Kamali Ardekani and B. Davvaz, Some notes on differential hyperrings, *Iran. J. Sci. Technol. (IJST)*, 39A1, 101–111 (2015).

[43] L. Kamali Ardekani and B. Davvaz, On (θ, σ)-derivations of two types of hyperrings, *J. Mult.- Valued Log. Soft Comput.*, 25(4–5), 491–510 (2015).

[44] H. Guan, E. Kaya, M. Bolat, S. Onar, B.A. Ersoy and K. Hila, $\phi - \delta$ primary hyperideals in Krasner hyperrings, *Math. Probl. Eng.* (2022), Article ID 1192684, 12 p.

[45] Y. Kemprasit, Multiplicative interval semigroups on $\mathbb{R}$ admitting hyperring structure, *Ital. J. Pure Appl. Math.*, 12, 205–212 (2002).

[46] Y. Kemprasit, Hyperring structures of some semigroups of linear transformations, *Ital. J. Pure Appl. Math.*, 13, 145–152 (2003).

[47] Y. Kemprasit and Y. Punkla, Transformation semigroups admitting hyperring structure, *Ital. J. Pure Appl. Math.*, 10, 213–218 (2001).

[48] N. Koblitz, *Introduction to Elliptic Curves and Modular Forms.* Graduate Texts in Mathematics, Vol. 97. Springer-Verlag, New York (1984).

[49] M. Koskas, Groupoids, demi-hypergroupes et hypergroupes, *J. Math. Pures Appl.*, 49(9), 155–192 (1970).

[50] M. Krasner, A class of hyperrings and hyperfields, *Int. J. Math. Math. Sci.*, 6, 307–312 (1983).

[51] N. Kuroki, Rough ideals in semigroups, *Inf. Sci.*, 100, 139–163 (1997).

[52] R. Lyndon, The representation of relational algebras, *Ann. Math.*, 2, 707–729 (1950).

[53] R. Lyndon, Relation algebras and projective property, *Mich. Math. J.*, 8, 21–28 (1961).

[54] F. Marty, Sur une généralization de la notion de groupe, in *8th Congress of Mathematicians*, Scandenaves, Stockholm, pp. 45–49 (1934).

[55] X. Ma, J. Zhan and B. Davvaz, Notes on approximations in hyperrings, *J. Mult. Valued Log. Soft Comput.*, 29(3–4), 389–394 (2017).

[56] J. McMullen and J. Price, Reversible hypergroups, *Rend. Sem. Mat. Fis. Milano*, 47, 67–85 (1977).

[57] J. McMullen and J. Price, Duality for finite abelian hypergroups over splitting fields, *Bull. Aust. Math. Soc.*, 20, 57–70 (1979).

[58] C.G. Massouros, On the theory of hyperrings and hyperfields, *Algebra Log.*, 24, 728–742 (1985).

[59] C.G. Massouros, Method of constructions of hyperfields, *Int. J. Math. Math. Sci.*, 8(4), 725–728 (1985).

[60] C.G. Massouros, Constructions of hyperfields, *Math. Balk.*, 5(3), 250–257 (1991).

[61] G.G. Massouros, C. Massouros and I.D. Mittas, Fortified join hypergroups, *Ann. Math. Blaise Pascal*, 3(2), 155–169 (1996).

[62] S. Mirvakili, S.M. Anvariyeh and B. Davvaz, On α-relation and transitivity condition of α, *Commun. Algebra*, 36(4), 1695–1703 (2008).

[63] S. Mirvakili and B. Davvaz, Applications of the α^*-relation to Krasner hyperrings, *J. Algebra*, 362, 145–156 (2012).

[64] J. Mittas, Hypergroupes canoniques (Fr), *Math. Balk.*, 2, 165–179 (1972).

[65] J. Mittas, Hypergroupes canoniques values et hypervalues-hypergroupes fortement et suprieurement canoniques (Fr), *Bull. Greek Math. Soc.*, 23, 55–88 (1982).

[66] J. Mittas, Certaines remarques sur les hypergroupes canoniques hypervaluables et fortement canoniques, *Riv. Mat. Pura Appl.*, 9, 61–67 (1991).

[67] J. Mittas, Sur la valuations stricte des hypergroupes polysymetriques canoniques (Fr), *Ratio Math.*, 13, 5–28 (1999).

[68] J. Mittas, Hypergroupes canoniques, *Math. Balk.*, 2, 165–179 (1972).

[69] T. Nakano, A theorem on lattice ordered groups and its applications to the evalution theory, *Math. Z.*, 83, 140–146 (1964).

[70] A. Nakassis, Expository and survey article of recent results in hyperring and hyperfield theory, *Int. J. Math. Math. Sci.*, 11, 209–220 (1988).

[71] S. Nenthein, S. Pianskool and Y. Kemprasit, Krasner hyperrings constructed from rings and the relationship between their algebraic structures, *East-West J. Math.* (Special Volume), 33–40 (2004).

[72] M. Norouzi, A. Mohammadi and V. Leoreanu-Fotea, Modeling of complete combustion reaction of room temperature gaseous alkanes via join spaces, *Bull. Comput. Appl. Math.*, 8(1), 81–94 (2021).

[73] S. Omidi, B. Davvaz and J. Zhan, Some properties of n-hyperideals in commutative hyperrings, *J. Algebraic Hyperstructures Log. Algebras*, 1(2), 23–30 (2020).

[74] S. Omidi, B. Davvaz and P. Corsini, Operations on hyperideals in ordered Krasner hyperrings, *Analele Univ. "Ovidius", Math. Ser.*, 24(3), 275–293 (2016).

[75] S. Omidi and B. Davvaz, On ordered Krasner hyperrings, *Iran. J. Math. Sci. Inf.*, 12(2), 35–49 (2017).

[76] S. Omidi and B. Davvaz, Hyperideal theory in ordered Krasner hyperrings, *Analele Univ. "Ovidius" din Constanta, Math. Ser.*, 27(1), 193–210 (2019).

[77] S. Omidi and B. Davvaz, Characterizations of bi-hyperideals and prime bi-hyperideals in ordered Krasner hyperrings, *TWMS J. Pure Appl. Math.*, 8(1), 64–82 (2017).

[78] S. Omidi and B. Davvaz, Study on clean ordered rings derived from clean ordered Krasner hyperrings, *J. Korean Soc. Math. Educ. Ser. B: Pure Appl. Math.*, 25(2), 115–125 (2018).

[79] S. Oğuz Unal, Permuting tri-derivations on hyperrings, *Palest. J. Math.*, 9(1), 105–111(2020).

[80] Z. Pawlak, Rough sets, *Int. J. Inf. Comp. Sci.*, 11, 341–356 (1982).

[81] Z. Pawlak, *Rough Sets — Theoretical Aspects of Reasoning about Data*. Kluwer Academic Publishing, Dordrecht (1991).

[82] W. Phanthawimol, Y. Punkla, K. Kwakpatoon and Y. Kemprasit, On homomorphisms of Krasner hyperrings, *An. Ştiinţ. Univ. Al. I. Cuza Iaşi. Mat. (N.S.)*, 57(2), 239–246 (2011).

[83] G. Pinotsis, The condition of isomorphism of a class of hyperfields with quotient hyperfields, *Ital. J. Pure Appl.*, 3, 49–54 (1998).

[84] E. Posner, Derivations in prime rings, *Proc. Am. Math. Soc.*, 8, 1093–1100 (1957).

[85] R. Prasad, L.A. Kennedy and E. Ruckenstein, Catalytic combustion, *Catal. Rev. Sci. Eng.*, 26, 1–58 (1984).

[86] W. Prenowitz and J. Jantosciak, *Join Geometries*. Springer-Verlag, UTM (1979).

[87] H. Rabinowitz and S. Vogel, Chapter 10: Style and usage for organic chemistry, in *The Manual of Scientific Style*. Academic Press, London, pp. 399–425 (2009).

[88] E. Stauffer, J.A. Dolan and R. Newman, *Review of Basic Organic Chemistry. Fire Debris Analysis*. Academic Press, Burlington (2008).

[89] N. Ramaruban, Commutative Hyperalgebra, Ph.D. Thesis, University of Cincinnati, USA, 2014.

[90] R. Roth, Character and conjugacy class hypergroups of a finite group, *Ann. Mat. Pura Appl.*, 105(4), 295–311 (1975).

[91] R. Roth, On derived canonical hypergroups, *Riv. Mat. Pura Appl.*, 3, 81–85 (1988).

[92] K. Serafimidis, Sur les L-Hyperideaux des hypergroupes canoniques strictement réticules (Fr), *Rend. Del Circ. Matem. Palermo* (Palermo, Italy), Serie II, Tomo XXXV, 411–419 (1986).

[93] K. Serafimidis, M. Konstantinidou and J. Mittas, Sur les hypergroupes canoniques strictement réticules, *Riv. Mat. Pura Appl.*, 2, 21–35 (1987).

[94] J.H. Silverman and I.T. Tate, *Rational Points on Elliptic Curves*. Springer-Verlag, New York (2015).

[95] A. Sonea and B. Davvaz, The Euler's totient function in canonical hypergroups, *Indian J. Pure Appl. Math.*, 53, 683–695 (2022).

[96] M. Ştefănescu, Constructions of hyperfields and hyperrings, *Stud. Cercet. Ştiinţ. Ser. Mat. Univ. Bacău*, 16 (suppl.), 563–571 (2006).

[97] V. Vahedi, M. Jafarpour, H. Aghabozorgi and I. Cristea, Extension of elliptic curves on Krasner hyperfields, *Commun. Algebra*, 47, 4806–4823 (2019).

[98] V. Vahedi, M. Jafarpour and I. Cristea, Hyperhomographies on Krasner hyperfields, *Symmetry* (Symmetry in Classical and Fuzzy Algebraic Hypercompositional Structures), 11(12), 1442 (2019).

[99] V. Vahedi, M. Jafarpour, S. Hoskova-Mayerova, H. Aghabozorgi, V. Leoreanu-Fotea and S. Bekesiene, Derived hyperstructures from hyperconics, *Mathematics*, 8(3), 42 (2020).

[100] M. Velrajan and A. Asokkumar, Note on isomorphism theorems of hyperrings, *Int. J. Math. Math. Sci.* (2010), Art. ID 376985, 12 p.

[101] T. Vougiouklis, The fundamental relation in hyperrings. The general hyperfield, in *Algebraic Hyperstructures and Applications*, Xanthi, 1990. World Science Publishing, Teaneck, NJ, pp. 203–211 (1991).

[102] T. Vougiouklis, *Hyperstructures and Their Representations*. Hadronic Press, Inc., Palm Harbor, USA (1994).

[103] P. Xu, M. Bolat, E. Kaya, S. Onar, B.A. Ersoy and K. Hila, r-hyperideals and generalizations of r-hyperideals in Krasner hyperrings, *Math. Prob. Eng.*, 2022 (2022), Article ID 7862425, 10 p.

[104] D. Yilmaz and H. Yazali, Semi-derivations on hyperrings, *Bull. Int. Math. Virtual Inst.*, 12(2), 309–319 (2022).

[105] M.M. Zahedi and R. Ameri, A hyperversion of Nakayama's lemma, in *Proceedings of the 27th Annual Iranian Mathematics Conference, Shiraz, 1996, Shiraz University, Shiraz*, pp. 331–347 (1996).

[106] M.F. Zwinkels, S.G. Jaras, P.G. Menon and T.A. Griffin, Catalytic materials for high-temperature combustion, *Catal. Rev. Sci. Eng.*, 35, 319–358 (1993).

Index